OPTIMIZATION AND CONTROL
OF BILINEAR SYSTEMS

Springer Optimization and Its Applications

VOLUME 11

Aims and Scope
Optimization has been expanding in all directions at an astonishing rate during the last few decades. New algorithmic and theoretical techniques have been developed, the diffusion into other disciplines has proceeded at a rapid pace, and our knowledge of all aspects of the field has grown even more profound. At the same time, one of the most striking trends in optimization is the constantly increasing emphasis on the interdisciplinary nature of the field. Optimization has been a basic tool in all areas of applied mathematics, engineering, medicine, economics and other sciences.

The series *Optimization and Its Applications* publishes undergraduate and graduate textbooks, monographs and state-of-the-art expository works that focus on algorithms for solving optimization problems and also study applications involving such problems. Some of the topics covered include nonlinear optimization (convex and nonconvex), network flow problems, stochastic optimization, optimal control, discrete optimization, multi-objective programming, description of software packages, approximation techniques and heuristic approaches.

OPTIMIZATION AND CONTROL OF BILINEAR SYSTEMS

Theory, Algorithms, and Applications

By

PANOS M. PARDALOS
University of Florida, Gainesville, FL

VITALIY YATSENKO
Space Research Institute of NASU-NASU, Kyiv, Ukraine

 Springer

Panos M. Pardalos
Department of Industrial
and Systems Engineering
University of Florida
303 Weil Hall
P.O.Box 116595
Gainesville FL 32611-6595
USA
pardalos@ufl.edu

Vitaliy Yatsenko
Space Research Institute
of NASU-NASU
Glushkov Avenue, 40
Kyiv 03680
Ukraine
vyatsenko@gmail.com

ISBN: 978-1-4419-4468-9   e-ISBN: 978-0-387-73669-3
DOI: 10.1007/978-0-387-73669-3

AMS Subject Classifications: 93C10, 93B29, 93B52, 93E10, 81P68

Printed on acid-free paper

9 8 7 6 5 4 3 2 1

springer.com

This book is dedicated to our families

Contents

Preface

The present book is based on results of scientific investigations and on the materials of special courses, offered for graduate and undergraduate students. The purpose of this book is to acquaint the reader with the developments in bilinear systems theory and its applications. Particular attention is paid to control of open physical processes functioning in a nonequilibrium mode.

The text consists of eight chapters. Chapter 1 is concerned with the problems of systems analysis of bilinear processes. Chapter 2 solves the problem of optimal control of bilinear systems on the basis of differential geometry methods. Chapter 3 deals with the progress made in an adaptive estimation technique. Chapter 4 is devoted to the application of the Yang–Mills fields to investigation of nonlinear control problems. Chapter 5 considers intelligent sensors, used to examine weak signals. This chapter also describes and analyzes bilinear models of intelligent sensing elements. Chapter 6 illustrates control problems of a quantum system. Chapter 7 discusses the problems of control and identification in systems with chaotic dynamics. Finally, Chapter 8 examines the controlled processes running in biomolecular systems.

This book is directed to students, postgraduate students, and specialists engaged in the fields of control of physical processes, quantum and molecular computing, biophysics, and physical information science.

University of Florida, Gainesville   *P.M. Pardalos*
Institute of Space Research, Kiev, Ukraine   *V.A. Yatsenko*

Thinking well is the greatest excellence; and wisdom is to act and speak what is true, perceiving things according to their nature.

—Heraclitus   (540 BC–480 BC)

Acknowledgments

Our thanks go to the members of the Industrial and System Engineering Department at the University of Florida in Gainesville for support during the writing of this book. We benefited from the intellectual intensity of the university and the freedom to focus on our own work. Research has been funded partially by NIH, STCU, and Air Force grants.

Foreword

Bilinear systems are special kinds of nonlinear systems capable of representing a variety of important physical processes. A great deal of literature related to the control problems of such systems has been developed over the past decades. Some results were concerned with bilinear systems with only multiplicative control. For bilinear systems with both additive and multiplicative control inputs, some control designs, such as bang-bang control or optimal control, obtain global asymptotic stability under the assumption that an open-loop system is either stable or neutrally stable. When the open-loop system is unstable, it is difficult to obtain global asymptotic stability except when independent additive and multiplicative control inputs exist.

This book proposes new algorithms motivated by bilinear models, which can approximate a wide class of nonlinear control systems. They can either be represented as a state space model or with an input-output equation. For the last form, different optimization algorithms have been derived. They all require the computationally complex evaluation of an optimality criterion for each update step. The goal of this book is to describe new methods, heuristics, and optimality criteria, which are less demanding in computational complexity than the exact criteria and which result in robust adaptive algorithms. This book has been written in cooperation with the Center for Applied Optimization at the University of Florida.

Notation

BM — bilinear model
BS — bilinear system
L — L-system
CDL — coupled dynamical lattice
CM — coupled maps
DITS — dynamic information-transforming system
DS — dynamical system
DSS — decision support system
DMZ — Dunkan–Martensen–Zakai equation
ECG — electrocardiograms
EEG — electroetinograms
LLPR — local linear polynomial regression
LM — lattice model
MCG — magnetocardiogram
MIMO — multiinput-output object
MS — multisensor system
MWE – low-intensity microwaves
NS — nonlinear system
NW — Nadaraya–Watson regression
PMP — Pontryagin Maximum Principle
QDS — quantum dynamical system
SE — sensitive element
SVD — singular value decomposition
SW — spike and wave activity
TS — T-signal

Introduction

Bilinear systems are one of the simplest nonlinear systems and therefore particularly applicable to analysis of much more complicated nonlinear systems. They can be used to represent a wide range of physical, chemical, biological, and social systems, as well as manufacturing processes that cannot be effectively modeled under the assumption of linearity.

A great deal of literature related to the control problems of such systems has been developed over the past decades. Some control problems of bilinear processes were solved for plasma, quantum devices, particle accelerators, nuclear power plants, and biomedicine (Bruni, 1971; Mohler, 1973; Brockett, 1973, 1975, 1976; Andreev, 1982; Butkovskiy and Samoilenko, 1990; Aganović and Gajić, 1995; Jurdjevic, 1998; Bacciotti, 2004).

The concept of bilinear systems was introduced in the 1960s. In the theory of automatic control these are control systems whose dynamics are jointly linear in the state and control variables. Their theory can be developed from the theories of time-variant linear systems and of matrix Lie groups; they have been applied to many areas of science and technology. The purpose of this book is to discuss the development of these ideas.

We emphasize the role of three disciplines that modified our outlook on bilinear system theory. The first one is modern differential geometry. The second discipline is the modern theory of control dynamical systems. The third discipline is optimization theory. Bilinear systems can approximate a wide class of nonlinear control systems. They can be represented as state space models or as systems of input-output (Kučera, 1966; Krener, 1973, 1975, 1998; Isidory and Krener, 1984; D'Alessandro, Isidori, and Ruberti, 1974; Semenov and Yatsenko, 1981; Yatsenko, 1984; Yatsenko and Knopov, 1992). For the last form, different adaptive algorithms were derived. They all require a stability criterion for each update step (Slemrod, 1978; Sontag, 1990; Willems, 1998; Pardalos et al., 2003).

The properties and behaviors of bilinear systems are being investigated and a number of useful results have been derived (Haynes and Hermes, 1970; Bailleul, 1978; Brockett, 1972, 1975, 1976, 1981; D'Alessandro, Isidory, and Ruberti, 1974; Isidory and Krener, 1984; Isidory, 1995; Krener, 1975; Kućera, 1966; Lo, 1975; Mohler, 1973; Pavlovskii and Elkin, 1988; Sontag, 1988, 1990). A number of practical algorithms were developed and this book is oriented to improve their performance (Yatsenko, 1996; Pardalos et al., 2001; Butkovskiy and Samoilenko, 1990).

The new methods developed in this context lead to a better understanding of quantum controlled processes. The problem of control of quantum states has existed since the beginning of quantum mechanics. Many experimental facts of quantum mechanics were established using macroscopic fields acting on quantum ensembles, which from the modern point of view can be considered as control. As the technology of experiments improved, new problems in controlling bilinear quantum systems arose (Butkovskiy and Samoilenko, 1990; Yatsenko, 1993, Yatsenko, Titarenko, and Kolesnik, 1994), and their solution required special methods (Andreev, 1982; Brockett, 1979; Butkovskiy and Samoilenko, 1990; Yatsenko, 1995).

The objective of bilinear control of open quantum systems can be formulated in terms of statistics for large numbers of quantum states, which brings this branch of the physical theory of control closer to the well-developed theory of statistical process control. Here quantum features reveal themselves in discrete energy spectra and in the symmetry or antisymmetry of wave functions for multiparticle systems (Butkovskiy and Samoilenko, 1990; Yatsenko, 1993, 1995; Pardalos, Sackellares, and Yatsenko, 2002).

It is an exciting prospect to produce novel quantum-based elements for control systems, quantum holography, and microscopy, for example, in connection with the increasing need for accuracy of measurement in experiments with test bodies (Yatsenko, 1987, 1989; Yatsenko and Knopov, 1992; Pardalos et al., 2001). A number of quantum effects, such as the Josephson effect and the phenomena of induced emission, are employed in controlled units and devices in computer technology. However, so far the problems of controlling quantum processes have been considered mainly in the space of averaged physical variables rather than in the space of quantum-mechanical states. This case is only valid if the characteristic control time is much bigger than the relaxation time of the underlying processes, so that the state variables and controlled processes are identical with the common macroscopical ones, and the quantum origin of the modeling equations is only exhibited in the non-linear dependencies involved. We do not discuss these cases further but focus our attention on considering the possibility of control of quantum states in both pure and mixed ensembles.

Bilinear models may also simulate a large class of communication systems. Methods of coherent optical communication consider the quantum features of the photon channels. The quantum theory of test of hypotheses was considered by Helstrom (1976), which includes a detailed bibliography on the subject. The Heisenberg uncertainty relation was suggested for use in estimating the limits of the possibility of control of microscopic processes and quantum ensembles (Krasovskiy, 1990; Petrov et al., 1982; Hirota and Ikehara, 1982; Belavkin, Hirota and Hudson, 1995) from the viewpoint of information theory. It should be pointed out that such estimates can be proven to be too sharp for the special classes of states and operators of observables. The theory of an optimal quantum estimation provides optimal observations for various estimation problems for the unknown parameters in the state of the system (Helstrom, 1976; Bendjaballah, Hirota, and Reynaud, 1991; Belavkin, Hirota and Hudson, 1995).

The wide spectrum of the above-mentioned problems can be represented by the following theoretical schemes.

1 Construction of a set of states accessible from a given initial state

2 Identification of the set of controls steering the system from a given initial state to a desired accessible state with the greatest or specified probability

3 Stability analysis for adaptive bilinear systems

4 Identification of a control that is optimal with respect to a given criterion, for example, the response time or the minimum of switches (in bang-bang control)

5 Control and optimization of biological systems

6 Construction of a system of a microscopic feedback providing for the possibility of control with accumulation of data

In this book we describe new principles of control and optimization of a large class of nonlinear objects including physical systems. Nonlinear physics and bilinear control are two rich and well-developed theories. Their efficient unification requires joint efforts of specialists in

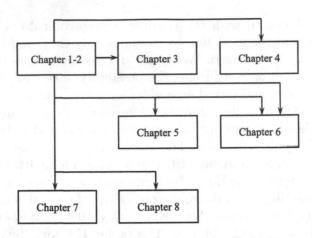

Figure 0.1. Chart of the book.

both fields. When two such abundant theories are joined, the effect is multiplicative rather than additive because they amplify each other's potential in proportion to their range of development. The chart shown in Figure 0.1 indicates the relationships among the chapters.

Chapter 1

SYSTEM-THEORETICAL DESCRIPTION OF OPEN PHYSICAL PROCESSES

One of the difficulties in designing controllers for complex physical systems is the problem of modeling. Very often, we have mathematical models that include highly nonlinear differential equations, for which control design techniques are a complicated problem. An example is given by the formation flight; accurate models are known for fluid dynamics, and for aircraft motion, that can be accurately simulated together. However, the equations of motion consist of partial and ordinary differential equations coupled via their boundary conditions, a model that offers little to the control designer. It is therefore a problem of considerable interest in developing explicit low-order models; once a control design has been constructed using such a low-order model, it can be tested by comparing with a full high-order simulation (Hunt, Su, and Meyer, 1983).

Bilinear models (BM) can approximate a wide class of nonlinear systems. They are used to model nonlinear processes in signal and image processing and communication systems modeling. In particular, they arise in areas such as channel equalization, echo cancellation, nonlinear tracking, multiplicative disturbance tracking, and many other areas of engineering, socioeconomics, and biology. BM represent a mathematically tractable structure over Volterra models for a nonlinear system. Also, a bilinear model can obviously represent the dynamics of a nonlinear system more accurately than a linear model. Hence, modeling and control of nonlinear systems in a bilinear framework are fundamental problems in engineering.

This chapter proposes new methodologies for analysis and optimal design of control systems using algebraic and geometric methods (Figure 1.1). These can be subdivided into methods that attempt to

P.M. Pardalos, V. Yatsenko, *Optimization and Control of Bilinear Systems*, doi: 10.1007/978-0-387-73669-3,
© Springer Science+Business Media, LLC 2008

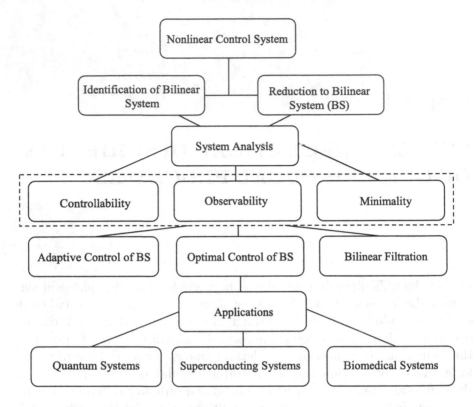

Figure 1.1. Control design methodology for nonlinear systems.

treat the system as a bilinear system in a limited range of operation and use bili-near design methods for each region. The most important aspect of these methodologies is transformation of a nonlinear control system into a bilinear system (Krener, 1975; Lo, 1975; Sontag, 1988; Elkin, 1999; Varga, 2001).

A global change in coordinates for transforming the system are used for finding a lower-order nonlinear subsystem. A constructive system analysis of such systems on the base of geometric and algebraic methods is conducted. The specific examples of nonlinear systems reduction to bilinear systems (BS) and dynamical systems (DS) with known physical properties are given. It is also shown that every nonlinear realization can be locally approximated by a bilinear realization, with an error that grows as a function of time t.

The controllability, observability, and invertibility of nonlinear control systems using Lie algebras of vector fields are considered. The study of this type of systems was initiated by R. Brockett (1972). In this chapter, Brockett's observability results are generalized and necessary and sufficient conditions for observability are presented. Effective algorithms are proposed to verify such conditions.

Local and global bilinear realizations of nonlinear control systems were studied in the literature (Krener, 1975; Lo, 1975; Sontag, 1988; Hammouri and Gauthier, 1988). For a controlled nonlinear system with control appearing linearly, there exist necessary and sufficient conditions for the existence of a dynamically equivalent bilinear system (Lo, 1975). It was also shown that every nonlinear realization can be approximated by abilinear realization (Krener, 1975).

Necessary and sufficient conditions for the invertibility of a class of nonlinear systems, which includes matrix bilinear systems, were also obtained. Lie algebraic invertibility criteria are obtained for bilinear systems in $\mathbb{R}^n$, which generalize standard tests for single input linear systems. These results are used to construct nonlinear systems that act as left-inverses for bilinear systems.

1. Reduction of Nonlinear Control Systems to Bilinear Realization

1.1 Equivalence of Control Systems

In this section, we consider the following nonlinear control system

$$\dot{x} = f(x, t, u), \quad z = r(x), \quad x(0) = x_0,$$
$$u(t) \in U, \quad t \in T, \quad x \in G, \tag{1.1}$$

where $x^T = (x_1, \ldots, x_n)$; $f^T = (f_1, \ldots, f_n)$ are C^∞ n-dimensional vector fields; u is an m-dimensional vector-valued function; G is a smooth manifold; T is a finite or infinite time interval; U is an m-dimensional domain in Euclidean space $\mathbb{R}^m$; $r(x)$ is a C^∞ output nonlinear function. The function $f(x, t, u)$ is defined on a smooth manifold $G_1 = G \times T \times U$.

An equivalent realization of an input-output map is

$$\dot{y} = g(y, u), \quad k = l(y), \quad y(0) = y_0,$$
$$u(t) \in U, \quad t \in T, \quad y \in M, \tag{1.2}$$

where M is a smooth manifold; $l(y)$ is an output nonlinear map; $g(y, u)$ is a vector-valued function (a comparison function of $f(x, u)$).

This chapter provides a necessary and sufficient condition for the existence of transformation $y = \gamma(x)$ of the state of (1.1) which provides a nonlinear realization (1.2).

As a corollary of using a geometric technique, we obtain conditions for the existence of an equivalent realization (1.2), such that $r(x) = l(\gamma(x))$.

1.2   Lie Algebras, Lie Groups, and Representations

In mathematics, aLie group is an analytic real or complex manifold that is also a group such that the group operations multiplication and inversion are analytic maps. Lie groups are important in mathematical analysis, physics, and geometry because they serve to describe the symmetry of analytical structures. They were introduced by Sophus Lie in 1870 in order to study symmetries of differential equations.

For most of this book, the C^∞-manifold that is of interest is $\mathbb{R}^n$ (which is covered by a single coordinate system). A vector space $\mathcal{L}$ over $\mathbb{R}$ is a real Lie algebra, if in addition to its vector space structure it possesses a product $\mathcal{L} \times \mathcal{L} \to \mathcal{L} : (X, Y) \to [X, Y]$ which has the following properties.

(a) It is bilinear over $\mathbb{R}$

(b) It is skew commutative: $[X, Y] + [Y, X] = 0$

(c) It satisfies the Jacobi identity

$$[X, [Y, Z]] + [Y, [Z, X]] + [Z, [X, Y]] = 0, \quad \text{where } X, Y, Z \in \mathcal{L}.$$

Example 1.1. Let $M_n(\mathbb{R})$ be the algebra of $n \times n$ matrices over $\mathbb{R}$. If we denote $[X, Y]$ by $XY - YX$, where XY is the usual matrix product, then this commutator defines aLie algebra structure on $M_n(\mathbb{R})$.

Example 1.2. Let $\mathcal{X}(M)$ denote the C^∞ vector fields on a C^∞ manifold M. $\mathcal{X}(M)$ is a vector space over $\mathbb{R}$ and a $C^\infty(M)$ module. (Recall, a vector field X on M is a mapping: $M \to T_p(M) : p \to x$, where $p \in M$ and $T_p(M)$ is the tangent space to the point p in M.) We can give a Lie algebra structure to $\mathcal{X}(M)$ by defining

$$\mathcal{F}_p(f) = (XY - YX)_p f = X_p(Yf) - Y_p(Xf), \quad f \in C^\infty(p)$$

with the C^∞ functions in a neighborhood of p and $[X, Y] = XY - YX$.

Let $\mathcal{L}$ be a Lie algebra over $\mathbb{R}$ and let $\{X_1, \ldots, X_n\}$ be a basis of $\mathcal{L}$ (as a vector space). There are uniquely determined constants $c_{rsp} \in \mathbb{R}$ $(1 \le r, s, p \le n)$ such that

$$[X_r, Y_s] = \sum_{1 \le p \le n} c_{rsp} X_p.$$

The c_{rsp} are called the structure constants of $\mathcal{L}$ relative to the basis $\{X_1, \ldots, X_n\}$. From the definition of a Lie algebra,

(a) $c_{rsp} + c_{srp} = 0 \quad (1 \le r, s, p \le n)$

(b) $\displaystyle\sum_{1 \le p \le n} (c_{rsp}c_{ptu} + c_{stp}c_{pru} + c_{trp}c_{psu}) = 0 \quad (1 \le r, s, t, u \le n).$

Let $\mathcal{L}$ be a Lie algebra over $\mathbb{R}$. Given two linear subspaces M, N of $\mathcal{L}$ we denote by $[M, N]$ the linear space spanned by $[X, Y]$, $X \in M$ and $Y \in N$. A linear subspace K of $\mathcal{L}$ is called a subalgebra if $[K, K] \subseteq K$, and ideal if $[\mathcal{L}, K] \subseteq K$.

If $\mathcal{L}$ and $\mathcal{L}'$ are Lie algebras over $\mathbb{R}$ and $\pi : \mathcal{L} \to \mathcal{L}' : X \to \pi(X)$, a linear map, π is called a homomorphism if it preserves brackets:

$$[\pi(X), \pi(Y)] = \pi([X, Y]), \quad (X, Y \in \mathcal{L}).$$

In that case $\pi(\mathcal{L})$ is a subalgebra of $\mathcal{L}'$ and $\ker \pi$ is an ideal in $\mathcal{L}$. Conversely, let $\mathcal{L}$ be a Lie algebra over $\mathbb{R}$ and K an ideal of $\mathcal{L}$. Let $\mathcal{L}' = \mathcal{L}/K$ be the quotient vector space and $\pi : \mathcal{L} \to \mathcal{L}'$ the canonical linear map. For $X' = \pi(X)$ and $Y' = \pi(Y)$, let

$$[X', Y'] = \pi([X, Y]).$$

This mapping is well defined and makes $\mathcal{L}'$ a Lie algebra over $\mathbb{R}$ and π is then a homomorphism of $\mathcal{L}$ into $\mathcal{L}'$ with K as the kernel. $\mathcal{L}' = \mathcal{L}/K$ is called the quotient of $\mathcal{L}$ by K.

Let $\mathcal{U}$ be any algebra over $\mathbb{R}$ whose multiplication is bilinear but not necessarily associative. An endomorphism D of $\mathcal{U}$ (considered as a vector space) is called a derivation if

$$D(ab) = (Da)b + a(Db), \quad a, b \in \mathcal{U}.$$

If D_1 and D_2 are derivations so is

$$[D_1, D_2] = D_1 D_2 - D_2 D_1.$$

The set of all derivations on $\mathcal{U}$ (assumed finite-dimensional) is a subalgebra of $gl(\mathcal{U})$ (the Lie algebra of all endomorphisms of $\mathcal{U}$).

The notion of a representation of a Lie algebra is very important.

Let $\mathcal{L}$ be a Lie algebra over $\mathbb{R}$ and $\mathcal{V}$ a vector space over $\mathbb{R}$. By a representation of $\mathcal{L}$ in $\mathcal{V}$ we mean a map

$\pi : X \to \pi(X) : \mathcal{L} + gl(\mathcal{V})$ (all endomorphisms of $\mathcal{V}$) such that

(a) π is linear.

(b) $\pi([X, Y]) = \pi(X)\pi(Y) - \pi(Y)\pi(X).$

For any $X \in \mathcal{L}$ let adX denote the endomorphism of $\mathcal{L}$

$$\text{ad} X : Y \rightarrow [X, Y] \quad (Y \in \mathcal{L}).$$

An endomorphism adX is also a derivation of $\mathcal{L}$ and $X \rightarrow \text{ad}X$ is a representation of $\mathcal{L}$ in $\mathcal{L}$, called the adjoint representation.

Let G be a topological group and at the same time a differentiable manifold. G is a Lie group if the mapping

$$(x, y) \rightarrow xy : G \times G \rightarrow G$$

and the mapping $x \rightarrow x^{-1} : G \rightarrow G$ are both C^{∞} mappings.

Given a Lie group G, there is an essentially unique way to define its Lie algebra. Conversely, every finite-dimensional Lie algebra is the Lie algebra of some simply connected Lie group.

In filtering theory, some special Lie algebras seem to arise. We give the basic definitions for three such Lie algebras.

A Lie algebra $\mathcal{L}$ over $\mathbb{R}$ is said to be nilpotent if ad X is a nilpotent endomorphism of $\mathcal{L}$, $\forall X \in \mathcal{L}$. Let the dimension of $\mathcal{L}$ be m. Then there are ideals $\mathcal{P}_j$ of $\mathcal{L}$ such that

(a) $\dim \mathcal{P}_j = m - j, \quad 0 \leq j \leq m,$

(b) $\mathcal{P}_0 = \mathcal{L} \supseteq \mathcal{P}_1 \supseteq \cdots \mathcal{P}_m,$

(c) $[\mathcal{L}, \mathcal{P}_j] \subset \mathcal{P}_{j+1}, \quad 0 \leq j \leq m - 1.$

Let g be a Lie algebra of finite-dimension over $\mathbb{R}$ and write $\mathcal{D}g = [g, g]$. $\mathcal{D}g$ is a subalgebra of g called the derived algebra. Define $\mathcal{D}^p g$ $(p \geq 0)$ inductively by

$$\mathcal{D}^0 g = g,$$
$$\mathcal{D}^p g = \mathcal{D}(\mathcal{D}^{p-1}g) \quad (p \geq 1).$$

1.3   Selection of Mathematical Models

In recent years there has been considerable interest in bilinear and L-systems (L). We consider BS and L-systems because there is a suitable theory that we can realize in the form of software. It is possible to consider many important nonlinear systems as L-systems (Yakovenko, 1972). Bilinear models describe the dynamics of many control objects. Therefore it is very important to derive the conditions under which the nonlinear control system (NS) (1.1) can be transformed to a bilinear model (BM) or L-system. In this case it is necessary to select comparison systems and estimate unknown parameters.

Definition 1.1. *System (1.2) is said to be real-analytical, if the manifold M is a real-analytical manifold, and g is an analytical function (Isidory, 1995).*

Definition 1.2. *System (1.1) is said to be symmetric, if for any control u from U there exists ũ from U such that $f(x, u) = -f(x, ũ)$.*

Definition 1.3. *The nonlinear system (1.1) is said to be complete if for any admissible control u the vector field X_u of the system is full, and the solution of $\dot{x} = f(x, u)$ is defined under any t.*

In particular, a system is said to be complete if the Lie algebra is finite-dimensional, and any fixed control is generated as a complete field (Andreev, 1982). In particular, BS are complete systems. A Lie algebra of these systems has dimension $\leq n^2$, and vector fields with fixed control are complete, because the solution of the linear equation $\dot{x} = Ax$ is defined for any $t \in \mathbb{R}$.

The linear-analytical systems form a very important subclass of the real-analytical systems. This class is defined by the equations

$$\dot{y} = f(x) + u(t)g(x), \quad y(t) = h(x). \tag{1.3}$$

Here f, g, h are analytical functions, and the state space is an analytical manifold. Some special behavior of such systems has been studied by many authors (Brockett, 1979; Bailleul, 1978; Crouch, 1984). If f, g, h are polynomial, then system (1.3) is said to be polynomial. Some results about these systems can be found in Andreev (1982).

A special case of linear-analytical systems is right-invariant systems. A right-invariant system on a manifold M has a structure of the Lie group, and the system is defined by right-invariant vector fields on the control group (Hirschorn, 1977)

$$\dot{x}(t) = A(x(t)) + \sum_{i=1}^{m} u_i(t)B_i(x(t)), \quad x(0) = x_0 \in M,$$

$$y(t) = Kx(t), \tag{1.4}$$

where M is a Lie group; K is the subgroup of M (Hirschorn, 1977; Pardalos et al., 2001). Let $\mathcal{K}$ and $\mathcal{H}$ be Lie algebras of K, M, respectively. The vector fields $A(x(t))$ and $B_i(x(t))$ belong to $\mathcal{H}$ (i.e., the Lie algebra of right-invariant vector fields on the group M). The right-invariance means that an area of the right group action transforming M does not change a control system (1.4). A control belongs to a class of real piecewise analytical functions on $(0, \infty)$.

Bilinear systems. Bilinear systems are defined by

$$\dot{x}(t) = \left(A + \sum_{i=1}^{m} u_i(t) B_i \right) x(t), \quad y(t) = Cx(t), \tag{1.5}$$

where $x \in \mathbb{R}_0^n = \mathbb{R}^n - \{0\}$; C is a constant $q \times n$ matrix; $u_i(t)$ is a scalar function of time; and A, B_i are constant $n \times n$ dimensional matrices. The algebra L of the system (1.5) is isomorphic to the matrix algebra Lie of a minimal dimension, which contains the matrices $\{A, B_i\}$. We can introduce a control semigroup $S(U)$ for BS, which assumes some representation in the form of the matrix semigroup. There is a connection between $S(U)$ and its representation in the form of semigroup one-to-one continuous maps $\mathbb{R}^n$ into $\mathbb{R}^n$ with a group operation of map composition. This composition is determined by means of a BS, in which the vector $x(t)$ corresponds to the matrix $X(t)$, and the initial condition $X(0) = I$. If we have the initial state x_0, then the set of states, accessible from x_0, will be identified by the set of points $x(t) = X(t)x_0$, where $X(t)$ is an element of the semigroup representation $S(U)$(i.e., a transition map of the BS under fixed control). The group G corresponding to the algebra L is constructed as a minimal group of the diffeomorphism of $\mathbb{R}_0^n$, which contains all one-parameter subgroups $\exp(At)$, where $A \in L$. The representation of the input semigroup does not realize all transformations of this group. If BS is parametric homogeneous and has the form $\dot{x}(t) = u_1(t)A + u_2(t)B$, and a piecewise control satisfies the condition $|u_i(t)| \leq \delta$, then the Lie algebra, which includes matrices A and B, corresponds to the Lie group. The transformations of this group are defined by the matrix $X(t)$, which is accessible from $X(0) = I$ by the system $\dot{X}(t) = (u_1(t)A + u_2(t)B)X(t)$. A homogeneous BS with bounded $|u_i(t)| \leq \delta$ assumes its dynamics with inverse time (Hirschorn, 1977), therefore an accessibility set of such system has a group structure.

L-systems. This class of dynamical systems has been introduced in the context of control problems; see Yakovenko (1972). Every L-system corresponds to a local finite-dimensional Lie group G of state space transforms and some Lie algebra with structural constants C_{ij}^k. The L-system defines a set of structural constants of nonsingular transformation of variables in the state space and the control space.

Definition 1.4. *An L-system is said to be an autonomous nonsingular system in the form*

$$\dot{x}^i(t) = f_j^i(x)u^j, \quad i, j = 1, \ldots, n, \ u \in U \subset \mathbb{R}^n, \tag{1.6}$$

where $[F_i, F_j] = C_{ij}^k F_k$; $F_j = f_j^i(x)(\partial/\partial x^i)$; $C_{ij}^k = \text{const}$; $|f_j^i| \neq 0$; $f_j^i(x)$ is a twice continuous differentiable function in the region $M \subset \mathbb{R}^n$.

An algebra L of the L-system is defined by the tensor C_{ij}^k. In the case of an L-system the equations can be obtained for a transformation group G. Let us assume $|u^j(t)| \leq 1$. Suppose that $x(t)$ is a solution of the system (1.6) under some fixed u_j and the initial condition $x^i(0) = x_0^i$. The solution has the form: $x^i = \Phi^i(x_0, v)$. If $u^j = v^j(t)$, then x^i is a parametric family of a state space transform $x^i = \Phi^i(x_0, v)$ with parameters v^j. This is a transform of a local n-parametric group G, which corresponds to an algebra L.

In particular, such a system describes ion channels of a biological membrane (Chinarov et al., 1990; see also Chapter 7). In this case, ion concentrations and electric potentials play the role of a control. We note these linear systems are a special case of L-systems and BS, and BS can be locally studied as an $n^2 + n$ dimensional L-system. BS can be transformed to right-invariant systems on a matrix group. The systems with a finite-dimensional algebra L can often be reduced to right-invariant or L-systems.

By this means, the usefulness of differential-geometric methods in the theory of nonlinear control stipulates the possibility of a strong mathematical formulation and solution of the main problems. The primary motivations for a Lie group approach are the generality of the analysis and synthesis techniques and the success of such an approach in the area of system identification. The theory of Lie algebras plays the same role as linear algebra in the theory of linear control systems.

Control systems and fiber bundles. If the control system (CS) (1.1) is defined on a smooth manifold M of C^r ($r \geq 1$) class, then a geometric structure of the CS is defined by a smooth manifold M, the set $\widetilde{B}$ with the map π: $\widetilde{B}$ in M, and the fiber $\pi^{-1}(x)$ of the set $\widetilde{B}$ at the point x. In the capacity of the fiber, $\widetilde{B}_x$ can be used as a vector space, which is defined by a group G of a matrix BS. Let M be a compact manifold; $\widehat{S}$ is a covering of a compact by sets M_i, and the system (1.1) is equivalent to a BS on each set. Then, the discrete-continuous structure of a CS is defined by a finite cycle of a local BS. The transition from one to another BS is realized by a discrete manner (a finite automaton (Hopcroft and Ullman, 1979)). As this takes place, a covering is considered as a discrete space, which can be identified by a set M_i with points of a discrete partially ordered set. This discrete space represents a base of fiber bundles with a Lie group of BS vector fields on a tangent space.

We define on a space state D of an automaton the skew-symmetric functions $f^{\tilde{r}}$ $(p_0, \ldots, p_{\tilde{r}})$ of $(\tilde{r} + 1)$ arguments, which pass through a set of various elements in D. The values of these functions belong to the Abelian group $\mathfrak{F}$, which is defined as an algebraic r-dimensional complex of a discrete space D. Let us suppose that a value of the function $f^{\tilde{r}}(p_0, \ldots, p_{\tilde{r}})$ is different from zero if and only if $p_0, p_1, \ldots, p_{\tilde{r}}$ represents an element of some ordered subset of a partially ordered set D. The boundary $g_u f^{\tilde{r}}$ of an algebraic complex $f^{\tilde{r}}$ is defined by the $(r-1)$-dimension algebraic complex $f^{\tilde{r}-1}$:

$$g_u f^{\tilde{r}} \equiv f^{\tilde{r}-1}(p_0, \ldots, p_{\tilde{r}-1}) = \sum_p f^{\tilde{r}}(p_0, \ldots, p_{\tilde{r}}),$$

where the summation is taken by all $p \in D$.

We can introduce for CS other algebra-geometric behaviors of cycles, in particular, homology, Betti group, and so on. These constructions are very important for the system analysis of equation (1.1). We point out that the compact set M can be reflected by ε-shift into a barycentric subpartition of any multiplicative ε-covering (open or closed), which is considered as some polytope. Hence, the system (1.1) can be simulated by a discrete-continuous system with a polytope structure of the state space, and the geometry of the system (1.1) describes the collection of a simplicial complex and a Lie group of a matrix BS, which act on fibers connected with the compact set M.

1.4 Bilinear Logic-Dynamical Realization of Nonlinear Control Systems

Consider the system

$$\dot{y}(t) = b_0(y) + \sum_{i=1}^{h} u_i(t) b_i(y),$$

$$z(t) = f(y(t)), \quad y(0) = y^0, \quad u(t) \in \Omega, \quad y \in Y, \qquad (1.7)$$

where $y = (y_1, \ldots, y_n)$ is a state vector; $z = (z_1, \ldots, z_n)$ is a vector of sensor outputs; $b_0(y), \ldots, b_n(y)$ are analytical vector fields; f is an infinite differentiable $\mathbb{R}^1$ vector-function; Y is a compact manifold, and $u(t) \in \Omega = \{u : |u_i| \leq 1, \ i = 1, \ldots, h\}$.

By using coordinate transformations we want to construct a logic-dynamical system (i.e., a system describing the processes evolving according to continuous dynamics, discrete dynamics, and logic rules).

Consider the system

$$\dot{x}(t) = \sum_{j=1}^{r} L_j \left[A_{0j} + \sum_{i=1}^{h} u_1(t) A_{ij} \right] x(t),$$

$$\omega(t) = \sum_{j=1}^{r} L_j C_j x(t), \quad x(0) = x, \quad u(t) \in \Omega, \qquad (1.8)$$

where L_j is a logic function, which can be realized by a finite automaton; $x = (x_1, \ldots, x_m)$ is a state vector; $\omega = (\omega_1, \ldots, \omega_l)$ is an output function; $A_{0j}, \ldots, A_{hj}$ are known real system $(m \times m)$-matrices; C_j is an $l \times m$-matrix; $u(t) = (u_1(t), \ldots, u_h(t))$ is a control function; $u(t) \in \Omega\{u, |u_i| \leq 1, \ i = 1, \ldots, h\}$. The system (1.8) satisfies the condition $\omega(t) = z(t)$, $t \in [0, T]$ with a fixed control function $u(t)$.

Consider the matrix equation

$$\dot{X}(t) = \left(A_0 + \sum_{i=1}^{h} u_i(t) A_i \right) X(t),$$

$$W(t) = CX(t), \quad X(0) = I, \quad u(t) \in \Omega, \qquad (1.9)$$

where $X(t)$ is a matrix, which evolves in $Gl(m, \mathbb{R})$, of invertible $(m \times m)$ matrices. Each column of this equation is a system in the form (1.7).

The Lie algebra of the group $Gl(m, \mathbb{R})$ is finite-dimensional over the real field $\mathbb{R}$. There is a closed Lie subgroup G of $Gl(m, \mathbb{R})$ which corresponds to the subalgebra g of the algebra $gl(m, \mathbb{R})$. This algebra is defined by the Lie bracket and the matrices $\{A_0, \ldots, A_h\}$ are characterized by the solution of the equation

$$\dot{X}(t) = \left(\sum_{i=1}^{h} u_i(t) A_i \right) X(t),$$

$$(X(0) = I, \quad |u_i| \leq 1, \quad i = 0, \ldots, h).$$

The group G contains the set of all accessible matrices of (1.9). The set of accessible matrices of the system is a subset of G with nonempty interior in the relative topology of G, hence G is the smallest subgroup of $Gl(m, \mathbb{R})$ containing all accessible matrices of (1.9).

Let S_j be some neighborhood of the point y_j^0; then $W_j(S_j)$ is a minimal subalgebra of the Lie algebra C^∞ of all vector fields on S_j over $\mathbb{R}$ containing $\{b_0, \ldots, b_h\}$, and a submanifold Y_j containing y_j^0, is an integral manifold $\widetilde{W}_j(S_j)$, whereas the dimension of Y_j is equal to the rank $\widetilde{W}_j(S_j)$ at the y_j^0. Then, according to Chow's theorem, the set of all points tY_j is accessible by the system (1.7) from y_j^0.

Because Y is a compact manifold, there exist submanifolds Y_j', such that $Y = \cup_{j=1}^{r} Y_j'$. If the subalgebra $\widetilde{W}_j(Y_j')$ is finite-dimensional, then there exists a Lie subalgebra g_j of the algebra $gl_j(m_j, \mathbb{R})$ for some m_j, and according to the Ado's theorem (Ado, 1947), an isomorphism of Lie algebras $\varphi_j \colon \widetilde{W}_j(Y_j') \mapsto g_j$. We define the matrix bilinear system (1.9) by the map $A_{ij} = \varphi_j(b_i)$. Let l_j be the map

$$l_j : W_j(Y_j') \mapsto \widetilde{W}_j(y_j^0),$$

such that $l_j(c) = c(y_j^0)$ for $c \in \widetilde{W}_j(Y_j')$. Then the linear map $l_j' = l_j \circ \varphi_j^{-1}$ satisfies the condition

$$l_j' = ([A_{i_1 j} \dots [A_{i_{\nu-1} j}, A_{i_\nu j}] \dots]) = [b_{i_1 j} \dots [b_{i_{\nu-1} j}, b_{i_\nu j}] \dots](y_j^0)$$

for any ν_i, $0 \le i_1, \dots, i_\nu \le h$. By Krener's theorem (Krener, 1975), there exists a neighborhood M of I and maps $\lambda_j \colon M_j \mapsto Y_j'$, that preserve the solutions.

By Brockett's theorem (Brockett, 1972), we can find the following result. If the equation (1.7) satisfies the above-stated conditions and the map $f \circ \lambda_j \colon X \mapsto Z$ is polynomial, then there exists a logic-dynamical realization (1.8) of $u(t) \mapsto \omega(t)$ and a constant $T \ge 0$, such that for any input $u(t)$, the corresponding outputs satisfy $\omega(t) = z(t)$ for $t \in [0, T]$.

Remark 1.1. The dimension of a state space of LDS is the maximal dimension of Euclidean space, corresponding to some submanifold M_j.

We define a logic variable L_j for each integral submanifold Y_j' of the compact space state Y by the following,

$$L_j = \begin{cases} 0, & \text{if } y \in Y_j', \quad j = 1, \dots, r, \\ 1, & \text{if } y \notin Y_j' \quad \text{otherwise.} \end{cases} \tag{1.10}$$

We suppose that the logic function L_j can be realized by a finite automaton. For each value $z_i \in Z$, $i = 1, \dots, r$ we can find a submanifold Y_t by the map $\gamma_t \colon T \times Y \mapsto Z$. This map satisfies the condition

$$\gamma_t(Y_j') = z_j, \quad Y_i' \cap Y_j' = \phi, \quad i \ne j.$$

If the system (1.7) satisfies the above hypothesis, then there exists a logic-dynamical system (1.8), such that for any input $u(t)$, the corresponding outputs satisfy $z(t) = \omega(t)$, $t \in [0.T]$.

2.   Global Bilinearization of Nonlinear Systems

In this section, we are concerned with the following nonlinear control system, with control (u, v) appearing linearly: for $t \in [0, T], T \subset \mathbb{R}$,

$$\begin{aligned} \dot{x}(t) &= f(x(t)) + G(x(t))u(t), \\ z(t) &= h(x(t)) + Q(x(t))v(t), \end{aligned} \tag{1.11}$$

where x (state), z (output), u, and v (inputs) are n, p, m, and q-dimensional vector-valued functions of time, respectively; f and h are n- and p-dimensional vector-valued functions of $x(t)$, respectively; and G and Q are matrix-valued of $x(t)$ of appropriate sizes.

The main result of this section is a necessary and sufficient condition for this system to have a dynamically equivalent bilinear system of the following form,

$$\dot{y}(t) = \left(A + \sum_{i=1}^{m} B_i u_i(t) \right) y(t),$$

$$z(t) = \left(C + \sum_{i=1}^{q} D_i v_i(t) \right) y(t), \tag{1.12}$$

where A, B_i, C, D_i, are constant matrices of appropriate sizes, and for some positive integers M_i, $i = 0, \dots, q$,

$$\text{rank}\,[C', A'C', \dots, (A')^{M_0-1}C', D_1', A'D_1', \dots, (A')^{M_1-1}D_1,$$
$$\dots, D_q', A'D_q', \dots, (A')^{M_q-1}D_q'] = \dim A. \tag{1.13}$$

Bilinear systems have been extensively studied in recent years for three primary reasons. First, it has been shown that bilinear systems are feasible mathematical models for large classes of problems of practical importance. Second, bilinear systems provide higher-order approximations to nonlinear systems than do linear systems. Third, bilinear systems have rich geometric and algebraic structures that promise a fruitful field of research.

As a final remark, we note that some necessary and sufficient conditions were given in Krener (1973) for two systems of the form (1.1) with $z = x$ to be locally equivalent.

Some of the results on stochastic systems, which are analogous to some of those given here, can be found in Lo (1975).

A necessary and sufficient condition. In a paper by Brockett (1976), it was reported that bilinear systems are capable of representing a wide variety of highly nonlinear models. Motivated by his results, in this section we derive a necessary and sufficient condition for a nonlinear system (1.11) to have the same input-output mapping as does a bilinear system (1.11) and (1.12). When the condition is satisfied, a procedure to construct such a bilinear system is provided in the proof of the sufficiency part of the condition.

To initiate the mathematical development, several definitions are in order. The reader is referred to Kalman et al. and Brockett (Kalman and Bucy, 1961; Kalman, Falb, and Arbib, 1969) for terminologies unspecified here.

Definition 1.5. *Given an initial state* $x(0) = x_0$ *and an input function* (u, v) *on* T, *the control system (1.11) produces a corresponding output function z on* T. *Thus for an initial state x_0, the system (1.11) defines an* input-output mapping. *Two control systems are said to be* dynamically equivalent *if, for appropriate initial states, they have the same input-output mapping.*

Definition 1.6. *Let L be the differentiation operator defined by*

$$L(g(x)) = \sum_{i=1}^{n} f_i(x) \frac{\partial}{\partial x_i}(g(x)) = g_0(x)f(x),$$

for a differentiable scalar function $g(x)$, where $g_0(x)$ is the gradient (a row vector) of g. When g is a vector function, $L(g)$ denotes the gradient of each component of g. We note that $L(g)$ is a vector function and $L(g_x)$ is a matrix function.

If h is infinitely differentiable, the set of functions $\{h(x(t)), Lh(x(t)), L^2h(x(t)), \ldots\}, \cup[\cup_{i=1}^{q}\{Q_i(x(t)), LQ_i(x(t)), L^2Q_i(x(t)), \ldots\}]$, where Q_i denotes the ith column of Q, is called the *sensor orbit* of the system (1.11) at the time t.

Remark 1.2. The notion of sensor orbits for stochastic systems was introduced in Bucy and Joseph (1968) to deduce a set of suboptimal filtering equations. Its application to optimal estimation was discussed in Lo (1973a, 1975). In defining sensor orbits for stochastic systems, the differentiation operator corresponding to L above, is a Kolmogorov backward operator.

Definition 1.7. *The system (1.11) is said to have a finite-dimensional sensor orbit, if there exist integers M_i, $i = 0, \ldots, q$, such that for $k = 1, \ldots, q$, and all state trajectories $x(t)$, $t \in T$,*

$$L^{M_0}h(x(t)) = \sum_{i=0}^{M_0-1} A(0,0,i+1)L^ih(x(t))$$

$$+ \sum_{j=1}^{q} \sum_{i=0}^{M_q-1} A(0,j,i+1)L^iQ_j(x(t)),$$

$$L^{M_k}Q_k(x(t)) = \sum_{i=0}^{M_0-1} A(k,0,i+1)L^ih(x(t))$$

$$+ \sum_{j=1}^{q} \sum_{i=0}^{M_q-1} A(k,j,i+1)L^iQ_j(x(t)),$$

where $A(i,j,k)$ are constant $p \times p$ matrices, and every column of $(L^ih(x(t)))_x \ G(x(t))$ and $(L^iQ_j(x(t)))_x \ G(x(t))$, $i = 0, \ldots, (M_j - 1)$, $j = 1, \ldots, q$, lies on the sensor orbit; that is, for $k = 1, \ldots, m$, and all state trajectories $x(t)$, $t \in T$, the kth column of

$$(L^i h(x(t)))_x G(x(t)) = \sum_{j=0}^{M_q} B_0(k,i,0,j) L^{j-1} h(x(t))$$

$$+ \sum_{j=1}^{q} \sum_{i=0}^{M_q-1} B_0(k,i,j,l) L^{l-1} Q_j(x(t)),$$

is the kth column of

$$(L^i Q_j(x(t)))_x G(x(t)) = \sum_{l=1}^{M_0} B_j(k,i,0,l) L^{l-1} h(x(t))$$

$$+ \sum_{r=1}^{q} \sum_{l=1}^{M_q} B_j(k,i,r,l) L^{l-1} Q_r(x(t)).$$

We are now in a position to state our main result.

Theorem 1.1. *The nonlinear system (1.11), with control appearing linearly, is dynamically equivalent to the bilinear system (1.12) and (1.13), if and only if the nonlinear system has a finite-dimensional sensor orbit.*

When this condition is satisfied, a procedure for constructing a dynamically equivalent bilinear system in an observability canonical form is suggested in the following proof of sufficiency.

Proof: *Sufficiency.* Set $y_{0i}(t) = L^{i-1} h(x(t))$ and

$$y_{ji}(t) = L^{i-1} Q_j(x(t)).$$

By the chain rule of differentiation, for $i = 1, \ldots, M_j$, $j = 1, \ldots, q$,

$$\dot{y}_{ji}(t) = y_{ji+1}(t) + (L^{i-1} Q_j(x(t)), G(x(t)) u(t)),$$
$$\dot{y}_{0i}(t) = y_{0i+1}(t) + (L^{i-1} h(x(t))_x G(x(t)) u(t)).$$

Because the system (1.11) has a finite-dimensional sensor orbit for $j = 0, \ldots, q$,

$$\dot{y}_{jM_j} = \sum_{k=0}^{q} \sum_{i=1}^{M_q} A(j,k,i) y_{ki}(t) \sum_{i=1}^{m} \sum_{k=0}^{q} \sum_{l=1}^{M_k} B_j(i, M_j, k, l) y_{kl}(t) u_i(t)$$

and, for $r = 1, \ldots, M_j - 1$, $j = 0, \ldots, q$,

$$\dot{y}_{jr} = y_{jr+1}(t) dt + \sum_{i=1}^{m} \sum_{k=0}^{q} \sum_{l=1}^{M_k} B_j(i, r, k, l) y_{kl}(t) u_i(t),$$

where $A(j, k, i)$ and $B_j(i, r, k, l)$ are constant matrices. Let
$$\dot{y} = [y'_{01}, \ldots, y'_{0M_0}, \ldots, y'_{q1}, \ldots, y'_{qM_0}]'.$$

Simple observation yields the first equation of (1.12), where

$$A = [A_{ij}],$$

$$A_{ii} = \begin{bmatrix} 0 & & I & & \\ & \ddots & & \ddots & \\ & & 0 & & I \\ A(i, i, l) & A(i, i, 2) & \cdots & & A(i, i, M_i) \end{bmatrix} \text{ for } i = 0, \ldots, q,$$

$$A_{ij} = \begin{bmatrix} 0 \\ A(i, j, 1) & A(i, j, 2) & \cdots & A(i, j, M) \end{bmatrix} \text{ for } i, j = 0, \ldots, q \text{ and } i \neq j,$$

$$B_i = \begin{bmatrix} B_0(i, 0) & B_0(i, 1) & \cdots & B_0(i, q) \\ B_1(i, 0) & B_1(i, 1) & \cdots & B_1(i, q) \\ \vdots & & & \vdots \\ B_q(i, 0) & \cdots\cdots & \cdots & B_q(i, q) \end{bmatrix},$$

$$B_j(i, k) = \begin{bmatrix} B_j(i, 1, k, 1) & B_j(i, 1, k, 2) & \cdots & B_j(i, 1, k, M_i) \\ B_j(i, 2, k, 1) & \cdots\cdots & \cdots & B_j(i, 2, k, M_i) \\ \vdots & & & \vdots \\ B_j(i, M_k, k, 1) & \cdots\cdots & \cdots & B_j(i, M_k, k, M_k) \end{bmatrix}.$$

It follows immediately from the second equation of (1.11) that the second equation of (1.12) holds, where
$$C = [I \quad 0 \quad \cdots \quad 0],$$
$$D = [\underbrace{0 \quad \cdots \quad 0}_{(\sum_{j=0}^{i-1} M_i)p} \quad I \quad 0 \quad \cdots \quad 0].$$

Straightforward calculation shows that (1.13) is true. This completes the proof of sufficiency.

Necessity. As the bilinear and nonlinear systems are dynamically equivalent, we have, for a control function v,
$$z(t) = h(x(t)) + Q(x(t))v(t) = \left(C + \sum_{i=1}^{p} D_i v_i(t)\right) y(t).$$

Setting $v = 0$, we have for all state trajectories $x(t)$, $t \in T$,
$$h(x(t)) = Cy(t). \tag{1.14}$$

Setting $v_i = 1$, we have for all state trajectories $x(t)$, $t \in T$,
$$Q_i(x(t)) = D_i y(t), \quad i = 1, \ldots, q. \tag{1.15}$$

Differentiating (1.14) and (1.15) with respect to time, we obtain, in view of (1.12), for $i = 2, 3, \ldots,$ and $k = 1, \ldots, q,$

$$L^i h(x(t)) = CA^i y(t),$$

$$(L^{i-1} h(x(t)))_x G(x(t)) = CA^{i-1}[B_1 y(t), \cdots, B_m y(t)].$$

$$L^i Q_i(x(t)) = D_k A^i y(t),$$

$$(L^{i-1} Q_k(x(t)))_x G(x(t)) = D_k A^{i-1}[B_1 y(t), \cdots, B_m y(t)].$$

By the Cayley–Hamilton theorem (Gradshteyn and Ryzhik, 2000), there exist constants c_i, $i = 0, \ldots, N$, such that $c_N \neq 0$ and $\sum_{i=0}^{N} c_i A^i = 0$. Hence for $k = 1, \ldots, q,$

$$\sum_{i=0}^{N} c_i L^i h(x(t)) = \sum_{i=0}^{N} c_i CA^i y(t) = 0,$$

$$\sum_{i=0}^{N} c_i L^i Q_k(x(t)) = \sum_{i=1}^{N} c_i D_k A^i y(t) = 0. \qquad (1.16)$$

Consider the jth column, say $CA^i B_j y(t)$, of $(L^i h(x(t)))_k G(x(t))$. Because of (1.13), the kth column of $B_j'(A')^i C'$ can be expressed as

$$\sum_{l=1}^{M_0} (A')^{l-1} C' B_0'(j, i, 0, l, k) + \sum_{r=1}^{q} \sum_{l=1}^{M_0} (A')^{l-1} D_r' B_0'(j, i, r, l, k)$$

for some constant p-vectors $B_0'(j, i, r, l, k)$. Hence

$$B_j'(A')^i C' = \sum_{l=1}^{M_0} (A')^{l-1} C' B_0'(j, i, 0, l) + \sum_{r=1}^{q} \sum_{l=1}^{M_0} (A')^{l-1} D_r' B_0'(j, i, r, l),$$

where $B_0(j, l, r, i) = [B_0'(j, i, r, l, 1), \ldots, B_0'(j, i, r, l, p)]$ and

$$CA^i B_j y(t) = \sum_{l=1}^{M_0} B_0(j, i, 0, l) L^{l-1} h(x(t))$$

$$+ \sum_{r=1}^{q} \sum_{l=1}^{M_r} B_0(j, i, r, l) L^{l-1} Q_r(x(t)). \qquad (1.17)$$

Similarly, the kth column of $(L^i Q_j(x(t)))_x$, $G(x(t))$,

$$D_j A^{i-1} B_k y(t) = \sum_{l=1}^{M_0} B_j(k, i, 0, l) L^{l-1} h(x(t))$$

$$+ \sum_{r=1}^{q} \sum_{l=1}^{M_q} B_j(k, i, r, l) L^{l-1} Q_r(x(t)),$$

for some constant matrices $B_j(k, i, r, l)$. This together with (1.16) and (1.17) completes the proof of necessity.

In the following, we look at an interesting special case of this theorem for which the proof follows directly from the theorem and is omitted.

Let $V_1, \ldots, V_r$ and $W_1, \ldots, W_s$ be vector spaces over the same field, and let the maps

$$\varphi_r : V_1 \times \cdots \times V_r \to W_1,$$
$$\Phi_r : V_1 \times \cdots \times V_r \to W_1 \times \cdots \times W_s$$

be r-linear; that is, for all α and β in the field, and $i = 1, 2, \ldots, r$,

$$\varphi_r(v_1, \ldots, \alpha v_i + \beta v_i^*, \ldots, v_r)$$
$$= \alpha \varphi_r(v_1, \ldots, v_i, \ldots, v_r) + \beta \varphi_r(v_1, \ldots, v_i^*, \ldots, v_r),$$
$$\Phi_r(v_1, \ldots, \alpha v_i + \beta v_i^*, \ldots, v_r)$$
$$= \alpha \Phi_r(v_1, \ldots, v_i, \ldots, v_r) + \beta \Phi_r(v_1, \ldots, v_i^*, \ldots, v_r),$$

respectively.

Corollary 1.1. *The nonlinear system,*

$$\dot{x}(t) = \left(F + \sum_{i=1}^{m} G_i u_i(t) \right) x(t),$$

$$z(t) = \sum_{r=1}^{p} \varphi_r(x(t), \ldots, x(t)) + \sum_{r=1}^{q} \Phi_r(x(t)) v(t),$$

where F and G_i are constant matrices, has a finite-dimensional sensor orbit and is dynamically equivalent to the bilinear system,

$$\dot{y}(t) = \left(A + \sum_{i=1}^{m} B_i u_i(t) \right) y(t),$$

$$z(t) = \left(C + \sum_{i=1}^{s} D_i v_i(t) \right) y(t), \tag{1.18}$$

where A, B, C, D_i are some constant matrices of appropriate sizes and, for some positive integers M_i, $i = 0, \ldots, q$, (1.13) holds.

3.   Identification of Bilinear Control Systems

Due to the widespread use of bilinear models, there is strong motivation to develop identification algorithms for such systems given noisy observations (Fnaiech, Ljung, and Fliess, 1987; Dang Van Mien and Norman-Cyrot, 1984; Krishnaprasad and Marcus, 1982). Fnaiech, Ljung and Fliess's paper (1987) presents methods for parameter identification of bilinear systems. These methods are directly transferred from linear system identification methods, such as least squares and recursive prediction error methods. A conjugate gradient method for identification of bilinear systems has been developed by Bose and Chen (1995). Most studies of the identification problem of bilinear systems have assumed an input-output formulation. Standard methods such as recursive least squares, extended least squares, recursive auxiliary variable, and recursive prediction error algorithms, have been applied to identifying bilinear systems.

In this section we describe an identification method based on the expansion of signal processes over an orthogonal basis. Using this methodology we can obtain a system of linear algebraic equations, which is used to determine the coefficients of the bilinear model. By means of the least squares method we obtain estimates of the unknown parameters of the model. The computational algorithm obtained has quite good accuracy. An algorithm for identification of the bilinear discrete models is described in Section 7 of this chapter. It is based on a discrete approximation of the input-output map of a nonlinear object (Yatsenko, 1984).

Consider the bilinear model

$$\dot{x}(t) = Ax(t) + Lu(t) + \sum_{j=1}^{n} B_j x(t) u_j(t), \qquad (1.19)$$

where A, L, and B_j are unknown parameters to be estimated; u is a control. By the generalized product of orthogonal series we mean

$$u_j(t) = \sum_{t=0}^{m-1} u_{jl} t^l,$$

$$x(t)u_j(t) = \sum_{l=0}^{m-1} u_{jl} X t^l \Pi(t) = \sum_{l=0}^{m-1} u_{jl} X R_l \Pi(t).$$

The integration of (1.19) gives

$$x(t) - x(0) = A \int_0^t x(t')dt' + L \int_0^t u(t')dt'$$

$$+ \sum_{j=1}^{n} L_j \int_0^t x(t')u_j(t')u_j(t')dt'. \qquad (1.20)$$

Using this result, we obtain

$$X\Pi - X(0)\Pi = AXE\Pi LUE\Pi + \sum_{j=1}^{n} B_j X \left[\sum_{t=0}^{m-1} u_{jl} R_l\right] E\Pi. \qquad (1.21)$$

Substituting the expression for Θ into (1.21) gives (for more details one should refer to Section 3.1.5 in Chapter 3)

$$XG\Pi - \sum_{j=1}^{n} X(0)G\Pi = AXEG\Pi + LUEG\Pi$$

$$+ \sum_{j=1}^{n} B_j X[\sum_{j=1}^{m-1} u_j R_j]EG\Pi(t)$$

or

$$XG - X(0)G = AXEG + LUEG + \sum_{j=1}^{n} B_j X[\sum_{l=0}^{m-2} u_{jl} R_l]EG,$$

$$ZS = (X - X(0))G, \qquad (1.22)$$

where Z is the parameter vector; that is,

$$Z = [ALB_1 B_2 \ldots B_n]. \qquad (1.23)$$

4.   Bilinear and Nonlinear Realizations of Input-Output Maps

4.1   Systems on Lie Groups

Instead of considering (1.5), it is useful to consider the matrix bilinear system

$$\dot{X}(t) = \left(A_0 + \sum_{i=1}^{h} u_i A_i\right) X(t),$$

$$W(t) = CX(t), \quad X(0) = I, \quad u(t) \in \Omega, \qquad (1.24)$$

where $X(t)$ takes values in the group $Gl(m, \mathbb{R})$ of all invertible $m \times m$ matrices.

Each column of the matrix equation (1.24) is a system of the form (1.5). Therefore instead of considering the problems of replacing or approximating (1.7) by (1.1), we study the equivalent problem of replacing or approximating (1.7) by (1.24).

The advantage of considering (1.24) over (1.5) is that $Gl(m, \mathbb{R})$ is a Lie group and each A_j defines a right-invariant vector field $A_j X$ on this group, hence a member of the associated Lie algebra $gl(m, \mathbb{R})$ of all $m \times m$ real matrices. This algebra is finite-dimensional over the field $\mathbb{R}$ and the multiplication is defined by the Lie bracket

$$[A_i, A_j] = A_j A_i - A_i A_j.$$

This is a noncommutative and nonassociative operation that satisfies the skew-symmetry and Jacobi relations,

$$[A_i, A_j] = -[A_j, A_i],$$

and

$$[A_i, [A_j, A_k]] = [[A_i, A_j], A_k] + [A_j, [A_i, A_k]].$$

For further discussion of Lie groups and algebras we refer the reader to Brockett (1973).

There is a unique subalgebra g of $gl(m, \mathbb{R})$ generated by $\{A_0, \ldots, A_n\}$ under bracketing and corresponding to this is a closed Lie subgroup G of $Gl(m, \mathbb{R})$. This subgroup is the set of all products of the form

$$\exp(t_{i_1} A_{i_1}) \cdots \exp(t_{i_k} A_{i_k})$$

for all $k \geq 0$ and $t_{i_j} \in \mathbb{R}$. Another characterization of G is that it is the set of all accessible matrices of

$$\dot{X}(t) = \left(\sum_{i=0}^{k} u_i(t) A_i \right) X(t),$$

$$X(0) = I, \quad |u_i| \leqq 1, \quad i = 0, \ldots, h.$$

This follows from the theorem of Chow (1939).

The dimension of G as a submanifold of $Gl(m, \mathbb{R})$ is precisely the dimension of the Lie subalgebra g. Furthermore, the set of accessible matrices of (1.24) is a subset of G with nonempty interior in the relative topology of G; hence G is the smallest subgroup of $Gl(m, \mathbb{R})$ containing all accessible matrices of (1.24).

The corresponding situation for (1.7) is more complicated because of the nonlinearity. We restrict our discussion of this system to some neighborhood $\mathcal{V}$ of y_0 in $\mathbb{R}$. If $b_i(y)$, $b_j(y)$ are C^∞-vector fields defined on $\mathcal{V}$, then the Lie bracket $[b_i, b_j](y)$ is another C^∞-vector field defined on $\mathcal{V}$ by

$$[b_i, b_j](y) = \frac{\partial b_j}{\partial y}(y) b_i(y) - \frac{\partial b_i}{\partial y}(y) b_j(y).$$

Once again the skew symmetry and Jacobi relations hold.

The set $V(\mathcal{V})$ of all C^∞-vector fields on $\mathcal{V}$ becomes a Lie algebra over $\mathbb{R}$ with this definition, however, in general, it is infinite-dimensional. Let $W(\mathcal{V})$ denote the smallest subalgebra of $V(\mathcal{V})$ containing $\{b_0, \ldots, b_h\}$. In many cases, but not in general, there is a submanifold $\mathcal{N}$ of $\mathcal{V}$ corresponding to $W(\mathcal{V})$, and containing y^n. To be more precise, let $W(y)$ be the linear subspace of $\mathbb{R}^n$ formed by evaluating the vector fields of $W(\mathcal{V})$ at y. A submanifold $\mathcal{N}$ of $\mathcal{V}$ is an integral manifold of $W(\mathcal{V})$ if for every $y \in \mathcal{N}$, $W(y)$ is precisely the tangent space to $\mathcal{N}$ at y. We define the rank of $W(\mathcal{V})$ at y to be the dimension of $W(y)$. If the rank of $W(\mathcal{V})$ is constant or $b_0(y), \ldots, b_h(y)$ are analytic, then there exists an integral manifold $\mathcal{N}$ of $W(\mathcal{V})$ containing y_0.

Henceforth we assume that $\mathcal{N}$ exists, the dimension of $\mathcal{N}$ is the same as the rank of $W(\mathcal{V})$ at y^0, and by Chow's theorem is the set of all points in $\mathcal{V}$ accessible from y^0 under the system

$$\dot{y}(t) = \sum_{i=0}^{h} u_i(t) b_i(y),$$

$$y(0) = y^0, \quad |u_i| \leqq 1, \quad i = Q, \ldots, h.$$

The set of all points in $\mathcal{N}$ accessible from y^n by (1.7) is again a subset of $\mathcal{N}$ with nonempty relative interior.

4.2   Bilinear Realization of Nonlinear Systems

The problem of replacing a nonlinear realization by a bilinear one can be broken into two parts. The first is: when does a change of state exist that linearizes the vector fields $b_0(y), \ldots, b_h(y)$, resulting in a system with bilinear dynamics and nonlinear output map? The second is: given a realization of this hybrid type, when can it be converted into a bilinear realization?

As for the first question, it is known that a family of vector fields $b_0(y), \ldots, b_k(y)$ can be converted to linear vector fields $A_0 x, \ldots, A_h x$ by a change of coordinates $x = x(y)$ in some neighborhood of y^0 if the vector fields are analytic, all vanish at y^0, and generate a finite-dimensional semisimple Lie algebra. Hermann (1973) gave a formal power series construction of the change of coordinates. However, these results are not directly applicable to our questions, because if all the vector fields vanish at y^0, then the system (1.7) is trivial:

$$W(\mathcal{N}) \overset{e}{\to} W(y'),$$

$$\mathrm{ad}\,(W(\mathcal{N})) \subseteq gl(m, \mathbb{R}).$$

As for the second step in bilinearization, we have a theorem of Brockett (1972) which states that every realization with bilinear dynamics and a polynomial output map is equivalent to a realization with bilinear dynamics and linear output map. This results in the following.

Corollary 1.2. *Given: any nonlinear realization (1.7) of the input-output $u(t) \mapsto z(t)$ satisfying the hypothesis of Theorem 1.1. If the map $f = \lambda : X \to z$ is a polynomial, then there exists a bilinear realization (1.7) of $u(t) \mapsto w(t)$ and a constant $T > 0$ such that for any input, the corresponding outputs satisfy $w(t) = z(t)$ for $t \in [0, T]$. (Polynomial here means each component of z is a polynomial in the components of X.)*

4.3   Approximation of Nonlinear Systems by Bilinear Systems

If the Lie algebra $W(\mathcal{N})$ is not finite-dimensional, then Theorem 1.1 does not hold; however, we can ask whether (1.7) can be approximated by systems of type (1.24). To be more precise, given (1.24) carried locally by $\mathcal{M}$ and (1.7) carried locally by $\mathcal{M}$, a C^{∞}-map $\lambda : \mathcal{M} \to \mathcal{N}$ preserves solutions up to order μ if there exists a $T > 0$ and $K \geqq 0$ such that for any solution $X(t)$ and $y(t)$ of (1.24) and (1.7) we have

$$|\lambda(X(t)) - y(t)| \leqq Kt^{\mu+1}$$

for $t \in [0, T]$.

Theorem 1.2. *Suppose that $b_0(y), \ldots, b_h(y)$ of (1.7) are C^{∞} and the system is carried locally by $\mathcal{A}$. Then for any $\mu \geqq 0$ there exists a system (1.24) carried locally by $\mathcal{M}$ in $Gl(m, \mathbb{R})$ and a C^{∞}-map $\lambda : \mathcal{M} \to \mathcal{N}$ preserving solutions up to order μ.*

Proof: An abstract Lie algebra g is a vector space over $\mathbb{R}$ that a multiplication which satisfies the skew symmetry and Jacobi relations. Suppose $a_0, \ldots, a_h$ are elements of g; then we call $[a_{i_1} \cdots [a_{i_{v-1}}, a_{i_v}] \cdots]$ a bracket of order v of $a_0, \ldots, a_h$. One way to construct an abstract Lie algebra g is to consider $a_0, \ldots, a_h$ to be elements of the algebra and linearly independent over $\mathbb{R}$. Then treat all the brackets of these up to and including order v as new elements of g that are linearly independent except for those relations implied by the skew symmetry and Jacobi relations. All brackets of order greater than v are taken to be 0. The result is a finite-dimensional Lie algebra that we call the *canonical algebra of order v with $h + 1$ generators*.

By Ado's theorem, this algebra is isomorphic to a subalgebra of $gl(m, \mathbb{R})$ which we also denote g. Under this identification, each a becomes an $m \times m$ matrix A_i, and these matrices are used to construct (1.24). We call the resulting system *the canonical system of order μ with h controls.*

Next we define a linear map $l : g \to \mathbb{R}$ by setting

$$l\big([A_{i_1} \cdots [A_{i_{v-1}}, A_{i_v}] \cdots]\big) = [b_{i_1} \cdots [b_{i_{v-1}}, b_{i_v}] \cdots](y^0).$$

It then follows from a theorem in Krener (1973) that there exists a neighborhood $\mathcal{M}$ of I in the subgroup G of $Gl(m, \mathbb{R})$ carrying (1.24), a neighborhood $\mathcal{N}$ of y^0 in the manifold carrying (1.7), and a C^{inf} map λ: $\mathcal{M} \to \mathcal{N}$ that *preserves solutions* to order μ.

Theorem 1.3. *Suppose that $b_0(y), \ldots, b_h(y)$ of (1.7) are analytic and the system is carried locally by $\mathcal{N}$. There exists a system (1.24) carried locally by $\mathcal{M}$ in $Gl(m, \mathbb{R})$ and an analytic map $\lambda : \mathcal{M} \to \mathcal{N}$ preserving solutions if and only if the Lie algebra generated by $b_0(y), \ldots, b_h(y)$ is finite-dimensional when restricted to $\mathcal{N}$.*

    Proof: Assume $W(\mathcal{N})$ is finite-dimensional. Then by Ado's theorem (Ado, 1947) there exist a Lie subalgebra g of $gl(m, \mathbb{R})$ for some m and a Lie algebra isomorphism $\varphi : W(\mathcal{N}) \to g$. Define a system with matrix bilinear dynamics, (1.24), by letting $A_i = \varphi(b_i)$. Let e be the evaluation map $e \colon W(\mathcal{N}) \to W(y^0)$, defined by $e(c) = c(y^0)$ for $c \in W(\mathcal{N})$. Then the map $l = e \circ \varphi^{-1}$ satisfies the following,

$$l([A_{i_1} \cdots [A_{i_{v-1}}, A_{i_v}] \cdots]) = [b_{i_1} \cdots [b_{i_{v-1}}, b_{i_v}] \cdots](y^0)$$

for any v and $0 \leq i_1, \ldots, i_v \leq h$.

    It follows from a theorem in Krener (1973) that there exist a neighborhood $\mathcal{M}$ of I and a map $\lambda \colon \mathcal{M} \to \mathcal{N}$ preserving solutions.

Remark 1.3. In general, the map λ is locally a projection from $\mathcal{M}$ onto $\mathcal{N}$. However, if the evaluation map $e \colon W(\mathcal{N}) \to W(y^0)$ is a vector space isomorphism, then so is I and the above-mentioned theorem implies λ is a local diffeomorphism.

Corollary 1.3. *Given any nonlinear realization (1.7) of the input-output map $i(t) \mapsto z(t)$ and any integer $\mu \geq 0$, there exists a bilinear realization (1.5) of $u(t) \mapsto w(t)$ and constants M and $T > 0$ such that for any input $u(t)$ the corresponding outputs satisfy*

$$[w(t) - z(t)] \leq Mt^{\mu+1} \quad for \ t \in [0, T].$$

    Proof: Using Theorem 1.2, we construct, a system with the matrix bilinear dynamics and a map $\lambda \colon \mathcal{M} \to \mathcal{N}$ that preserves solutions to order μ. We define a polynomial output map ψ for this system by letting ψ be the power series expansion around I of $f \circ \lambda$ up to and including terms of order μ. Using Brockett's technique (Brockett, 1972), an equivalent system with bilinear dynamics and linear output map can always be constructed, so all we need to show is that our system with bilinear dynamics and polynomial output map approximates (1.7) as required.

5.   Controllability of Bilinear Systems

Bilinear systems evolve as natural models or as accurate approximations to numerous dynamical processes in engineering, economics, biology, ecology, and so on, and in other uses bilinear control may be implemented to improve controllability of an otherwise linear system (Goka, Tarn, and Zaborszky, 1973; Butkovskiy, 1991; Khapalov and Mohler, 1996). There are a large number of publications related to the reachability and controllability properties of the following BS,

$$\dot{x} = Ax + \sum_{j=1}^{m}(B_j x + b_i)u_j, \quad x(0) = x_0, \quad t > 0. \qquad (1.25)$$

Here A, B_j are $n \times n$ matrices, $b_i \in \mathbb{R}^n, j = 1, \ldots, m$, $u = (u_1, \ldots, u_m)$ is an m-dimensional control, and $u \in L^2(0, \inf, \mathbb{R}^n)$ There are a large number of publications related to this topic. Assuming the existence of a pair of "connected" stable and unstable equilibrium points at which the vector field is nonsingular, sufficient conditions for complete controllability of (1.25) by bounded controls were derived in Rink and Mohler (1968). The complete account of the qualitative controllability behavior of the planar BS was given in Koditschek and Narendra (1985). Geometrical properties of reachable sets and their time-evolution for equicontinuous controls were studied in Rink and Mohler (1971). Closedness and connectivity under the sole assumption that a given control set is compact were the subjects of interest in Susmann (1972). Sufficient conditions for convexity of reachable sets of the BS of special type were discussed in Brockett (1985). An extensive analysis of controllability of the systems described by the vector fields on manifolds (inspired by BS) by differential geometric methods was given in Koditschek and Narendra (1985), Boothby and Wilson (1979); Gauthier, Kupka, and Sallet (1984); Gauthier and Bornard (1982), and Jurdjevic and Sallet (1984). Reachable sets and controllability of bilinear time-invariant systems were studied in Khapalov and Mohler (1996).

In this section, we consider a bilinear control system

$$\dot{x} = Ax + uBx + vb, \quad u, v \in \mathbb{R}, \quad x \in \mathbb{R}^n. \qquad (1.26)$$

We are interested in when the system (1.26) is locally controllable at the origin (Susmann and Jurdjevic, 1989; Butkovskiy, 1991; Susmann, 1998); that is,

$$0 \in int\, \mathfrak{A}_0(t) \quad \forall t > 0,$$

where $\mathfrak{A}_0(t)$ is the attainable set at the time t. Negation of necessary conditions for geometric optimality gives sufficient conditions for local controllability. Now we apply second-order conditions (Agrachev and Sachkov, 2004) to our system. Suppose that

$$0 \in \partial \mathfrak{A}_0(t) \quad \text{for some } t > 0.$$

Then the reference trajectory $x(t) \equiv 0$ is geometrically optimal, thus it satisfies the Pontryagin Maximum Principle (PMP). The control-dependent Hamiltonian is

$$h_{u,v}(p,x) = pAx + upBx + upb, \quad \lambda = (p,x) \in T^*\mathbb{R}^n = \mathbb{R}^{n*} \times \mathbb{R}^n.$$

The vertical part of the Hamiltonian system along the reference trajectory $x(t)$ reads:

$$\dot{p} = -pA, \quad p \in \mathbb{R}^{n*}. \tag{1.27}$$

It follows from PMP that

$$p(\tau)b = p(0)e^{-A\tau}b = 0, \quad \tau \in [0,t]; \tag{1.28}$$

that is,

$$p(0)A^i B = 0, \quad i = 0, \dots, n-1, \tag{1.29}$$

for some covector $p(0) \neq 0$, thus

$$\operatorname{Span}(B, Ab, \dots, A^{n-1}b) \neq \mathbb{R}^n.$$

We pass to second-order conditions. The Legendre condition degenerates because the system is control affine, and the Goh condition (Agrachev and Sachkov, 2004) takes the form:

$$p(\lambda)Bb = 0, \quad \tau \in [0,t].$$

Differentiating this identity by virtue of Hamiltonian system (1.26), we obtain, in addition to (1.28), new restrictions on $p(0)$:

$$p(0)A^i Bb = 0, \quad i = 0, \dots, n-1.$$

The generalized Legendre condition degenerates.

Summing up, the inequality

$$\operatorname{Span}(b, Ab, \dots, A^{n-1}b, Bb, ABb, \dots, A^{n-1}Bb) \neq \mathbb{R}^n$$

is necessary for geometric optimality of the trajectory $x(t) \equiv 0$. In other words, the equality

$$\operatorname{Span}(b, Ab, \dots, A^{n-1}b, Bb, ABb, \dots, A^{n-1}Bb) = \mathbb{R}^n$$

is sufficient for local controllability of bilinear system (1.26) at the origin.

6.   Observability of Systems on Lie Groups

Most of the work in bilinear systems has concentrated on bilinear systems up to output injection. One of the first results on observers for bilinear systems was obtained in Williamson (1977). In the early 1980s, Derese and Noldus (1981), and Grasselli and Isidori (1981) obtained necessary and sufficient conditions for the existence of bilinear observers for bilinear systems. In the late 1990s, a renewed interest was seen in bilinear systems. In Kinnaert (1999), the design of a residual generator for robust fault detection in bilinear systems was considered, utilizing methods based on a linear time-invariant observer up to output injection and the so-called Kalman-like observer. Martinez Guerra (1996) utilized Fliess generalized observable canonical form and generalized controllable canonical form to derive an observer-based controller for bilinear systems. In Hanba and Yoshihiko (2001), an output-feedback stabilizing controller for bilinear systems was proposed utilizing a periodic switching of the controller and the use of a dead-beat observer. A separation principle was posited for a class of dissipative systems with bilinearities. The class of bilinear systems that we study in this section is motivated by nonlinear control problems. The solution of the control problem in a Lyapunov framework leads us into observer design for a class of multiple-output bilinear systems that are not transformable into the so-called nonlinear observable canonical form for which exponential observers exist (Isidory, 1995). Robust fault detection based on observers for bilinear systems were studied in Kinnaert (1999).

6.1   Observability and Lie Groups

To avoid unnecessary complexity, we assume throughout this chapter that the controls are piecewise-constant. In fact, this is not essential. For instance, if we replace the set of piecewise-constant functions by the set of the piecewise-continuous functions, all of the arguments remain valid.

Let $R(x, t)$ be the reachable set starting from x; that is, $R(x, t)$ is the set of points y such that there exist a piecewise-constant control u and a time $t \geq 0$, such that the solution of (1.4) satisfies $x(0) = x$, $x(T) = y$.

It is proved in Jurdjevic (1997) that for the right-invariant system (1.4), the reachable set of x is related to the reachable set of the identity e by

$$R(x) = R(e)x. \tag{1.30}$$

Using this fact, we prove the following elementary result, which shows that distinguishing two arbitrary points is equivalent to distinguishing a point from the identity.

Lemma 1.1. *Two points p and q are indistinguishable if and only if for each $r \in R(e)$*

$$\mathrm{Ad}\,(r)pq^{-1} \in C. \tag{1.31}$$

Proof: By the structure of the output (1.4) it is clear that p and q are indistinguishable if and only if for all t, $R(p,t)$ and $R(q,l)$ are in the same coset of C. From (1.30), it follows that

$$Crp = Crq \quad \text{for all} \quad r \in R(e); \tag{1.32}$$

that is,

$$rpq^{-1}r^{-1} = \mathrm{Ad}\,(r)pq^{-1} \in C.$$

Now we may define an unobservable state as follows. (It is similar to the linear case: x_1 and x_2 are indistinguishable if and only if $x_1 - x_2$ belongs to an unobservable subspace.)

Definition 1.8. *A point h is called unobservable if there exist p and q such that $pq^{-1} = h$ and p and q are indistinguishable.*

Remark 1.4. Let h be unobservable. Then it follows from Lemma 1.1 that for a pair (p^1, q^1) if $p^1(q^1) = h$, then p^1 and q^1 are indistinguishable.

Let

$$H = \{h \in G \mid h \text{ is unobservable}\}.$$

By the definition of unobservable states, it is clear that

$$H \subset C. \tag{1.33}$$

In fact, H has a subgroup structure that is shown in the following lemma.

Lemma 1.2. *Assume C is closed. Then the unobservable set H is a closed Lie subgroup of G.*

Proof: By definition and Lemma 1.1,

$$H = \{h \in G \mid rhr^{-1} \in C \quad \text{for all} \quad r \in R(e)\}. \tag{1.34}$$

Let h_1, $h_2 \in H$. Then,

$$rh_1h_2^{-1}r^{-1} = rh_1r^{-1}rh_2^{-1}r^{-1} = (rh_1r^{-1})(rh_2r^{-1})^{-1} \in C.$$

Thus, H is a subgroup of G.

Because C is closed, if for a sequence $\{h_n\} \subset H$, $h_n \to h$, as $n \to \infty$, then

$$rh_nr^{-1} \to rhr^{-1} \in C.$$

Thus, $h \in H$, and hence H is closed. Now the result follows from well-known facts.

If C is closed, the output mapping has an analytic structure that is described in the following well-known theorem.

Theorem 1.4. *Let G be a Lie group and C a closed subgroup of G. Then the quotient space $C\backslash G$ admits the structure of a real analytic manifold in such a way that the action of G on $C\backslash G$ is real analytic; that is, the mapping $G \times C\backslash G \to C\backslash G$, which maps (p, Cq) into Cpq, is real analytic. In particular, the projection $G \to C\backslash G$ is real analytic.*

Let $\{R(e)\}_G$ be the subgroup of G generated by $R(e)$ and let $\overline{\{R(e)\}_G}$ denote the closure of $\{R(e)\}_G$. For convenience denote the vector fields $A(x)$, $B_1(x), \ldots, B_m(x)$ by $A, B_1, \ldots, B_m$, respectively, where A and B are elements in $\mathcal{G}(G)$, the Lie algebra of G. Then we have the following lemma.

Lemma 1.3. *Assume $h \in H$. Then*

$$\text{Ad}\,(r)h \in H \quad \text{for all } r \in \overline{\{G(e)\}_G}. \tag{1.35}$$

Proof: First, we claim that

$$\text{Ad}\,(r)h \in H \quad \text{for all } r \in G(e). \tag{1.36}$$

Because $R(e)$ is a semigroup, for any $\tilde{r} \in R(e)$ we have $\tilde{r}r \in R(e)$. Thus,

$$(\tilde{r}r)h(\tilde{r}r)^{-1} = \tilde{r}(rhr^{-1})^1\tilde{r}^{-1} \in C \quad \text{for all } r \in R(e).$$

It follows that $rhr^{-1} \in H$.

From its defining properties, it is clear that

$$\{R(e)\}_G = \Bigg\{ \exp(t_s X_s) \mid t_i \in \mathbb{R}, \quad s \in Z^+,$$

$$X_i \in \Bigg\{ A + \sum_{j=1}^m u_j B_j \,\Bigg|\, u_j \in \mathbb{R} \Bigg\}, \quad i = 1, \ldots, s \Bigg\}. \tag{1.37}$$

Set

$$E_s = \{(t_1, \ldots, t_s) \in \mathbb{R}^s \mid \text{Ad}\,(\exp(t_s X_s) \cdots \exp(t_1 X_1))h \in C\}.$$

Then, to prove (1.35) for $r \in \{R(e)\}_G$ it is enough to show that $E_s = \mathbb{R}^s$, $s = 1, 2, \ldots$. We proceed by induction. For $s = 1$, if $\text{Ad}(\exp t_1 X_1)h \notin C$, then there exists $\tilde{t}$, such that

$$\frac{d}{dt_1}\bigg|_{\tilde{t}_1} \text{Ad}\,(\exp t_1 X_1)h \notin (R_p)_* \mathcal{G}(C),$$

where $p = \text{Ad}(\exp \tilde{t}_1 X_1)h$, $\mathcal{G}(C)$ is the Lie algebra of C, and R_p is the right translation (i.e., $R_p : G \to G$ is defined as $x \to xp$). In other words, there exists a right-invariant one-form $w(x)$ generated by $w \in (\mathcal{G}(C))$ such that

$$\left\langle w_{(p)}, \frac{d}{dt_1}\Big|_{\tilde{t}_1} \mathrm{Ad}\,(\exp t_1 X_1) h \right\rangle \neq 0. \tag{1.38}$$

By analyticity, (1.38) holds in an open dense subset of $\mathbb{R}$. But according to (1.36), for $\tilde{t}_1 \in \mathbb{R}_+ = \{t \in \mathbb{R},\ t \geq 0\}$ the left-hand side of (1.38) is zero; this leads to a contradiction. Now, assume that

$$\mathrm{Ad}\,(\exp(t_{s-1} X_{s-1}) \cdots \exp(t_1 X_1))\, h \in C, \quad t \in \mathbb{R}$$

and

$$\{\mathrm{Ad}\,(\exp(t_s X_s) \cdots \exp(t_1 X_1))\, h \mid (t_1, \ldots, t_s) \in \mathbb{R}^s\} \not\subset C.$$

Then there exists $\tilde{t} = (\tilde{t}_1, \ldots, \tilde{t}_s)$ such that

$$\frac{d}{dt}\Big|_{\tilde{t}_s} \mathrm{Ad}\,(\exp(t X_s) \exp(\tilde{t}_{s-1} X_{s-1}) \cdots \exp(\tilde{t}_1 X_1)) h \in \mathcal{G}(C).$$

Similar to the case when $s = 1$, we have a contradiction.

Thus, we have shown that (1.35) holds for all $r \in \{R(e)\}_G$. By continuity, it holds for all $r \in \overline{\{R(e)\}_G}$.

Remark 1.5. It is clear by (1.37) that $\{R(e)\}_G$ is a path-connected group, hence a Lie subgroup. Now because $\{R(e)\}_G$ is a connected Lie group, and $A, B_1, B_2, \ldots, B_m$, generate $\mathcal{G}(\{R(e)\}_G)$, then

$$\begin{aligned} \{R(e)\}_G = \{\exp(t_x X_s) \cdots \exp(t_1 X_1) \mid t_i \in \mathbb{R}, \quad s \in Z^+, \\ X_i \in \{A, B_1, \ldots, B_m\}, \quad i = 1, \ldots, s\}. \end{aligned} \tag{1.39}$$

Next, we investigate the relations among the Lie algebras $\mathcal{G}(H)$, $\mathcal{G}(C)$, and $\mathcal{G}(G)$, which are the Lie algebras of H, C, and G, respectively.

Let $\{X_1(x), \ldots, X_s(x)\}$ be a set of right-invariant vector fields generated by $X_i \in \mathcal{G}(G)$, $i = 1, \ldots, s$, respectively. Let A denote the subspace of $\mathcal{G}(G)$ spanned by $\{x_1, \ldots, x_n\}$. A subspace Δ of $\mathcal{G}(G)$ is called $Y \in \mathcal{G}(G)$ invariant if

$$\{[Y, X] \mid X \in \Delta\} \subset \Delta.$$

Likewise, for the right-invariant we form $w_1(x), \ldots, w_s(x)$ generated by $w_i \in \mathcal{G}^*(G)$, the cotangent space of G at the identity e; we have a right-invariant subspace

$$\Omega = \mathrm{span}\,\{w_1, \ldots, w_s\}.$$

Ω is Y invariant if

$$\{L_Y w \mid w \in \Omega\} \subset \Omega.$$

The following two lemmas are generalizations of Theorem 1.1.

Lemma 1.4. $\mathcal{G}(H)$ *is A and B_i $(i = 1, \ldots, m)$, invariant.*

Proof: Let $X = A$ or B_i $t \in \mathbb{R}$, $p = \exp(tX)$. According to Lemma 1.3 and (1.39), $(\text{Ad} \exp(tX))_* \mathcal{G}(H) \subset \mathcal{G}(H)$. Now let $Y \in \mathcal{G}(H)$. Then

$$[X, Y] = \frac{d}{dt}\bigg|_{t=0} \text{Ad} \exp(tX)_* Y \in \mathcal{G}(H).$$

Lemma 1.5. $\mathcal{G}(H)$ *is the largest invariant Lie subalgebra contained in* $\mathcal{J}(C)$.

Proof: We claim that

$$\mathcal{G}(H) = \bigcap_{X_1,\ldots,X_p \in \{A, B_1,\ldots,B_m\} p \in Z^+} \text{ad}_{X_i}^{-1} \cdots \text{ad}_{X_p}^{-1} \mathcal{G}(C). \tag{1.40}$$

First, we show that (1.40) implies $\mathcal{G}(H)$ is the largest A and B invariant Lie subalgebra contained in $\mathcal{G}(C)$. Assume $\mathcal{G}(\tilde{H}) \subset \mathcal{G}(C)$ is also A and B invariant. Then, for any $X_1, \ldots, X_p \in \{A, B_1, \ldots, B_m\}$,

$$\text{ad}_{X_1} \cdots \text{ad}_{X_p} \mathcal{G}(\tilde{H}) \subset \mathcal{G}(\tilde{H}) \subset \mathcal{G}(C).$$

Thus,

$$\mathcal{G}(\tilde{H}) \subset \text{ad}_{X_p}^{-1} \mathcal{G}(C).$$

Because $X_1, \ldots, X_p$ are chosen arbitrarily, we have that

$$\mathcal{G}(\tilde{H}) \subset \mathcal{G}(H).$$

Next, we prove (1.40).

Lemma 1.4 shows that $\mathcal{G}(H)$ is A and B invariant. The inclusion follows from an argument similar to the above.

Let

$$Y \in \bigcap_{X_1,\ldots,X_p \in \{A, B_1,\ldots,B_m\} p \in Z^+} \text{ad}_{X_1}^{-1} \cdots \text{ad}_{X_p}^{-1} \mathcal{G}(C).$$

To show that $Y \in \mathcal{G}(C)$, it is enough to show that

$$\exp(\tau Y) \in H \quad \text{for all} \quad \tau \in \mathbb{R}.$$

Using (1.39), it suffices to show that for any

$$X_1, \ldots, X_p \in \{A, B_1, \ldots, B_m\} \quad (t_1, \ldots, t_P) \in \mathbb{R}^n,$$
$$\text{Ad} (\exp(t_p X_p) \cdots \exp(t_1 X_1)) \exp \tau Y \in C. \tag{1.41}$$

Because $\text{Ad} (\exp(t_p X_p) \cdots \exp(t_1 X_1))$ is a diffeomorphism, we have

$$\text{Ad} (\exp(t_p X_p) \cdots \exp (t_1 X_1)) \exp \tau Y$$
$$= \exp (\text{Ad} (\exp(t_p X_p) \cdots \exp (t_1 X_1) \tau Y)).$$

Now to prove (1.41), it suffices to show that

$$\text{Ad}\,(\exp(t_p X_p) \cdots \exp(t_1 X_1)) \tau Y \in \mathcal{G}(C). \tag{1.42}$$

Let us denote the right-hand side of (1.40) by $\mathcal{J}$. Now because

$$\text{Ad}\,(\exp(t_p X_p) \cdots \exp(t_1 X_1) \tau Y)$$
$$= \text{Ad}\,(\exp(t_p X_p))\,\text{Ad}\,(\exp(t_{p-1} X_{p-1})) \cdots \text{Ad}\,(\exp(t_1 X_1)) \tau Y,$$

it suffices to prove that

$$\text{Ad}\,(\exp(t_1 X_1) \tau Y) \in \mathcal{J}.$$

But

$$\text{Ad}\,(\exp(t_1 X_1) \tau Y) = \exp\,(\text{ad}\,(t_1 x_1)) \tau y = \sum_{n=0}^{\infty} \frac{(\text{ad}\,(t_1 x_1))^n}{n!}(\tau y).$$

Therefore, we have $\text{Ad}\,(\exp(t_1 X_1) \tau y) \in \mathcal{J}$.

We are now ready to discuss the observability properties of (1.4).

Definition 1.9. *System (1.4) is locally observable at x if there exists a neighborhood V_x of x such that*

$$I_x \cap V_x = \{x\},$$

where I_x is the set of points that are indistinguishable from x. System (1.4) is locally observable if it is locally observable everywhere. System (1.4) is (globally) observable if

$$I_x = \{x\}.$$

In fact, Lemma 1.5 leads to the following local observability result, which is now obvious.

Theorem 1.5. *System (1.4) is locally observable if and only if the largest A and B_i, $i = 1, \ldots, m$, invariant subalgebra contained in $\mathcal{G}(C)$ is zero. Moreover, if V_e is a neighborhood of e such that $I_e \cap V_e = \{e\}$, then $V_x = R_x(V_x)$ is a neighborhood of x such that $I_x \cap V_x = \{x\}$.*

Let S be the centralizer of $\{R(e)\}_G$, that is,

$$S = \{x \in G \mid rx = xr \quad \text{for all } r \in \{R(e)\}_G\}. \tag{1.43}$$

According to (1.39), we may express S in an easily verifiable form as

$$S = \{x \in G \mid x\exp(tX) = \exp(tX)x, \quad X \in \{A, B_1, \cdots, B_m\}\}. \tag{1.44}$$

We use S to establish a global result.

It is obvious that S is a closed subgroup of G, and hence is a Lie subgroup. Moreover, by the construction of $\{R(e)\}^G$ we see that to verify that $x \in S$ it is enough to verify that

$$\text{Ad}\,(x)\,\exp(tY) = \exp tY$$

for

$$Y \in \{A, B_1, \ldots, B_m\}, \quad t \in \mathbb{R}.$$

Now we state our global observability theorem.

Theorem 1.6. *System (1.4) is globally observable if and only if the following two conditions are satisfied:*

(a) $\mathcal{G}(H) = \{0\}$,

(b) $S \cap C = \{e\}$.

Proof: *Necessity.* The necessity of (a) has been proved in Theorem 1.5. The necessity of (b) is obvious, because if $e \neq h \in S \cap C$, then $h \in H$; that is, h is indistinguishable from e.

Sufficiency. From (a) we see that H is a discrete subgroup of G. Now for each $h \in H$, we define a mapping $\phi : \{R(e)\}_G \to H$ as

$$\phi(r) = \text{Ad}\,(r)h.$$

According to Lemma 1.3, ϕ maps $\{R(e)\}_G$ into H. Now, because $\{R(e)\}_G$ is connected and ϕ is continuous, $\text{Ad}\,(r)h \mid r \in \{R(e)\}_G \subset H$ is connected, but because

$$h \in \{\text{Ad}\,(r)h \mid r \in \{R(e)\}_G\}$$

it follows that

$$\{h\} = \{\text{Ad}\,(r)h \mid r \in \{R(e)\}_G\};$$

that is,

$$\text{Ad}\,(r)h = h \quad \text{for all } r \in \{R(e)\}_G.$$

In other words, $h \in S$. Now using condition (b), we see that $h = e$; that is, $H = \{e\}$.

6.2   Algorithms of Observability

In the previous section, we saw that the Lie subalgebra $\mathcal{G}(H)$ of the unobservable Lie group H plays an important role in investigating the observability of the system (1.4). The following algorithm gives a method to compute it.

Algorithm 1 produces an increasing sequence of right-invariant subspaces. To see that it provides $\mathcal{G}(H)$, we need the following theorem. The proof may be found in Isidory (1995).

$$\Omega \triangleq \mathcal{G}(C)^{\perp}, \Omega_{k+1} \triangleq \Omega_k + L_A \Omega_k + \sum_{i=1}^{m} L_{B_i} \Omega_k, \quad k \geq 1.$$

Algorithm 1: Algorithm of observability.

Theorem 1.7. *In Algorithm 1 if $\Omega_{k^*+1} = \Omega_{k^*}$ then*

$$\mathcal{G}(H) = \Omega_{k^*}^{\perp}. \tag{1.45}$$

Note that the algorithm converges because the sequence of subspace $\{\Omega_k\}$ is increasing.

Because every Lie algebra over the field of real numbers $\mathbb{R}$ is isomorphic to some matrix algebra, we may consider further algorithmic details for the Lie algebras of groups of matrices.

First, let $w(x) \in V^*(G)$ be a right-invariant covector field (one-form) generated by $w \in \mathcal{G}^*(G)$, and let $A(x)$, $B(x) \in V(G)$ be the right-invariant vector fields generated by $A, B \in \mathcal{G}(G)$, respectively. Then,

$$\langle L_A \omega, B \rangle = \langle L_{A(x)} \omega(x), B(x) \rangle$$
$$= L_{A(x)} \langle \omega(x), B(x) \rangle - \langle \omega(x), [A(x), B(x)] \rangle. \tag{1.46}$$

Because $\langle \omega(x), B(x) \rangle$ is constant, the first term of the right-hand side of (1.46) is zero. Thus, we have

$$\langle L_A \omega, B \rangle = -\langle \omega, [A, B] \rangle. \tag{1.47}$$

Now we consider a group of matrices. Assume the group considered is $GL(n, \mathbb{R})$ (or a subgroup of it). Then, $A, B \in gl\,(n, \mathbb{R})$ may be considered as matrices $A = (a_{ij})$ and $B = (b_{ij})$, respectively. Let $\omega \in gl^*\,(n, \mathbb{R})$. We may assume ω is also expressed as a matrix $\omega = (\omega_{ij})$ and define

$$\langle \omega, A \rangle = \sum_{i=1}^{n} \sum_{j=1}^{n} \omega_{ij} a_{ij}. \tag{1.48}$$

Now,

$$\langle L_A \omega, B \rangle = -\langle \omega, [A, B] \rangle$$
$$= \sum_{i=1}^{n} \sum_{j=1}^{n} \sum_{k=1}^{n} (a_{kj} \omega_{ij} b_{ij} - a_{ik} \omega_{ik} b_{kj})$$
$$= \sum_{p=1}^{n} \sum_{q=1}^{n} \left(\sum_{k=1}^{n} \omega_{pk} a_{qk} - a_{kp} \omega_{kq} \right) b_{pq}.$$

Thus

$$L_A\omega = [\omega, A^\top] = \omega A^\top - A^\top\omega, \tag{1.49}$$

where $\top$ stands for transpose. To apply Algorithm 1, formula (1.47) is helpful.

Remark 1.6. As shown in Brockett (1975), a right-invariant vector field on $\mathcal{G}(n, \mathbb{R})$ may be written as

$$A(x) = Ax,$$

where $A = A(e)$ and $A(x) = (R_x)_* A(e) = Ax$. Similarly, a right-invariant covector field may be written as

$$\omega(x) = \omega(x^\top)^{-1},$$

where $\omega = \omega(e)$ and $\omega(x) = (R_{x^{-1}})^* \omega(e) = \omega(x^\top)^{-1}$.
   To see this, we only have to show that

$$\langle \omega(x), A(x) \rangle = \langle \omega, A \rangle.$$

In fact, if we denote $y = x^{-1}$, $x = (x_{ij})$, and $y = (y_{ij})$, then

$$
\begin{aligned}
\langle \omega(x), A(x) \rangle &= \sum_i \sum_j \left(\sum_p \omega_{ip} y_{jp} \right) \left(\sum_q a_{iq} x_{qj} \right) \\
&= \sum_i \sum_p \sum_q \omega_{ip} a_{iq} \left(\sum_j x_{qj} y_{jp} \right) \\
&= \sum_i \sum_p \sum_q \omega_{ip} a_{iq} \delta_{qp} \\
&= \sum_i \sum_p \omega_{ip} a_{ip} = \langle \omega, A \rangle.
\end{aligned}
$$

In fact, if we rewrite $A(x)$ in the "usual fashion" as a vector

$$A(e) = (a_{11}, \ldots, a_{1n}, a_{21}, \ldots, a_{2n}, \ldots, a_{n1}, \ldots, a_{nn})^\top,$$

then

$$A(x) = \underbrace{(x^\top \dotplus x^\top \dotplus \cdots \dotplus x^\top)}_{(n \text{ terms})} A(e).$$

Similarly,

$$\omega(x) = \omega(e)((x^\top)^{-1} \dotplus (x^\top)^{-1} \dotplus \cdots \dotplus (x^\top)^{-1}),$$

where "$\dotplus$" denotes the direct sum of matrices, and $(x^\perp \dotplus x^\perp \dotplus \cdots \dotplus x^\perp)$ and $((x^\perp)^{-1} \dotplus (x^\perp)^{-1} \dotplus \cdots \dotplus (x^\perp)^{-1})$ are the Jacobian matrices of R and $R_x - 1$, respectively.

6.3   Examples

In this section, we present some examples to demonstrate our results and algorithms.

Example 1.3. Consider a system

$$\dot{x} = uBx, \tag{1.50}$$

$$t = Cx, \tag{1.51}$$

where $x \in \text{GL}\,(3, \mathbb{R})$, $C \in SO(3)$, $SO(3)$ is a schematic of the group of rotations in three dimensions, and

$$B = \begin{bmatrix} 0 & 0 & 0 \\ 1 & 0 & 0 \\ 0 & 0 & 0 \end{bmatrix}.$$

Now $\mathcal{G}(C)$ is the following set of skew-symmetric matrices:

$$\mathcal{G}(C) = \text{span} \left\{ \begin{bmatrix} 0 & 1 & 0 \\ -1 & 0 & 0 \\ 0 & 0 & 0 \end{bmatrix}, \begin{bmatrix} 0 & 0 & 1 \\ 0 & 0 & 0 \\ -1 & 0 & 0 \end{bmatrix}, \begin{bmatrix} 0 & 0 & 0 \\ 0 & 0 & 1 \\ 0 & -1 & 0 \end{bmatrix} \right\}.$$

According to Algorithm 1, we set

$$\Omega_0 = \text{span} \left\{ \begin{bmatrix} 1 & 0 & 0 \\ 0 & 0 & 0 \\ 0 & 0 & 0 \end{bmatrix}, \begin{bmatrix} 0 & 0 & 0 \\ 0 & 1 & 0 \\ 0 & 0 & 0 \end{bmatrix}, \begin{bmatrix} 0 & 0 & 0 \\ 0 & 0 & 0 \\ 0 & 0 & 1 \end{bmatrix}, \begin{bmatrix} 0 & 1 & 0 \\ 1 & 0 & 0 \\ 0 & 0 & 0 \end{bmatrix}, \right.$$

$$\left. \begin{bmatrix} 0 & 0 & 1 \\ 0 & 0 & 0 \\ 1 & 0 & 0 \end{bmatrix}, \begin{bmatrix} 0 & 0 & 0 \\ 0 & 0 & 1 \\ 0 & 1 & 0 \end{bmatrix} \right\} \triangleq \text{span}\{\omega_1, \omega_2, \omega_3, \omega_4, \omega_5, \omega_6\}.$$

Using formula (1.49), we see that

$$L_B\omega_1 = \begin{bmatrix} 0 & 1 & 0 \\ 0 & 0 & 0 \\ 0 & 0 & 0 \end{bmatrix}, \quad L_B\omega_5 = \begin{bmatrix} 0 & 0 & 0 \\ 0 & 0 & 0 \\ 0 & 1 & 0 \end{bmatrix}, \quad L_B\omega_6 = \begin{bmatrix} 0 & 0 & -1 \\ 0 & 0 & 0 \\ 0 & 0 & 0 \end{bmatrix}.$$

Thus, $\Omega_1 = \mathcal{G}(G)$ and $k_* = 1$. Therefore,

$$\mathcal{G}(H) = \Omega_1^\top = \{0\}.$$

Next, let us consider

$$S \cap C = \{xeC \mid x \exp tB = \exp(tB)x, \quad \text{for all } t \in \mathbb{R}\}.$$

Let $x = (x_{ij}) \in C$. Because

$$\exp tB = \begin{bmatrix} 1 & 0 & 0 \\ t & 1 & 0 \\ 0 & 0 & 1 \end{bmatrix},$$

we set

$$x \exp tB = \begin{bmatrix} x_{11} + tx_{12} & x_{12} & x_{13} \\ x_{21} + tx_{22} & x_{22} & x_{23} \\ x_{31} + tx_{32} & x_{32} & x_{33} \end{bmatrix}$$

$$= \exp(tB)x = \begin{bmatrix} x_{11} & x_{12} & x_{13} \\ tx_{11} + x_{21} & tx_{12} + x_{22} & tx_{13} + x_{23} \\ x_{31} & x_{32} & x_{33} \end{bmatrix}.$$

It follows that

$$x_{12} = 0, \quad x_{11} = x_{22}, \quad x_{13} = 0, \quad x_{32} = 0. \tag{1.52}$$

Because $x \in SO(3)$, the only solutions of (1.52) are

$$x_1 = e, \quad x_2 = \begin{bmatrix} -1 & 0 & 0 \\ 0 & -1 & 0 \\ 0 & 0 & 1 \end{bmatrix}. \tag{1.53}$$

According to Theorem 1.3, system (1.50), (1.51) is not globally observable.

Example 1.4. Consider the following system:

$$\dot{x} = Ax + uBx, \tag{1.54}$$
$$y = Cx, \tag{1.55}$$

where B and C are as in Example 1.3, and

$$A = \begin{bmatrix} 0 & 0 & 1 \\ 0 & 0 & 0 \\ 0 & 0 & 0 \end{bmatrix}.$$

As in the previous example, we see that $\mathcal{G}(H) = \{0\}$. Hence, the system is locally observable. Now

$$e^{At} = \begin{bmatrix} 1 & 0 & t \\ 0 & 1 & 0 \\ 0 & 0 & 1 \end{bmatrix}.$$

According to (1.44), we have only to check the commutativity of both x_1 and x_2 of (1.53) with $\exp(tA)$. For x_2 the answer is "no". Therefore, $x_1 = e = I_3$ is the only element in $S \cap C$. It follows that system (1.54), (1.55) is globally observable.

Remark 1.7. In Example 1.4, if we consider e^{At}, e^{Bt}, $e^{-Bt}e^{-At}$ $e^{Bt}e^{At}$ and their products, it is easy to see that

$$\{R_{(e)}\}_G = \left\{ \begin{pmatrix} 1 & 0 & a \\ b & 1 & c \\ 0 & 0 & 1 \end{pmatrix} \middle| a, b, c \in \mathbb{R} \right\}.$$

It follows that

$$S = \left\{ \begin{pmatrix} x & 0 & 0 \\ 0 & x & y \\ 0 & 0 & x \end{pmatrix} \middle| \; x, y \in \mathbb{R}, \;\; x \neq 0 \right\}$$

and therefore,

$$S \cap C = I_3.$$

But in general, it is difficult to calculate $\{R(e)\}_G$ and S. In fact, Example 1.4 shows that to use Theorem 1.3 it is not necessary to construct $\{R(e)\}_G$ and S directly. We may check the global observability by the following rule, which may be considered as a corollary of Theorem 1.3.

Corollary 1.4. *System (1.4) is globally observable, if and only if,*

(a) $\mathcal{G}(H) = \{0\}$,

(b) $\{x \in C \mid \exp(tX)x = x \exp(tX),$

$$X \in \{A, B_1, \dots, B_m\}, \; t \in \mathbb{R}\} = \{e\}.$$

6.4   Decoupling Problems

As an application, we consider a decoupling problem. To keep the right-invariance of $A(x)$ and $B(x)$, we consider only a constant feedback

$$u = \alpha + \beta u, \tag{1.56}$$

where $a \in \mathbb{R}^m$ and $\beta \in \mathrm{GL}\,(m, \mathbb{R})$.

Now consider

$$\dot{x} = A(x) + \sum_{i=1}^{m} u_i B_i(x) + \omega W(x),$$
$$y = Cx,$$

where ω is a disturbance.

Lemma 1.6. *The disturbance ω does not affect the output y if and only if*

$$W \in \mathcal{G}(H). \tag{1.57}$$

Proof: In fact, we may choose a local coordinate (ϕ, U) around e, say $x = (x_1, x_2)$, such that

$$C \cap U = \{p \in U \mid x_2^p = 0\}.$$

Thus,

$$y = x_2(q), \quad q \in U.$$

Now, it is easy to see that on U, $\mathcal{G}(H)$ is the largest A and B invariant distribution contained in the $\ker(y_*)$. Note that constant feedback does not affect $\mathcal{G}(H)$. Thus, the canonical decoupling result shows that (Isidory, 1995) is a necessary and sufficient condition that ω does not affect y on V. By the analyticity, it is also true globally.

Next, we consider the input-output decoupling problem. Assume $C_1, \ldots, C_k$ are Lie subgroups of G. Let $C = C_1 \cap \cdots \cap C_k$. Then it is easy to see that

$$y = Cx \tag{1.58}$$

is equivalent to

$$y_1 = C_1 x$$
$$\vdots \tag{1.59}$$
$$y_k = C_k x$$

in the sense that any points p and q are indistinguishable in (1.58) if and only if they are indistinguishable in (1.59).

Let $\mathcal{G}(H^i)$ be the largest A and B_i invariant Lie subalgebra contained in $\mathcal{G}(C_1 \cap \cdots \cap C_{i-1} \cap C_{i+1} \cap \cdots \cap C_k)$. Consider the system

$$\dot{x} = A(x) + \sum_{i=1}^{m} u_i B_i(x),$$
$$y_j = C_j x, \quad j = 1, \ldots, k. \tag{1.60}$$

We say that the input-output decoupling problem is solvable if there exists $\beta = (\beta_{ij}) \in \mathrm{GL}\,(m, \mathbb{R})$, such that for

$$u = v\beta$$

there exists a partition of v, namely $v = (v^1, \ldots, v^k)$, such that v^i affects only the corresponding y_i, $i = 1, \ldots, k$.

Theorem 1.8. *For the system described by (1.60) the input-output decoupling problem is solvable if and only if*

$$B = B \cap \mathcal{G}(H^1) + \cdots + B \cap \mathcal{G}(H^k),$$

where $B = \mathrm{span}\,\{B_1, \ldots, B_m\}$. Moreover, if the system (1.60) satisfies the controllability rank condition (i.e., $\mathcal{G}(\{R_e\}_G) = \mathcal{G}(G)$), then v^i controls y^i completely.

Proof: The proof is immediate from Lemma 1.6 and the well-known decoupling results (Nijmeijer and Van der Schaft, 1985; Nijmeijer and Respondek, 1988).

We have considered a system defined on a Lie group with outputs in a coset space as described in Brockett (1975). The main results of this chapter are two observability theorems, Theorems 1.2 and 1.3, that give necessary and sufficient conditions for local and global observability, respectively. Algorithm 1 calculates the A and B invariant Lie subalgebra contained in a given Lie subalgebra, which makes the condition in the above two theorems computationally verifiable. Some examples are included. Finally, we have briefly discussed the input-output decoupling problem of a system on a Lie group with output in a coset space.

7.   Invertibility of Control Systems
7.1   Right-Invariant Control Systems

In this section we review some basic results and definitions that are used in this chapter.

The purpose of this section is to show that the linear results on invertibility can be extended to a large class of systems.

Let $\mathbf{H}$ be a Lie group. The right multiplication mapping $R_x\colon y \to yx$ from $\mathbf{H} \to \mathbf{H}$ has differential dR_x. A vector field X on $\mathbf{H}$ is called *right-invariant* if $dR_x X(y) = X(yx)$ for all $y \in \mathbf{H}$. The collection of right-invariant vector fields, $\mathbf{H}$ is called the Lie algebra of $\mathbf{H}$ (Hirschorn, 1977).

A single–input–single–output bilinear system is a control system of the form

$$
\begin{aligned}
\dot{x}(t) &= Ax(t) + u(t)Bx(t), \quad x(0) = x_0, \\
y(t) &= cx(t),
\end{aligned}
\tag{1.61}
$$

where the state $x \in \mathbb{R}^n$; A and B are $n \times n$ matrices over $\mathbb{R}$, c is a $1 \times n$ matrix over $\mathbb{R}$, and $u \in \mathcal{U}$, the class of piecewise real analytic functions on $(0, \infty)$.

It is often convenient to express the solution to (1.61) as $x(t) = X(t)x_0$ where $X(t)$ is an $n \times n$ matrix valued function of t which is the trajectory of the corresponding matrix bilinear system.

A *single-input matrix bilinear system* is a system of the form

$$
\begin{aligned}
\dot{X}(t) &= AX(t) + u(t)BX(t), \quad X(0) = X_0, \\
Y(t) &= CX(t),
\end{aligned}
\tag{1.62}
$$

where A, B, X are $n \times n$ matrices over $\mathbb{R}$, $u \in \mathcal{U}$, and C is an $r \times n$ matrix over $\mathbb{R}$. We assume that X_0 is invertible so that $X(t)$ evolves in $GL(n, \mathbb{R})$, the Lie group of invertable $n \times n$ real matrices (Brockett, 1972; Jurdjevic, 1997).

The matrix system (1.62) is a special case of the more general class of right-invariant systems studied in Jurdjevic (1997).

Definition 1.10. *A right-invariant system is a system of the form*

$$\dot{x}(t) = A(x(t)) + \sum_{i=1}^{m} u_i(t) B_i(x(t)), \quad x(0) = x_0 \in \mathbf{H},$$

$$y(t) = \mathbf{K} x(t),$$

(1.63)

where $u_1, \ldots, u_m \in \mathcal{U}$, $\mathbf{H}$ *is a Lie group,* $\mathbf{K}$ *is a Lie subgroup of* $\mathbf{H}$ *with Lie algebra* $\mathcal{K}$, *and* $A, B_1, \ldots, B_m \in \mathcal{H}$, *the Lie algebra of right-invariant vector fields on* $\mathbf{H}$.

We remark that the coset output $y(r) = \mathbf{K} x(t)$ generalizes the output $Y(t) = CX(t)$ in (1.63). In particular one could set $\mathbf{K} = \{X \in GL(n, \mathbb{R}): CX = C\}$ and $\mathbf{H} = GL(n, \mathbb{R})$.

A *single-input right invariant system* is a system of the form

$$\dot{x}(t) = A(x(t)) + u(t) B(x(t)), \quad x(0) = x_0 \in \mathbf{H},$$

$$y(t) = \mathbf{K} x(t),$$

(1.64)

where A, B, $\mathbf{K}$ are defined as above.

Definition 1.11. *The right-invariant system (1.63) is said to be invertible if the output* $\tau \to y(\tau)$ *on any interval* $0 \leq \tau < t$ *uniquely determines the input* $\tau \to u(\tau)$ *for* $0 \leq \tau < t$. *That is, distinct inputs produce distinct outputs. Invertibility for systems (1.61), (1.62) and (1.64) are defined in an analogous manner.*

The properties of a right-invariant system are related to the structure of the Lie algebra $\mathcal{H}$. The Lie algebra $\mathcal{H}$ is a vector space with a nonassociative "multiplication" defined as follows.

For $X, Y \in \mathcal{H}$ the *Lie bracket of X and Y* is

$$[X, Y](m) = X(m)Y - Y(m)X.$$

We define $\mathrm{ad}_X^n Y$ inductively as follows:

$$\mathrm{ad}_X^0 Y = Y, \quad \mathrm{ad}_X^k Y] = [X, \mathrm{ad}_X^{k-1} Y].$$

For matrix bilinear systems with X, $Y \in \mathcal{H}$ right-invariance it means that $X(M) = XM$ and

$$[X, Y](M) = (YX - XY)M.$$

Let $\mathcal{J}$ be a subset of the Lie algebra $\mathcal{H}$. We define $\{\mathcal{J}\}_{\mathrm{LA}}$ to be the *Lie algebra generated in $\mathcal{J}$ in $\mathcal{H}$.* Thus $\{\mathcal{J}\}_{\mathrm{LA}}$ is the smallest Lie subalgebra of $\mathcal{H}$ containing $\mathcal{J}$. For each $x \in \mathbf{H}$ let $\mathcal{J} = \{L(x): L \in \mathcal{J}\}$.

It is known that the structure of the reachable set for (1.63) is related to the structure of the Lie algebras:

$$\mathcal{L} = \{A, B_1, B_2, \ldots, B_m\}_{\text{LA}},$$
$$\mathcal{L}_0 = \{\text{ad}_A^k B_i : k = 0, 1, \ldots \text{ and } i = 1, \ldots, m\}_{\text{LA}},$$
$$\mathcal{B} = \{B_1, \ldots, B_m\}_{\text{LA}}.$$

Thus each right-invariant system has associated with it the chain of Lie algebras

$$\mathcal{H} \supset \mathcal{L} \supset \mathcal{L}_0 \supset \mathcal{B}.$$

If $\exp : \mathcal{H} \to \mathbf{H}$ is the standard exponential map in Lie theory then $\exp \mathcal{L} = \{\exp L: L \in \mathcal{L}\} \supset \mathbf{H}$ and the group generated by $\mathcal{L}$, $\{\exp \mathcal{J}\}_G$, is a Lie subgroup of $\mathbf{H}$. Set

$$\mathbf{G} = \{\exp \mathcal{L}\}, \quad \mathbf{G}_0 = \{\exp \mathcal{L}_0\}_G$$

and

$$\mathbf{B} = \{\exp \mathcal{B}\}_G.$$

Thus each right-invariant system gives rise to the chain of Lie groups

$$\mathbf{H} = \mathbf{G} \supset \mathbf{G}_0 \supset \mathbf{B}.$$

Because $\mathcal{L}_0$ is an ideal $\mathcal{L}$ (that is, for each $L_0 \in \mathcal{L}_0$, $L \in \mathcal{L}$, $[L_0, L] \in \mathcal{L}_0$) we know that $\mathbf{G}_0$ is a normal subgroup of $\mathbf{G}$. The following theorem relates the structure of the trajectories of a bilinear system to the above group decompositions.

Theorem 1.9. (Jurdjevic, 1997). *Consider the right-invariant system (1.63) where the state x evolves in the Lie group $\mathbf{H}$ and $A, B_1, \ldots, B_m \in \mathcal{H}$. Associated with this system is the chain of Lie groups $\mathbf{H} \supset \mathbf{G} \supset \mathbf{G}_0 \supset \mathbf{B}$. Then for any set of controls $u_1, \ldots, u_m \in \mathcal{U}$ with corresponding trajectory $t \to x(t)$ we have $x(t) \in (\exp tA) \cdot \mathbf{G}_0 \cdot x_0$ for all $t \geq 0$, where $(\exp tA) \cdot \mathbf{G}_0 \cdot x_0 = \{\exp tA \cdot g \cdot x_0: g \in \mathbf{G}_0\}$.*

We conclude this section by presenting two formulae that are used in the next section. The mapping $L_x : y \to xy$ from $\mathbf{H} \to \mathbf{H}$ is called the left multiplication map. Suppose that $x = \exp X$ where $X \in \mathbf{H}$. The mapping $A_x = L_x \circ R_{x-1}: y \to xyx^{-1}$ of $\mathbf{H} \to \mathbf{H}$ has differential $dA_x = \text{Ad}(x): \mathcal{H} \to \mathcal{H}$. The *Campbell–Baker–Hausdorff* formula for right-invariant vector fields asserts that

$$\text{Ad}(x)(Y) = Y - \text{ad}_X Y + \frac{1}{2!}\text{ad}_X^2 Y - \frac{1}{3!}\text{ad}_X^3 Y + \cdots.$$

The exp mapping of $\mathcal{H} \to \mathbf{H}$ has a differential $X \in \mathcal{H}$, $d \exp_X : \mathcal{H} \to \mathbf{H}$ where

$$d \exp_X Y(e) = (dR_{\exp X})_e \circ \frac{1 - e^{\mathrm{ad}_X}}{-\mathrm{ad}_X} Y(e)$$

$$= Y(\exp X) + \frac{1}{2!}\mathrm{ad}_X Y(\exp X) + \frac{1}{3!}\mathrm{ad}_X^2 Y(\exp X) + \cdots .$$

7.2   Invertibility of Right-Invariant Systems

In this section we derive necessary and sufficient conditions for the invertibility of right-invariant systems. The main result in this Section is the following theorem (Hirschorn, 1977; Pardalos et.al., 2001).

Theorem 1.10. *The right-invariant system (1.64) is invertible if and only if* $\mathrm{ad}_A^k B \notin \mathcal{K}$ *for some integer* $k \in \{0, 1, \ldots, n-1\}$, *where n is the dimension of $\mathcal{L}$ and $\mathcal{K}$ is the Lie algebra of* $\mathbf{K}$.

Corollary 1.5. *Consider the right-invariant system (1.64) with output* $y(t) = c\,(x(t))$ *where* $c: \mathbf{H} \to \mathbf{J}$ *is a Lie group homomorphism and* $c_* : \mathcal{H} \to \mathcal{J}$ *is the differential of c.*

This system is invertible if and only if $c_* \, \mathrm{ad}_A^k B \neq 0$ *for some positive integer* $k \in \{0, 1, \ldots, n-1\}$.

Corollary 1.6. *The matrix bilinear system (1.62) is invertible if and only if*

$$C \, \mathrm{ad}_A^k B \neq 0$$

for some positive integer $k \in \{0, 1, \ldots, n^2 - 1\}$ *where A and B are $n \times n$ matrices.*

Corollary 1.7. *The matrix bilinear system (1.62) fails to be invertible if and only if every control gives rise to the same output function.*

The similarity between the standard linear invertibility results and the above conditions is striking. The single-input-single-output linear system $x = Ax + bu$; $y = cx$ is invertible if and only if $cA^k b \neq 0$ for some positive integer k. The relative order α of the system is the least positive integer k such that $cA^{k-1}b \neq 0$ (or an infinity). Rewriting this system in bilinear form (Brockett, 1972), we have

$$\dot{z} = A_1 z + u B_1 z,$$
$$y = C_1 z,$$

where

$$z = \begin{pmatrix} x \\ 1 \end{pmatrix}, \quad A_1 = \begin{pmatrix} A & 0 \\ 0 & 0 \end{pmatrix}, \quad B_1 = \begin{pmatrix} 0 & b \\ 0 & 0 \end{pmatrix}, \quad C_1 = (c \quad 0).$$

The state transition matrix for this system is the state for the corresponding matrix bilinear system. Corollary 1.7 above asserts that this matrix system is invertible if and only if $C_1 \operatorname{ad}_{A_1}^k B_1 \neq 0$ for some k. Because

$$C_1 \operatorname{ad}_{A_1}^k B_1 = \begin{pmatrix} 0 & cA^k b \\ 0 & 0 \end{pmatrix},$$

and invertibility in the linear case is independent of the initial state, the well-known linear result follows from the more general bilinear result, Corollary 1.7. This motivates the following definition.

Definition 1.12. *The relative order α of the matrix bilinear system (1.62) is the least positive integer k such that $C_1 \operatorname{ad}_{A_1}^k B \neq 0$ or $\alpha = \infty$ if $C \operatorname{ad}_A^k B = 0$ for all $k > 0$.*

As in the linear case, a matrix bilinear system is invertible if and only if the relative order $\alpha < \infty$. The remainder of this section is devoted to proving this result.

In studying the invertibility of matrix bilinear systems one is tempted to repeat the approach that is successful in the linear case: differentiate the output until the control $u(t)$ appears, and solve for $u(t)$ in terms of the derivatives of the output. Unfortunately the bilinear dependence of the control on the state greatly complicates the situation and little insight is obtained. Instead we use the fact that the trajectory evolves in a Lie group. We begin by looking for a sufficient condition for invertibility for the right-invariant system (1.64). Suppose that this system is not invertible. This means that there are two different controls u_1 and u_2 which give rise to outputs y_1 and y_2, respectively, where $y_1 \equiv y_2$. Let $t \to x_1(t)$ and $t \to x_2(t)$ denote the trajectories corresponding to u_1 and u_2. Then

$$y_1(t) = \mathbf{K} \cdot x_1(t) \equiv \mathbf{K} \cdot x_2(t) = y_2(t)$$

and

$$\mathbf{K} \cdot x_1(t) x_2(t) = \mathbf{K} \quad \text{for all } t \geq 0.$$

In particular the curve

$$t \to a(t) = x_1(t) x_2(t)^{-1}$$

is contained in the Lie subgroup K for all $t \geq 0$, and for each positive time t, the derivative $d(t)$ is contained in the tangent space to $\mathbf{K}$ at $a(t)$, $\mathcal{K}(a(t))$. The following lemma establishes some of the basic properties of $a(t)$ and $d(t)$.

Lemma 1.7. *Consider the right-invariant system (1.64). Suppose that $u_1, u_2 \in \mathcal{U}$ are controls which give rise to trajectories $t \to x_1(t)$ and $t \to x_2(t)$, respectively. The curve*

$$t \to a(t) = x_1(t)x_2(t)^{-1}$$

is contained in $\mathbf{G}_0$ *and*

$$\dot{a}(t) = (A + u_1(t)B)(a(t)) + dL_{a(t)}(A + u_2(t)B)(e),$$

where e is the identity element of $\mathbf{G}_0$. *In particular,*

$$dR_{a(t)} - 1(\dot{a}(t)) = (A + u_1(t)B)(e) - \text{Ad}\,(a(t))(A + u_2(t)B)(e)$$

is contained in $\mathcal{L}_0(e)$ *for all real t.*

Proof: Let $t \to x_1(t)$ and $t \to x_2(t)$ be smooth trajectories corresponding to controls u_1 and u_2. By Theorem 1.11 $x_1(t), x_2(t) \in (\exp tA) \cdot \mathbf{G}_0 \cdot x_0$ for all $t \in \mathbf{R}$. It follows that $x_i(t) = (\exp tA) \cdot P_i(t) \cdot x_0$ where $P_i(t)$ is a smooth curve in $\mathbf{G}_0$, for $i = 1, 2$. Thus

$$a(t) = x_i(t)x_2(t)^{-1} = (\exp tA)P_1(t)P_2(t)^{-1}\exp(-tA),$$

and because $\exp tA \in \mathbf{G}$, $P_1(t)P_2(t)^{-1} \in \mathbf{G}_0$, and $\mathbf{G}_0$ is a normal subgroup of $\mathbf{G}$, we see that $a(t) \in \mathbf{G}_0$ for all $t \in \mathbb{R}$.

The product rule for differentiation implies that

$$\dot{a}(t) = \frac{d}{dt}(x_1(t)x_2(t)^{-1}) = \frac{\partial}{\partial t}x_1(t)x_2(s)^{-1}\bigg| + \frac{\partial}{\partial s}x_1(t)x_2(s)^{-1}\bigg|_{s=t}.$$

Because A and B are right-invariant vector fields and $\dot{x}_i(t) = (A + u_i(t)B)x_i(t)$ for $i = 1, 2$, we have

$$\frac{\partial}{\partial t}x_1(t)x_2^{-1}(s)\bigg|_{s=t} = (A + u_1(t)B)(x_1(t)x_2(t)^{-1}) = (A + u_1(t)B)(a(t)).$$

To obtain an expression for $\dot{x}_2(s)^{-1}$ we observe that $x_2(s)x_2(s)^{-1} = e$ for all $s \in \mathbb{R}$. Differentiating both sides of this equality results in the equation

$$dL_{x_2(t)}(\dot{x}_2(t)^{-1}) + dR_{x_2(t)-1}\dot{x}_2(t) = 0$$

or

$$(x_2(\dot{t})^{-1}) = -dL_{x_2(t)-1}dR_{x_2(t)-1}\dot{x}_2(t) = -dL_{x_2(t)-1}(A + u_2(t)B)(e).$$

Thus

$$\frac{\partial}{\partial s}x_1(t)x_2^{-1}(s) = dL_{x_2(t)} \circ (-dL_{x_2(t)}(A + u_2(s)B)(e))$$

and using the chain rule we conclude that

$$\dot{a}(t) = (A + u_1(t)B)(a(t)) - dL_{a(t)}(A + u_2(t)B)(e).$$

To complete the proof we identify $\mathcal{L}_0(g)$ with $T_e(\mathbf{G}_0)$ for all $g \in \mathbf{G}_0$. Then $\dot{a}(t)$ is identified with $dL_{a(t)^{-1}}(\dot{a}(t))$ for all real t. We observe that the mapping $C_x: g \to xgx^{-1}$ of $\mathbf{G}_0 \to \mathbf{G}_0$ can be written as the composition $R_x^{-1} \circ L_x$ and thus $dR_{x^{-1}} \circ dL_x = dC_x = \mathrm{Ad}\,(x)$ and $dR_{a(t)^{-1}}(\dot{a}(t)) = dR_{a(t)^{-1}}(A + u_1(t)B)(a(t)) - dR_{a(t)^{-1}} \circ dL_{a(t)}dL_{a(t)}(A + u_2(t)B)(e) = (A + u_1(t)B)(e) - \mathrm{Ad}\,(a(t))(A + u_2(t)B)(e)$. This completes the proof.

We have observed that if system (1.61) fails to be invertible then the curve $t \to a(t)$ is contained in the Lie Group $\mathbf{K}$ and $\dot{a}(t) \in \mathcal{K}(a(t))$. Thus $dR_{a(t)^{-1}}(\dot{a}(t)) \in \mathcal{K}(e)$ for all t in $\mathbb{R}$ and if we set

$$Q(t) = (A + u_1(t)B)(e) - \mathrm{Ad}\,(a(t))(A + u_2(t)B)(e)$$

then the curve $t \to Q(t)$ is contained in $\mathcal{K}(e)$ by Lemma 1.7. If we identify $\mathcal{K}(e)$ with $\mathcal{K}$ then $Q(t)$ and its derivatives with respect to t of all orders are contained in $\mathcal{K}$. In particular $(d^n Q(t)/dt^n)\,|_{t=0} = Q^{(n)} \in \mathcal{K}$ for $n = 0, 1, \dots$ and the Lie algebra generated by these tangent vectors is contained in $\mathcal{K}$.

In proving Theorem 1.12 we show that the Lie algebra generated by the derivatives $Q(0)$, $Q^{(1)}(0)$, $Q^{(2)}(0), \dots$ is the Lie algebra $\mathcal{L}_0$ of $\mathbf{G}_0$. Thus a sufficient condition for invertibility is that $\mathcal{L}_0 \not\subset \mathcal{K}$. The next lemma examines the relationship between the curve $t \to a(t)$ and the Lie algebra $\mathcal{L}_0$.

Lemma 1.8. *Consider the right-invariant system (1.64). Suppose that $u_1, u_2 \in \mathcal{U}$ are distinct controls and $t \to x_1(t)$, $t \to x_2(t)$ are the corresponding trajectories. Then there exist $\varepsilon > 0$ and a real analytic curve $t \to L(t)$ in $\mathcal{L}_0$, defined for $|t| < \varepsilon$, such that $x_1(t)x_2(t)^{-1} = \exp L(t)$ for $|t| < \varepsilon$ and $\mathcal{L}_0 = \{L(t): |t| < \varepsilon\}_{LA}$.*

Proof: The curve $t \to a(t) = x_1(t)x_2(t)^{-1}$ is contained in $\mathbf{G}_0$ as a consequence of Lemma 1.7. It is well known that $\exp \mathcal{L}_0 \to \mathbf{G}_0$ is a local diffeomorphism in some neighborhood $\mathcal{N}$ of 0 in $\mathcal{L}_0$ [10]. Thus there exists an $\varepsilon > 0$ and a real analytic curve $t \to L(t)$ in $\mathcal{L}_0$ such that $\exp L(t) = a(t)$ and $L(t)$ has a Taylor series expansion $L(t) = \sum_{i=0}^{\infty} t^i L_i$ for $|t| < \varepsilon$. Because $a(0) = x_1(0)x_2(0)^{-1} = ee^{-1} = e$ and $a(0) = \exp L(0)$, we have $L(0) = 0$ and $L_0 = 0$. Clearly $L_1, L_2, \dots \in \mathcal{L}$ and $\{L(t) : |t| < \varepsilon\}_{LA} = \{L_i: i = 1, 2, \dots\}_{LA}$. We set $\widehat{\mathcal{L}}_0 = \{L_i: 1, 2, \dots\}_{LA}$. It follows that $\widehat{\mathcal{L}}_0 \subset \mathcal{L}_0$ and the proof is complete if we can show that $\widehat{\mathcal{L}}_0 = \mathcal{L}_0$.

Set $Q(t) = (A + u_1(t)B)(e) - \mathrm{Ad}\,(a(t))(A + u_2(t)B)(e)$ for all real t. Lemma 1.7 asserts that

$$Q(t) = dR_{a(t)^{-1}}(\dot{a}(t)) = dR_{a(t)^{-1}}\left(\frac{d}{dt}\exp L(t)\right).$$

We use this equality to study $\widehat{\mathcal{L}}_0$. We begin by noting that $\mathrm{Ad}\,(a(t)) = \mathrm{Ad}(\exp L(t))$. Using the Campbell–Baker–Hausdorff formula we have

$$\mathrm{Ad}\,(a(t))(A + u_2(t)B)(e) = \sum_{k=0}^{\infty} \left((-1)^k/k!\right)\,\mathrm{ad}_{L(t)}^k(A + u_2(t)B)(e),$$

and so

$$Q(t) = (u_1(t) - u_2(t))B(e) - \sum_{k=1}^{\infty}\left((-1)^k/k!\right)\,\mathrm{ad}_{L(t)}^k(A + u_2(t)B)(e),$$

Choosing e smaller if necessary we can assume that $u_1(t) = \sum_{i=0}^{\infty} a_i t^i$ and $u_2(t) = \sum_{i=0}^{\infty} b_i t^i$ for $|t| < \varepsilon$. Setting $c_i = a_i - b_i$, we have

$$Q(t) = \sum_{i=0}^{\infty} c_i t^i B(e) - \sum_{k=1}^{\infty} ((-1)^k/k!)\,\mathrm{ad}_{L(t)}^k A(e)$$

$$= \sum_{k=1}^{\infty}\sum_{i=0}^{\infty} ((-1)^k/k!)\,\mathrm{ad}_{L(t)}^k B(e).$$

Expressing $L(t)$ as $\sum_{j=1}^{\infty} t^j L_j$ we can collect like powers of t and write

$$Q(t) = c_0 B(e) + \sum_{k=1}^{\infty} t^k F_k(e),$$

where $F_k \in \mathcal{L}_0$ for all k. A straightforward induction argument shows that

$$F_k = c_k B_k + (-1)^k \mathrm{ad}_A L_k + R_k + S_k, \qquad (1.65)$$

where R_k is a linear combination of terms of the form $\mathrm{ad}_{L_{k_1}} \mathrm{ad}_{L_{k_2}} \cdots \mathrm{ad}_{L_{k_p}} A$ with $p \geq 1$, $k_i < k$ for $i = 1, 2, \ldots, p$ and S_k is a linear combination of terms of the form $\mathrm{ad}_{L_{k_1}} \cdots \mathrm{ad}_{L_{k_q}} B$ with $q \geq 1$, $k_i \leq k$ for $i = 1, 2, \ldots, q$.

A second expression for $Q(t)$ comes from the identity

$$Q(t) = dR_{a(t)^{-1}}\left(\frac{d}{dt}\exp L(t)\right).$$

Using the formula for exp and the Taylor series expansion for $L(t)$ it is easy to verify that

$$Q(t) = L_1(e) + \sum_{k=1}^{\infty} t^k((k+1)L_{k+1}(e) + M_k(e)),$$

where M_k is contained in $\{L_1, \ldots, L_k\}_{\mathrm{LA}}$. Combining these two expressions for $Q(t)$ we find that $L_1 = c_0 B$ and

$$(k+1)L_{k+1} + M_k = F_k \qquad (1.66)$$

for $k = 1, 2, \ldots$. To complete the proof we use these relations to show that $\mathrm{ad}_A^k B \in \widehat{\mathcal{L}}_0$ for $k = 0, 1, \ldots$, which implies that $\widehat{\mathcal{L}}_0 = \mathcal{L}_0$.

Because $u_1 \neq u_2$, we have $c_k \neq 0$ for some k. Let p be the smallest positive integer k such that $c_k \neq 0$.

Claim. $L_1, L_2, \ldots, L_p = 0$ and $(p+1)L_{p+1} = c_p B$: If $p = 0$ then this is the case, because $L_1 = c_0 B$. If $p > 0$ then $c_0 = 0$ and $L_1 = 0$. Suppose that $L_1, L_2, \ldots, L_k = 0$ for $0 \leq k < p$. Combining (1.65) and (1.66) we find that $F_k = (k+1)L_{k+1} + M_k = c_k B + (-1)^k \, \mathrm{ad}_A L_k + R_k + S_k$. Because $c_1 = c_2 = \cdots = c_k = 0$ and $L_1, L_2, \ldots, L_k = 0$ it follows from the definitions that $M_k = c_k B = \mathrm{ad}_A L_k = R_k = S_k = 0$. This induction argument proves that $L_1, L_2, \ldots, L_p = 0$ and hence $M_p = \mathrm{ad}_A L_p = R_p = S_p = 0$. Thus $(p+1)L_{p+1} = F_p = C_p B_p$, which proves the assertion.

Claim. $\widehat{\mathcal{L}}_0$ *is an* ad_A-*invariant subspace of* $\mathcal{L}_0$: Let p be chosen as above. Then $L_1 = L_2 = \cdots = L_p = 0$ and it suffices to show that $(-1)^k \, \mathrm{ad}_A L_k - (k+1)L_{k+1} \in \{L_i \colon p < i \leq k\}_{\mathrm{LA}}$ for all $k > p$, Because this implies that $\mathrm{ad}_A L_k \in \widehat{\mathcal{L}}_0$ for all k. The proof uses induction on k. We have shown that $(p+1)L_{p+1} = c_p B$, $(p+2)L_{p+2} + M_{p+1} = F_{p+1}$ from (1.66), and $F_{p+1} = c_{p+1}B + (-1)^{p+1} \, \mathrm{ad}_A L_{p+1} + R_{p+1} + S_{p+1}$ from (1.65). Because $L_1 = \cdots = L_p = 0$, we have $R_{p+1} = S_{p+1} = 0$ and because $M_{p+1} \in \{L_1, \ldots, L_{p+1}\}_{\mathrm{LA}}$, $M_{p+1} = \alpha B$ for some real number a. Combining these results we have

$$(p+2)L_{p+2} = (-1)^{p+1}(c_p/(p+1)) \, \mathrm{ad}_A B + (c_{p+1} - \alpha)B.$$

If $k = p + 1$ then $(-1)^k \, \mathrm{ad}_A L_k - (k+1)L_{k+1} = (-1)^k \, \mathrm{ad}_A((c_{k-1}/k)B) - (-1)^k(c_{k-1}/k) \, \mathrm{ad}_A B - (c_k - \alpha)B = (\alpha - c_k)B \in \{L_k\}_{\mathrm{LA}}$. Now assume that $(-1)^k \, \mathrm{ad}_A L_k - (k+1)L_{k+1} \in \{L_i \colon p << i \leq k\}_{\mathrm{LA}}$ for $p < k < n$ where n is a positive integer greater than $p + 1$. For $k = n$ we have

$$(n+1)L_{n+1} = F_n - M_n = c_n B + (-1)^n \, \mathrm{ad}_A L_n + R_n + S_n - M_n,$$

from (1.65) and (1.66). Thus

$$(-1)^n \mathrm{ad}_A L_n - (n+1)L_{n+1} = M_n - c_n B - R_n - S_n,$$

and the induction will be completed if we can show that the right-hand side of the above inequality is contained in $\{L_{p+1}, L_{p+2}, \ldots, L_n\}_{\mathrm{LA}}$. Set $\mathcal{K} = \{L_{p+1}, \ldots, L_n\}_{\mathrm{LA}}$. Now $M_n \in \mathcal{K}$ by definition, and because $L_{p+1} = c_p B \in \mathcal{K}$ and $c_p \neq 0$, both B and $c_n B$ are in $\mathcal{K}$. Recall that S_n is a linear combination of terms of the form $\mathrm{ad}_{L_{k_1}} \mathrm{ad}_{L_{k_2}} \cdots \mathrm{ad}_{L_{k_q}} B$

where $k_i \leqq n$, hence $S_n \in \mathcal{K}$. Finally, R_n is a linear combination of terms of the form $\mathrm{ad}_{L_{k_1}} \cdots \mathrm{ad}_{L_{k_q}} A$ where $k_q < n$. By the induction hypothesis $(-1)^n \mathrm{ad}_A L_{n-1} = (-1)^{n-1} \mathrm{ad}_{L_{n-1}} A = -n L_n + L$ where $L \in \{L_{p+1}, \ldots, L_{n-1}\}_{\mathrm{LA}}$ and it follows that R_n is also contained in $\mathcal{K}$. This completes the induction.

Because $c_{pk} \neq 0$, $B \in \widehat{\mathcal{L}}_0$ and $\widehat{\mathcal{L}}_0$ is an ad_A-invariant subspace of $\mathcal{L}_0 \, \mathrm{ad}_A^k \in \widehat{\mathcal{L}}_0$ for all $k \geqq 0$, which completes the proof of this lemma.

Proof of Theorem 1.10: First we suppose that the system (1.64) is invertible but $\mathrm{ad}_A^k B \in \mathcal{K}$ for $k = 0, 1, \ldots$. For each control $u \in \mathcal{U}$ the corresponding trajectory is $x(t) = \exp tA \cdot P(t) \cdot x_0$, where $P(t) \in G_0$, as a consequence of Theorem 1.11. Because $\mathcal{L}_0$ is the Lie algebra generated by $\{\mathrm{ad}_A^k B : k = 0, 1, \ldots\}$, a subset of the Lie algebra $\mathcal{K}$, we are assuming that $\mathcal{L}_0 \subset \mathcal{K}$, and thus $G_0 \subset \mathbf{K}$. If $u_1, u_2 \in \mathcal{U}$ are two controls producing trajectories $x_1(t) = \exp tA \cdot P_1(t) \cdot x_0$ and $x_2(t) = \exp tA \cdot P_2(t) \cdot x_0$, then $x_1(t) x_2(t)^{-1} = \exp tA \cdot P_1(t) \cdot x_0 x_0^{-1} P_2^{-1}(t) (\exp tA)^{-1} = (\exp tA)^{-1}$. Because $\exp tA \in \mathbf{G}$, $P_1(t) P_2^{-1}(t) \in \mathbf{G}_0$, and $\mathbf{G}_0$ is a normal subgroup of $\mathbf{G}$, $x_1(t) x_2(t)^{-1} \in \mathbf{G}_0 \subset \mathbf{K}$. Thus $\mathbf{K} x_1(t) x_2(t)^{-1} = \mathbf{K}$ and $\mathbf{K} x_1(t) = \mathbf{K} x_2(t)$ for all $t > 0$. In other words $u_1(t)$ and $u_2(t)$ produce the same outputs. Clearly this system is not invertible, a contradiction. Thus invertibility implies that $\mathrm{ad}_A^k B \notin \mathcal{K}$ for some positive integer k. Because ad_A is a linear operator on the n dimensional Lie algebra $\mathcal{L}$, a necessary condition for invertibility is that $\mathrm{ad}_A^k B \notin \mathcal{K}$ for some $k \in \{0, 1, \ldots, n-1\}$, by the Cayley–Hamilton theorem.

To show that this Lie algebraic condition implies invertibility it suffices to show that if two different controls result in the same output then $\mathcal{L}_0 \subset \mathcal{K}$. Suppose that $u_1, u_2 \in \mathcal{U}$ are distinct controls producing the same outputs, $y_1(t) = \mathbf{K} \cdot x_1(t) = y_2(t) = \mathbf{K} \cdot x_2(t)$. Then for t sufficiently small the real analytic curve $t \to a(t) = x_1(t) x_2(t)^{-1}$ is contained in $\mathbf{K}$. From Lemma 1.8 we know that there exists an $\varepsilon > 0$ and a real analytic curve $t \to L(t)$ in $\mathcal{L}_0$ such that $a(t) = \exp L(t)$ for $|t| < \varepsilon$ and $\widehat{\mathcal{L}}_0 = \{L(t) : |t| < \varepsilon\} = \mathcal{L}_0$. Decreasing ε if necessary we can express $L(t)$ by the Taylor expansion $L(t) = \sum_{i=1}^{\infty} t^i L_i$, where $\{L_i = i = 1, 2, \ldots\}_{\mathrm{LA}} = \widehat{\mathcal{L}}_0$. Because $a(t) = \exp L(t) \in \mathbf{K}$ for $|t| < \varepsilon$ we know that $L(t) \in \mathcal{K}$ for $|t| < \infty$. Thus $\{L_i : i = 1, 2, \cdots\}_{\mathrm{LA}} \subset \mathcal{K}$ and $\widehat{\mathcal{L}}_0 = \mathcal{L}_0 \subset \mathcal{K}$. This completes the proof.

7.3 Left-Inverses for Bilinear Systems

Suppose that a given control system is invertible: that is, the output uniquely determines the control. One then faces the practical problem of determining the input given only the output record of the system. In the linear case this problem has been solved in a very elegant manner.

A second linear system, called a left-inverse system, can be constructed. This *left-inverse* system, when driven by appropriate derivatives of the output of the original system, produces as an output $u(t)$, the input to the original system (Hirschorn, 1977). In this section we construct nonlinear systems that are left-inverses for bilinear systems.

Consider the bilinear system (1.61). As in the matrix case the *relative order* α of this bilinear system is the least positive integer k such that $c \operatorname{ad}_A^{k-1} B \neq 0$ or $\alpha = \infty$ if $c \operatorname{ad}_A^k B = 0$ for all $k > 0$.

In contrast with the linear case it is not yet known whether an invertible bilinear system has a bilinear left-inverse system. We look for a left-inverse in the class of nonlinear systems of the form

$$\hat{x}(t) = a(\hat{x}(t)) + \hat{u}(t)b(\hat{x}(t)), \quad \hat{x}(0) = \hat{x}_0 \in M,$$
$$\hat{y}(t) = d(\hat{x}(t)) + \hat{u}(t)e(\hat{u}(t)), \tag{1.67}$$

where $\hat{x} \in M$, a differentiable manifold, $\hat{x} \in \mathcal{U}$, $a(\,\cdot\,)$ and $b(\,\cdot\,)$ are smooth vector fields on M, and d, e are smooth functions on M.

Definition 1.13. *The system (1.67) is called a left-inverse for the bilinear system (1.61) if $\hat{u}(t) = y^\alpha(t)$ implies that $\hat{y}(t) = u(t)$.*

The following theorem generalizes the well known linear result on left-inverses to the bilinear case.

Theorem 1.11. *If the bilinear system (1.61) is invertible then its relative order is $\alpha < \infty$. If $\alpha < \infty$ and $c \operatorname{ad}_A^{\alpha-1} B \neq 0$ then the bilinear system is invertible with left-inverse (1.67), where $M = \mathbb{R}^n \sim (cA^{\alpha-1}B)^\perp$, $\hat{x}_0 = x_0$, and*

$$a(\hat{x}) = A\hat{x} - (cA^\alpha \hat{x}/cA^{\alpha-1}B\hat{x})B\hat{x},$$
$$b(\hat{x}) = (1/cA^{\alpha-1}B\hat{x})B\hat{x},$$
$$d(\hat{x}) = -(cA^\alpha \hat{x}/cA^{\alpha-1}B\hat{x})$$

and

$$e(\hat{x}) = (1/cA^{\alpha-1}B\hat{x}).$$

If $\hat{u}(t) = y^\alpha(t)$ then $\hat{y}(t) = u(t)$.

Proof: We begin by showing that an invertible bilinear system has relative order $\alpha < \infty$. If a is infinite then the corresponding matrix bilinear system (1.62) is not invertible by Theorem 1.10, Corollary 1.5. Choose distinct controls u_1, $u_2 \in \mathcal{U}$ which produce identical outputs for the matrix bilinear system. The output of the bilinear system (1.61) is $t \to Y(t)x_0$, where $t \to Y(t)$ is the output of the corresponding matrix

system, thus the bilinear system (1.61) is not invertible. This completes
the first part of the proof.

Suppose that $\alpha < \infty$ and $c \, \mathrm{ad}_A^{\alpha-1} Bx_0 \neq 0$. Differentiate the output
$y(t) = cx(t)$ to obtain

$$\dot{y}(t) = c\dot{x}(t) = c\,Ax(t) + u(t)c\,Bx(t).$$

If $\alpha > 1$ then $cB = 0$, and differentiating $\dot{y}(t)$ we find that

$$y^{(2)}(t) = cA\dot{x}(t) = cA^2 x(t) + u(t)cABx(t).$$

If $\alpha > 2$ then $c\,\mathrm{ad}_A = 0$ and so $cAB - cBA = 0$. Because $cB = 0$ we
have $cAB = 0$ and

$$y^{(3)}(t) = cA^3 x(t) + u(t)cA^2 Bx(t).$$

Continuing this procedure we find that

$$y^{(\alpha)}(t) = cA^\alpha x(t) + u(t)cA^{\alpha-1} Bx(t). \tag{1.68}$$

Because $cA^{\alpha-1} Bx_0 \neq 0$ by assumption, the scalar function $cA^{\alpha-1} Bx(t)$
is nonzero for t sufficiently small. The set of vectors x in $\mathbb{R}^n$ for which
$cA^{\alpha-1} Bx \neq 0$ is the differentiable manifold $M = \mathbb{R}^n \sim (cA^{\alpha-1})^\perp$. Consider the nonlinear system (1.67) described in the statement of this the-
orem, and set $\hat{u}(t) = y^{(\alpha)}(t)$. Then

$$\dot{\hat{x}}(t) = a(\hat{x}(t)) + (cA^\alpha x(t) + u(t)cA^{\alpha-1} Bx(t))b(\hat{x}); \quad \hat{x}(0) = x_0.$$

We claim that $\hat{x}(t) = x(t)$. Because $\hat{x}(0) = x(0)$ it suffices to verify
that both $\hat{x}(t)$ and $x(t)$ solve the same differential equation. Replacing $\hat{x}$
by x in the above differential equation in $\hat{x}$, and invoking the definitions
for $a(\,\cdot\,)$ and $b(\,\cdot\,)$, this equation reduces to the differential equation

$$\dot{x}(t) = Ax(t) + u(t)Bx(t).$$

Thus $\hat{x}$ and x satisfy the same differential equation when $\hat{u}(t) = y^{(\alpha)}(t)$.

The corresponding output is

$$\hat{y}(t) = d(x(t)) + \hat{u}(t)e(x(t))$$
$$= -(cA^\alpha x(t)/cA^{\alpha-1} Bx(t)) + y^{(\alpha)}(t)(1/cA^{\alpha-1} Bx(t)).$$

Substituting the expression (1.68) for $y^{(\alpha)}(t)$ we find that $\hat{y}(t) = u(t)$.
Because $x(t)$ involves in M for some interval of time and the controls
are piecewise real analytic functions, $u(t)$ is completely determined for
all $t > 0$. Thus the bilinear system is invertible and the given nonlinear
is a left-inverse system. This completes the proof.

We remark that in the proof of this theorem we show that when
$\hat{u}(t) = y^{(\alpha)}(t)$ the state $\hat{x}(t) = x(t)$ is the state of the original bilinear

system. Thus the left-inverse system acts as a state observer for the bilinear system, a result which itself is of some interest.

We also note that for certain bilinear systems the vector fields $a(x)$, $b(x)$ may not be complete. That is, the integral curves for these vector fields need not be defined for all time. Thus after a finite time has passed the trajectory $x(t)$ could leave M, and in this case $u(t)$ would only be recovered for t in some bounded interval. For a linear system in bilinear form $y(t)$ is defined for all t and the left-inverse system reduces to the standard linear left-inverse.

Theorem 1.11 presents a sufficient condition for inverting vector bilinear systems in the case where $\alpha < \infty$ but this condition is far from being necessary. In Example 1.4, $c\,\mathrm{ad}_A^{\alpha-1} Bx_0 = 0$ but $c\,\mathrm{ad}_A^{\alpha} Bx_0 \neq 0$ and the system is invertible. It seems reasonable to expect that a necessary and sufficient condition for invertibility must take into account the action of the matrix Lie group $\mathbf{G}$ on the state space $\mathbb{R}^n$.

Definition 1.14. *The initialized relative order $\alpha(x_0)$ for a bilinear system (1.61) is the least positive integer k such that $c\,\mathrm{ad}_A^{k-1}\,Bx_0 \neq 0$ or $\alpha(x_0) = \infty$ if $c\,\mathrm{ad}_A^k Bx_0 = 0$ for $k = 0, 1, 2, \ldots, n^2 - 1$.*

Theorem 1.12. *Consider the bilinear system (1.61) with associated Lie algebras $\mathcal{L} \supset \mathcal{L}_0 \supset \mathcal{B}$. If $\alpha(x_0) < \infty$ and*

$$\alpha(x) \geqq \alpha(x_0) \quad \text{for } x \in \mathbf{G} \cdot x_0 \tag{1.69}$$

then the system is invertible with left-inverse (1.67), where

$$\hat{x}_0 = x_0, \quad M = \mathbb{R}^n \sim (cA^{\alpha(x_0)-1}B)^{\perp},$$
$$a(\hat{x}) = A\hat{x} - (cA^{\alpha(x_0)-1}\hat{x}/cA^{\alpha(x_0)-1}B\hat{x})B\hat{x},$$
$$b(\hat{x}) = (1/cA^{\alpha(x_0)-1}B\hat{x})B\hat{x},$$
$$d(\hat{x}) = -(cA^{\alpha(x_0)}\hat{x}/cA^{\alpha(x_0)-1}B\hat{x}),$$

and

$$e(\hat{x}) = (1/cA^{\alpha(x_0)-1}B\hat{x}).$$

If $\hat{u}(t) = y^{(\alpha(x_0))}(t)$ then $\hat{y}(t) = u(t)$.

Proof: Suppose $\alpha(x_0) < \infty$ and condition (1.69) is satisfied. Condition (1.69) implies that $c\,\mathrm{ad}_A^k Bx = 0$ for $0 \leq k < \alpha(x_0)-1$ and $x \in \mathbf{G}\cdot x_0$. In particular $cBx = 0$ and $c\,\mathrm{ad}_A Bx = c(BA-AB)x = cBAx - cABx = 0$ for $\alpha(x_0) > 2$. Now (1.69) implies that $cB(\exp tA)x = 0$ for all real t, as $\exp tA \in \mathbf{G}$. Differentiating with respect to t and setting $t = 0$ shows that $cBAx = 0$. Combining this with the above expression for $c\,\mathrm{ad}_A Bx$

we see that $cABx = 0$. A similar argument proves that $cA^k Bx = 0$ for $0 \leq k \leq \alpha(x_0) - 1$ and for all $x \in \mathbf{G} \cdot x_0$. In particular, if $x(t)$ is the trajectory for the system (1.61) with $x(0) = x_0$, then $cA^k Bx(t) = 0$ for $0 \leq k < \alpha(x_0) - 1$ and

$$y^{(\alpha(x_0))}(t) = cA^{\alpha(x_0)}x(t) + u(t)A^{\alpha(x_0)-1}Bx(t).$$

In the proof of Theorem 1.11 we showed that this implies that $\hat{y}(t) = u(t)$ when $\hat{u}(t) = y^{(\alpha(x_0))}(t)$. This completes the proof.

Example 1.5. In this example we apply Theorem 1.11 to a linear system in bilinear form.

Consider the linear system $\dot{x}(t) = Ax(t) + bu(t)$; $x(0) = x_0$, with output $y(t) = cx(t)$. In bilinear form

$$\dot{z}(t) = Fz(t) + u(t)Gz(t), \quad z(0) = z_0,$$
$$y(t) = Hz(t),$$

where $z(t) = (x(t), 1)$, $z(0) = (x_0, 1)$,

$$F = \begin{pmatrix} A & 0 \\ 0 & 0 \end{pmatrix}, \quad G = \begin{pmatrix} 0 & b \\ 0 & 0 \end{pmatrix}, \quad H = (c \quad 0).$$

Here $H \, \mathrm{ad}_F^{\alpha-1} G = (0 \quad cA^{\alpha-1}b)$ and $H \, \mathrm{ad}_F^{\alpha-1} Gz_0 = cA^{\alpha-1}$. Thus Theorem 1.11 asserts that a linear system is invertible if and only if $\alpha < \infty$. If $\alpha < \infty$ the left-inverse described in Theorem 1.11 is of the form

$$\dot{\hat{z}}(t) = a(\hat{z}(t)) + \hat{u}(t)b(\hat{z}(t)), \quad \hat{z}(0) = z(0),$$
$$\hat{y}(t) = d(\hat{z}(t)) + \hat{u}(t)e(\hat{z}(t)),$$

where $\hat{z}(t) = (x(t), \alpha(t))$ with $x \in \mathbb{R}^n$ and $\alpha \in \mathbb{R}$,

$$a(\hat{z}) = (A\hat{x} - (cA^{\alpha}\hat{x}/\alpha \cdot cA^{\alpha-1}b)\alpha \cdot b, 0),$$
$$b(\hat{z}) = ((1/\alpha \cdot cA^{\alpha-1}b)\alpha \cdot b, 0).$$
$$d(\hat{z}) = -(cA^{\alpha}\hat{x}/\alpha \cdot cA^{\alpha-1}b),$$
$$e(\hat{z}) = (1/\alpha \cdot cA^{\alpha-1}b),$$

$M = \{(x, \alpha): x \in \mathbb{R}^n, \alpha \in \mathbb{R} \sim \{0\}\}$, and $z_0 = (x_0, 1)$. With z_0 given it follows that $z(t) = (x(t), 1)$ so that the above system of equations reduces to

$$a\dot{\hat{z}} = [A - (bcA^{\alpha}/cA^{\alpha-1}b)]\hat{x}(t) + (1/cA^{\alpha-1}b)b\hat{u}(t),$$
$$\hat{y}(t) = -(cA^{\alpha}/cA^{\alpha-1}b)\hat{x}(t) + (1/cA^{\alpha-1}b)\hat{u}(t),$$

which is the well-known linear left-inverse.

Example 1.6. Consider the matrix bilinear system

$$\dot{X}(t) = AX(t) + u(t)BX(t), \quad X(0) = I,$$
$$Y(t) = CX(t)$$

with

$$A = \begin{pmatrix} 0 & 0 & 0 & 0 \\ 0 & 1 & 0 & 0 \\ 0 & 0 & 0 & 0 \\ 0 & 0 & 0 & 0 \end{pmatrix}, \quad B = \begin{pmatrix} 1 & 0 & 0 & -1 \\ -1 & 0 & 0 & -1 \\ 0 & 0 & 0 & 0 \\ 0 & 0 & -1 & 0 \end{pmatrix},$$

and $C = (1\ 1\ 0\ 0)$. By direct computation we find that

$$[A, B] = BA - AB = \begin{pmatrix} 0 & 0 & 0 & 0 \\ 1 & 0 & 0 & -1 \\ 0 & 0 & 0 & 0 \\ 0 & 0 & 0 & 0 \end{pmatrix},$$

$$[B, \mathrm{ad}_A B] = [A, B]B - B[A, B] = \begin{pmatrix} 0 & 0 & 0 & 0 \\ 1 & 0 & 1 & -1 \\ 0 & 0 & 0 & 0 \\ 0 & 0 & 0 & 0 \end{pmatrix},$$

$$\mathrm{ad}_A^2 B = -\mathrm{ad}_A B, \quad \mathrm{ad}_A[B, \mathrm{ad}_A B] = -[B, \mathrm{ad}_A B],$$

and

$$\mathrm{ad}_A^2 \mathrm{ad}_A B = [B, \mathrm{ad}_A B].$$

Thus

$$\mathcal{L} \text{ has basis } \{B, A, \mathrm{ad}_A B, [B, \mathrm{ad}_A B]\},$$
$$\mathcal{L} \text{ has basis } \{B, \mathrm{ad}_A B, [B, \mathrm{ad}_A B]\},$$

and

$$\mathcal{B} \text{ has basis } \{B\}.$$

Here $CB = 0$, $C\,\mathrm{ad}_A B = (1, 0, 0, -1)$, and hence the relative order is $\alpha = 2$, and this system is invertible by Corollary 1.5 of Theorem 1.10.

Now we consider the corresponding bilinear system

$$\dot{x}(t) = Ax(t) + u(t)Bx(t), \quad x(0) = x_0,$$
$$y(t) = cx(t)$$

with A, B, c defined above and $x_0 = (1, 0, 0, 0)$. Because $c\,\mathrm{ad}_A Bx \neq 0$ Theorem 1.11 asserts that this system is invertible with left-inverse

$$\dot{\hat{x}} = a(x) + \hat{u}b(\hat{x}), \quad \hat{x}_0 = x_0,$$
$$\hat{y} = d(\hat{x}) + \hat{u}e(\hat{x}),$$

where

$$a(\hat{x}) = a(\hat{x}_1, \hat{x}_2, \hat{x}_3, \hat{x}_4) = (\hat{x}^2, 0, 0, \hat{x}_2\hat{x}_3/(\hat{x}_4 - \hat{x}_1)),$$
$$b(x) = (-1, 1, 0, \hat{x}_3/(\hat{x}_1 - \hat{x}_4)),$$
$$d(\hat{x}) = \hat{x}_2/(\hat{x}_1 - \hat{x}_4),$$
$$e(\hat{x}) = 1/(\hat{x}_4 - \hat{x}_1).$$

According to Theorem 1.11 we have $\hat{y}(t) = u(t)$ if $\hat{u}(t) = y^{(2)}(t)$. We now verify this fact directly. We know that

$$\dot{y}(t) = c\dot{x}(t) = cAx(t) + u(t)cBx(t) = cAx(t),$$

inasmuch as $cB = 0$, and

$$y^{(2)}(t) = cA(Ax(t) + u(t)Bx(t))$$
$$= cA^2x(t) + u(t)cABx(t) = x_2(t) + u(t)(x_4) - x_1(t)).$$

Thus if $\hat{u}(t) = y^{(2)}(t)$ then

$$\dot{\hat{x}} = \begin{pmatrix} \hat{x}_2 - x_2 \\ x_2 \\ 0 \\ [\hat{x}_2\hat{x}_3/(\hat{x}_4 - \hat{x}_1)] - [x_2x_3/(x_4 - x_1)] \end{pmatrix} + u \begin{pmatrix} x_1 - x_4 \\ x_4 - x_1 \\ 0 \\ -x_3 \end{pmatrix}.$$

But if we set $\hat{x}(t) = x(t)$ this equation is just

$$\dot{x}(t) = Ax(t) + u(t)Bx(t),$$

so when $\hat{u}(t) = y^{(2)}(t)$ we see that $\hat{x}(t) = x(t)$, and

$$\hat{y}(t) = x_2(t)/(x_1(t) - x_4(t)) + y^{(2)}(t)/(x_4(t) - x_1(t)) = u(t)$$

for all $t \geq 0$.

Example 1.7. The following system has $\alpha = 1$ and $c\,\mathrm{ad}_A A^{\alpha-1} Bx_0 = 0$, so Theorem 1.11 can't be used. Consider the bilinear system

$$\dot{x}(t) = Ax(t) + u(t)Bx(t), \quad x(0) = x_0,$$
$$y(t) = cx(t),$$

where

$$x_0 = (0, 0\,1), \quad c = (1\,0\,1), \quad A = \begin{pmatrix} 0 & 0 & 0 \\ 0 & 0 & 0 \\ 0 & 1 & 0 \end{pmatrix} \text{ and } B = \begin{pmatrix} 1 & 0 & 0 \\ 0 & 0 & 1 \\ 0 & 0 & 0 \end{pmatrix}.$$

Here $cB = (1\ 0\ 0)$, $cBx_0 = 0$, and $c\ \mathrm{ad}_A Bx_0 = -1$. Thus $\alpha = 1$, $\alpha(x_0) = 2$, and Theorem 1.13 does not apply because $c\ \mathrm{ad}_A^{\alpha-1} Bx_0 = 0$.

To apply Theorem 1.12 we must check that $\alpha(x) \geq \alpha(x_0)$ for all $x \in \mathbf{G} \cdot x_0$. In this case we must verify that $cBx = 0$ for all $x \in \mathbf{G} \cdot x_0$. By direct computation both $\exp tA$ and $\exp tB$ are matrices with first rows of the form $(b\ 0\ 0)$ with b real. Because $\mathbf{G} = \{\exp t_1 A, \exp t_2 B: t_1, t_2 \text{ real}\}_G$, $\mathbf{G} \cdot x_0$ consists of vectors whose first entries are zero. This means that $cBx = (1\ 0\ 0)x = 0$ for all $x \in \mathbf{G} \cdot x_0$.

Theorem 1.12 states that this system is invertible and provides a left-inverse. Here

$$a(\hat{x}) = A\hat{x}, \quad b(\hat{x}) = \begin{pmatrix} \hat{x}_1/\hat{x}_3 \\ 1 \\ 0 \end{pmatrix}, \quad d(\hat{x}) = 0, \quad e(\hat{x}) = (1/\hat{x}_3),$$

and $M = \{(a, b, c): a,\ b,\ c \in \mathbb{R},\ c \neq 0\}$. Condition (1.69) is satisfied, therefore we know that $cBx(t) \in cB\mathbf{G} \cdot x_0 = \{0\}$. Thus $\dot{y}(t) = cAx(t) + u(t)cBx(t) = cAx(t)$, $y^{(2)}(t) = y^{\alpha(x_0)-1}(t) = cA^2(t) + u(t)cABx(t) = u(t)x_3(t)$, and when $\hat{u}(t) = y^{(2)}(t)$, $\dot{\hat{x}}(t) = \dot{x}(t)$, $\hat{x}(t) = x(t)$, and

$$\hat{y}(t) = (\hat{u}(t)/\hat{x}_3(t)) = u(t)x_3(t)/x_3(t) = u(t).$$

Of course for certain controls $u(t)$ one could have $x_3(T) = 0$ for some $T > 0$, in which case $x(T) \notin M$ and $u(t)$ is recovered for some interval $0 \leq t < \varepsilon$ on which $x(t)$ exists.

8.   Invertibility of Discrete Bilinear Systems
8.1   Discrete Bilinear Systems and Invertability

Consider the multi input-output BS described by the equations

$$x(t+1) = Ax(t) + \sum_{i=1}^{m} B_i x(t)u_i(t), \tag{1.70}$$

$$y(t) = Cx(t), \tag{1.71}$$

where $x(t)$ is the n-dimensional state vector, $u_i(t)$, $i = 1, \ldots, m$ are the scalar controls, and $y(t)$ is the m-dimensional measurement vector. The matrices A, B_i $(i = 1, \ldots, m)$, C are real constant matrices of appropriate dimensions and

$$\operatorname{rank} B_i = 1, \quad i = 1, \ldots, m. \tag{1.72}$$

Consider another system in the form

$$\tilde{x}(t+1) = \widetilde{A}\tilde{x}(t) + \widetilde{B}\tilde{u}(t),$$
$$\tilde{y}(t) = H(\tilde{x}(t), \tilde{u}(t)), \tag{1.73}$$

where $\tilde{x}(t)$ is the n-dimensional state vector, $\tilde{u}(t)$ is the m-dimensional control vector, and $\tilde{y}(t)$ is the m-dimensional output vector, the matrices $\tilde{A}$, $\tilde{B}$ are constant matrices of appropriate dimensions, and H is a nonlinear map.

Definition 1.15. *The system (1.70) is called the inverse for BS (1.70)–(1.72) if, when driven by the output of the original system $y(t+1)$, it produces the input $u(t)$ of the original system as an output.*

The purpose here is to construct an inverse system for the BS (1.70)–(1.72), that is, to find the matrices $\tilde{A}$, $\tilde{B}$ and the form of the map H.

8.2   Construction of Inverse Systems

We now derive the equations for the Inverse system. If the unit shift operator is applied to equations (1.71) and equation (1.70) is substituted into it, the result is

$$y(t+1) = CAx(t) + \sum_{i=1}^{m} CB_i x(t) u_i(t). \tag{1.74}$$

Equations (1.70) and (1.74) taken together give a pair of algebraic equations relating $x(t+1)$, $x(t)$, $y(t+1)$, and $[u_1(t)\ldots u_m(t)]^T$:

$$\begin{bmatrix} I_n & -B_1 x(t) & \cdots & -B_m x(t) \\ 0 & CB_1 x(t) & \cdots & CB_m x(t) \end{bmatrix} \begin{bmatrix} x(t+1) \\ u(t) \end{bmatrix} = \begin{bmatrix} Ax(t) \\ y(t+1) - CAx(t) \end{bmatrix}. \tag{1.75}$$

The desired inverse system is a pair of equations for $x(t+1)$ and $u(t)$ in terms of $x(t)$ and $y(t+1)$. It is known that the equations (1.75) can be solved for $x(t+1)$, $u(t)$ uniquely if and only if the matrix

$$R = \begin{bmatrix} I_n & -B_1 x(t) & \cdots & -B_m x(t) \\ 0 & CB_1 x(t) & \cdots & CB_m x(t) \end{bmatrix} \tag{1.76}$$

has an inverse. Now consider the following fact from matrix theory.

Lemma 1.9. *Any n-dimensional square matrix of rank one can be uniquely (within a scalar factor) expressed as a product of a column and a row n-vector.*

Consequently, the matrices B_i can be expressed as

$$B_i = g_i h_i^T, \quad i = 1,\ldots,m.$$

Let us introduce notations

$$h_I^T x(t) = \gamma_i(x), \quad \Gamma(x) = \text{diag}\,\{\gamma_1(x)\ldots\}, \quad G = [g_1 \ldots g_m].$$

In these notations

$$R = \begin{bmatrix} I_n & -G\Gamma(x) \\ 0 & CG\Gamma(x) \end{bmatrix}.$$

The matrix R has an inverse if and only if the matrix CG has a full rank and

$$\gamma_i(x) = h_i^T x(t) \neq 0, \quad i = 1, \dots, m, \quad t = 0, 1, \dots .$$

If the matrix R is invertible, then applying the well-known matrix inversion lemma, we get

$$R^{-1} = \begin{bmatrix} I_n & G(C\Gamma)^{-1} \\ 0 & \Gamma^{-1}(CG)^{-1} \end{bmatrix}$$

and the solution of equations (1.75) is

$$x(t+1) = [I - G(CG)^{-1}C]Ax(t) + G(CG)^{-1}y(t+1),$$
$$u(t) = -\Gamma^{-1}(x)(CG)^{-1}CAx(t) + \Gamma^{-1}(x)(CG)^{-1}y(t+1). \qquad (1.77)$$

Now, denote $H_i = \{x \mid h_x^T = 0\}$, $X = R^n - \cup_{i=1}^m H_i$. For any $x(t_0) \in X$ introduce a set $\mathcal{U}_{x(t_0)}^{t_0,t_1}$, containing the sequences $\{u(t_0), u(t_0+1), \dots, u(t_1)\}$ that yield $x(t) \in X$, $t = t_0, \dots, t_1$.

We have proved the following theorem.

Theorem 1.13. *The BS (1.70)–(1.72) has the inverse on the time interval $[t_0, t_1]$ at $x(t_0) \in X$ with respect to $\{u(t_0), \dots, u(t_1)\} \in \mathcal{U}_{x(t_0)}^{t_0,t_1}$ if and only if the matrix CG has a full rank. If the BS (1.70)–(1.72) has an inverse, then the inverse system is a linear system with nonlinear output, defined by*

$$x(t+1) = \tilde{A}x(t) + \tilde{B}y(t+1),$$
$$u(t) = \tilde{C}x(t) + \tilde{D}y(t+1),$$

where

$$\tilde{A} = [I - G(CG)^{-1}C]A, \quad \tilde{B} = G(CG)^{-1},$$
$$\tilde{C} = -\Gamma^{-1}(x)(CG)^{-1}CA, \quad \tilde{D} = \Gamma^{-1}(x)(CG)^{-1}.$$

8.3   Controllability of Inverse Systems

A theorem is presented that gives sufficient conditions for the complete controllability of inverse systems.

Theorem 1.14. *If the BS (1.70)–(1.72) is completely controllable, then the inverse system (1.76) is completely controllable.*

Proof: The proof of this theorem is based on Lemma 2 by P. Hollis and D.N.P. Murthy (Kotta, 1983; Fliegner, Kotta, and Nijmeijer, 1996).

Lemma 1.10. *If the BS (1.70)–(1.72) is completely controllable, then* rank $[G\, AG \ldots A^{n-1}\, G] = n.$

The inverse system of (1.70)–(1.72), which is a time-invariant linear system by Theorem 1.13, is completely controllable if and only if rank $[\widetilde{B}\widetilde{A}\widetilde{B} \ldots \widetilde{A}^{n-1}\widetilde{B}] = n.$ *Using the relationship between $\widetilde{A}$, $\widetilde{B}$ and A, G, C given by Theorem 1.13, the matrix $[\widetilde{B}\ \widetilde{A}\widetilde{B} \ldots \widetilde{A}^{n-1}\ \widetilde{B}]$ can be expressed as*

$$[\widetilde{B}\widetilde{A}\widetilde{B} \ldots \widetilde{A}^{n-1}\widetilde{B}] = [GAG \ldots A^{n-1}G]T,$$

where T is an $nm \times nm$ upper triangular block Toeplitz matrix with the matrices $(CG)^{-1}$ on the main diagonal. It follows, therefore, that the matrix T has full rank and

$$\text{rank}\,[\widetilde{B}\widetilde{A}\widetilde{B} \ldots \widetilde{A}^{n-1}\widetilde{B}] = \text{rank}\,[GAG \ldots A^{n-1}G].$$

Applying Lemma 1.10, Theorem 1.13 follows.

9. Versal Models and Bilinear Systems
9.1 General Characteristics of Versal Models

Controlled dynamic systems are synthesized and analyzed in Udilov (1974) based on versal or universal models. The concept of a versal or universal mapping was introduced in Arnold (1968), however, the methods for calculation of the parameters of a versal or universal model using an initially given model of a time-varying system are important for engineering applications. In other words, the case in point is the construction of analytical dependence of parameters of a universal model as a function of parameters of an a priori given model (e.g., of its controlling part). This problem can also be interpreted as the problem of robust decomposition of sets of dynamic systems. It should be pointed out that each subsystem forming a part of the universal model contains a minimum admissible number of parameters from the point of view of completeness of consideration of possible variants of subsystem interaction in the initial model and admits an independent investigation. In this case, interaction between the subsystems in the initial model is reduced to parametric interaction (self-operation) in these subsystems. Interactions between the initial subsystems that cannot be removed in this way appear only in the cases where there are singularities in the initial subsystems (symmetry, close eigenfrequencies, singularity of the matrix of higher derivatives of differential equations in the initial model, and possibly some others). In addition to the circumstances mentioned above, selection of dimension of universal subsystems is determined by computing resources used for calculation of parameters of universal models from preset interaction coefficients and for investigation of the models themselves. Once such dependencies are obtained, investigation of a universal model becomes practically manageable and can be easily performed analytically.

Let us point out also that the construction of a universal model admits its extension by connecting new subsystems. In this case, algorithms for calculation of universal model parameters are arranged so that they allow us to refine the parameters of the initial universal model with regard to the presence of new subsystems and, at the same time, to determine parameters of the universal model of the connected subsystem as a function of the initial varied parameters of the whole system.

Methods for calculation of versal model parameters based on the Campbell–Hausdorff decomposition are well known.

9.2   Algorithms

Let $A = A_0 + B$, where A_0 is the constant principal matrix of the object, B is the matrix of the interaction constant or is analytically depending on the parameters. We apply to the matrix A, the homothetic transformation e^S parameterized by means of a matrix exponential curve and obtain

$$\widehat{A} = e^{-S} A e^S = e^{-S}(A_0 + B)e^S = A_0 + X.$$

The matrices S and X should be determined from the known matrix B.

Let us consider a formal expansion of S and X in terms of degrees of the matrix B:

$$S = S^1 + S^2 + \cdots, \quad X = X^1 + X^2 = \cdots,$$

where the superscript is the exponent of the expansion with respect to B. To obtain the component of this expansion, we expand the matrix A into the Campbell–Hausdorff series:

$$\widehat{A} = A_0 + X = e^s A e^s = A + [A, S] + \frac{1}{2!}[[AS]S] + \frac{1}{3!}[[[AS]S]S] + \cdots,$$

where $[A, S] = AS - SA$ is a Lie bracket. We substitute the expansions of the matrices S and X into this expansion and obtain an infinite system of relations by comparing the terms with equal indices of homogeneity:

$$[A_0 S^1 + B^1] = X^1, \quad B^1 \equiv B,$$

$$[A_0 S^2] + [B^1 S^1] + \frac{1}{2}[[A_0 S^{-1}]S^1] = X^2,$$

$$[A_0 S^3] + [B^1 S^2] + \frac{1}{2}[[A_0 S^2]S^1] + \frac{1}{2}[[B^1 S^1]S^1] + \frac{1}{2}[[A_0 S^1]S^2] = X^3.$$

The formal algorithm for the solution of these equations with respect to the homogeneous components S^i and X^i can be described as follows.

1. Select the first-degree component S^1 in such a way that a maximum number of terms of nonzero elements of the matrix B^1 are annihilated and then determine the first-degree component X^1; the known component $[B^1 S^1] + \frac{1}{2}[[A_0 S^1]S^1]$ appears in this case in the second-degree equations,

2. Select the component S^2 of the transformation so as to annihilate a maximum number of elements in the appeared component and then determine the second-degree component X^2.

The same method should be applied to the third-degree components by selecting S^3, and so on. The algorithm of the transformation e^s is reduced to compensation of as many as possible degrees of perturbation of B, and thus, to decrease its influence in the transformed matrix A. As a whole, this process turns out to be infinite. If we terminate it in N steps, then the terms of degree $N+1$ and higher with respect to B will remain in the transformed matrix, which symbolically can be written as

$$e^{-s}(A + B)e^s = A_0 + X \quad (\mathrm{mod}\ B^{N+1}).$$

A practical implementation of this algorithm is difficult, inasmuch as it is not clear how to perform its first step.

Based on the versal model theory, an alternate, more constructive algorithm can be proposed for calculation of the transformation e^s and the component X that is not annihilated in principle by this transformation.

Essentially it can be reduced to the solution of equations obtained from the Campbell–Hausdorff expansion, simultaneously for the matrices S and X, using the structure of these matrices known from the versal model theory. In other words, we search for the matrices S in the form of expansion in terms of the base $\{S\}$ from matrices transversal to the centralizer of the matrix A_0:

$$S = \sum_{i=1}^{m} \omega_i S_i \equiv S^1 + S^2 + \cdots.$$

The basic matrices S, for different types of the matrices A_0 can be constructed in an explicit form. We search for the matrices X in the form of $\{x_k\}$-base expansion of the normal to the orbit:

$$X = \sum_{k=1}^{p} \lambda_k X_k \equiv X^1 + X^2 + \cdots, \quad p = n^2 - m.$$

Let us point out that each matrix of the infinite sequences of the matrices $S^1, S^2, \ldots$ (or $X^1, X^2, \ldots$) can be decomposed in terms of a finite base $\{S_i\}$ or $\{x_i\}$, respectively.

If the matrix B is given numerically, then we have the following system of equations for determination of the homogeneous components S^i and X^i from the Campbell–Hausdorff expansion,

$$X^1 - [A_0 S^1] = B^1 \equiv B,$$

$$X^2 - [A_0 S^2] = B^2 = [B^1 S^1] + \frac{1}{2}[[A_0 S^{-1}]S^1],$$

$$X^3 - [A_0 S^3] = B^3 = [B^1 S^2] + \frac{1}{2}[[A_0 S^2]S^1]$$

$$+ \frac{1}{2}[[B^1 S^1]S^1] + \frac{1}{2}[[A_0 S^1]S^2],$$

which can be solved recurrently. With a given structure of the matrices S^i and X^i, each equation of this system is of the same type and they differ only by their right-hand sides. A solution of each equation can be obtained by parts using a block representation of the matrices A, S^i, and X^i. The required result is obtained through summation of a finite number of the matrices S^i and X^i with the selected degree N of homogeneity.

Let us consider the algorithm of construction of the solution in the form of an explicit dependence on varied parameters. Let the matrix B of dimension $n \times n$ be a linear function of parameters

$$B(\mu) = \sum_{i=1}^{S} \mu_i B_i, \quad S \le n^2,$$

where B_i are constant matrices.

We present homogeneous components of the matrices X and S in the form

$$X^1 = \sum_{j=1}^{s} \mu_j Y_j, \quad X^2 = \sum_{j,k=1}^{s} \mu_j \mu_k Y_{jk},$$

$$X^3 = \sum_{j,k,l=1}^{s} \mu_j \mu_k \mu_i Y_{j,k,i}, \dots,$$

$$S^1 = \sum_{j=1}^{s} \mu_j Q_j, \quad S^2 = \sum_{j,k=1}^{s} \mu_j \mu_k Q_{jk},$$

$$S^3 = \sum_{j,k,l=1}^{s} \mu_j \mu_k \mu_i Q_{j,k,i}, \dots, \tag{1.78}$$

where $Y_j Y_{jk} \dots, Q_j Q_{jk} \dots$ are two infinite sequences of matrices from finite-dimensional spaces $X = \{X_1 \dots X_p\}$ and $S = \{S_1 \dots S_m\}$.

Having substituted these expansions into the Campbell–Hausdorff expansion, we obtain the equations for determining the matrices Y_j, Q_j, Y_{jk}, and Q_{jk}:

$$Y_j - [A_0 Q_j] = B_j,$$

$$Y_{jk} - [A_0 Q_{jk}] = B_{jk} = [B_j Q_k] + \frac{1}{2}[[A_0 Q_j]Q_k],$$

$$Y_{jki} - [A_0 Q_{jkl}] = B_{jkl} = [B_j Q_{kl}] + \frac{1}{2}[[A_0 Q_{kl}]Q_j]$$

$$+ \frac{1}{2}[[B_j Q_k]Q_l] + \frac{1}{2}[[A_0 Q_l]Q_{kl}],$$

$$j, k, l = 1, \ldots, S.$$

Because the spaces of the matrices X and S are of finite dimension, each of the two infinite sequences of the matrices $\{Y_j, Y_{jk}, \ldots\}$ and $\{Q_j, Q_{jk}, \ldots\}$ is a finite-dimensional linear combination of the basic sequences:

$$Y_j = \sum_{q=1}^{p} a_{jq} X_q, \quad Y_{jk} = \sum_{q=1}^{p} a_{jkq} X_q, \quad Y_{jkl} = \sum_{q=1}^{p} a_{jklq} X_q,$$

$$Q_j = \sum_{r=1}^{m} b_{jr} S_r, \quad Q_{jk} = \sum_{r=1}^{m} b_{jkr} S_r, \quad Q_{jklr} = \sum_{r=1}^{m} b_{jklr} S_r, \qquad (1.79)$$

where $\{a_{iq} a_{jkq} \ldots\}$, $\{b_{ir} b_{jkr} \ldots\}$ are constant coefficients that can be calculated from the systems of linear algebraic equations of the type

$$\sum_{q=1}^{p} a_q Sp X_q X_{q'}^* = Y, \quad q' = 1, \ldots, p,$$

$$\sum_{r=1}^{m} b_r Sp S_r S_{r'}^* = Q, \quad r' = 1, \ldots, m,$$

after substitution of the matrices $\{Y_j Y_{jk} \ldots\}$ for the coefficients $\{a_j, a_{jq} \ldots\}$ and matrices $\{Q_j, Q_{jk} \ldots\}$ for the coefficients $\{b_j, b_{jk} \ldots\}$ into their right-hand sides.

Having substituted expansions (1.79) into expansion (1.78), we obtain expressions for the parameters of the universal model in the form of power series in parameters of the initial strain:

$$\omega_r(\mu) = \sum_{j=1}^{s} b_{jr}\mu_j + \sum_{j,k=1}^{s} b_{jkr}\mu_j\mu_k + \sum_{j,k,l=1}^{s} b_{jklr}\mu_j\mu_k\mu_l + \cdots,$$

$$\lambda_q(\mu) = \sum_{j=1}^{s} a_{jq}\mu_j + \sum_{j,k=1}^{s} a_{jkq}\mu_j\mu_k + \sum_{j,k,l=1}^{s} f_{jklq}\mu_j\mu_k\mu_l + \cdots.$$

If we restrict ourselves to the terms of the Nth degree in these series, then we can speak about universal models of the orders $1, 2 \ldots, M$. More results can be found in Udilov (1999.)

10.   Notes and Sources

The study of bilinear systems was initiated by Brockett (1972) and Mohler (1973). The material contained in Section 1.1 is a synthesis from the early papers of Yatsenko (1984, 1985). Theorem 1.1 and Corollary 1.1 are essentially taken from Lo (1975). The results presented in Section 1.4 are fundamentally inspired by the papers of Krener (1973, 1975) and also appear in the Andreev (1982) paper.

The idea of using Lie algebra as the basic object for bilinear systems is a synthesis from the papers of Brockett (1972, 1973, 1979) and Jurdjevic (1997, 1998). Controllability and observability of control systems can be traced to the early papers of Kuchtenko (1963), Butkovskiy (1990), Andreev (1982,) and Jurdjevic (1997).

The results presented in Sections 1.7,1.8 are inspired by the papers of Hirshorn (1977) and Kotta (1983) and also appear in the Willsky (1973) paper. These results are used to construct nonlinear systems that act as left-inverses for bilinear systems.

Chapter 2

CONTROL OF BILINEAR SYSTEMS

Bilinear systems are an important subclass of nonlinear systems with numerous applications in engineering, biology, and economics (Mohler, 1973; Espana and Landau, 1978; Brockett, 1979; Baillieul, 1978, 1998). Most papers consider time-invariant continuous bilinear systems with linear feedback. In almost all references stability problems are studied using a sufficient condition for the existence of a feedback control. These studies usually do not deal with stability problems with a large class of inputs as considered here. The linear-quadratic optimal control problem is apparently one of the simplest and most thoroughly studied. It is not surprising, therefore, that the nonlinear analogues of this problem have long since attracted the attention of control scientists (Andreev, 1982; Aganovic, 1995; Bloch and Crouch, 1996; Bloch, 1998; Jurdjevic, 1998; Agrachev and Sachkov, 2004). The symplectic structure of the linear-quadratic problem was also studied in Faibusovich (1988a, 1991), and has naturally led to the question of nonlinear generalizations.

In this chapter we consider optimal control problems for nonlinear systems. In Section 1 we demonstrate how to reduce this problem to the equivalent problem for bilinear systems. Then we investigate an optimal control problem for nonlinear systems through necessary conditions of optimality. We apply soliton methods and Pontryagin's principle to the optimal control problem. We give theoretical justification for the equations in the Lax form, which can be constructed from the Hamiltonian system arising through the expression of the necessary optimality conditions. The Lax forms we consider, extend and unify similar results obtained previously in the publications on optimal control of bilinear systems. The theoretical justification was done by V. Yatsenko (1984),

P.M. Pardalos, V. Yatsenko, *Optimization and Control of Bilinear Systems*, doi: 10.1007/978-0-387-73669-3,
© Springer Science+Business Media, LLC 2008

who applied the soliton method to optimal control problems. However, the approach is general and the specific Lie algebraic structure of bilinear systems is overlooked. This explains why here we obtain two equations in the Lax form instead of one. Moreover, we show that it is not difficult to extend the approach to nonlinear systems. Note that the proposed procedure is global and, in particular, it is applicable to controllable systems on manifolds. This degree of generality is dictated by potential applications (mechanical systems, robotics, multiagent systems, nonlinear circuits, control of quantum-mechanical processes) and by the fact that a number of geometrical problems that have recently been attracting considerable attention fit into the proposed framework. We also examine the selection of a performance criterion and present a number of examples. Sections 2 and 3 deal with stabilization problems for controlled physical objects. A variety of physical and biological systems are well modeled by coupled bilinear equations. In most cases such systems are capable of displaying several types of dynamical behavior: limit cycles, bistability, excitation, or chaos.

1.   Optimal Control of Bilinear Systems

1.1   Optimal Control Problem

In this section, we consider the following bilinear control systems:

$$\dot{x}(t) = f_0(x) + F(x)u(t), \qquad u(t) \in \Omega,$$
$$z(t) = h(x) + Q(x)v(t), \qquad v(t) \in \Gamma,$$
$$z(t) \in \mathbb{R}^r, \qquad x(t) \in \mathbb{R}^n, \tag{2.1}$$

where x is a state vector; z is an output signal; $u(t)$ and $v(t)$ are control functions of the time of dimensions m, q, respectively; $f_0(x)$ and $h(x)$ are given vector-valued functions of the $x(t)$ of dimensions n, p, respectively; and F and Q are matrix-valued functions of $x(t)$ of appropriate sizes.

At the initial time $t = t_0$, the initial condition for the system (2.1) is specified,

$$x(t_0) = x_0, \tag{2.2}$$

where x_0 is a given n-dimensional vector.

We consider the following objective function,

$$\eta = \int_0^T \sum_{i,j=1}^m q_{ij} u_i u_j dt, \tag{2.3}$$

where $\tilde{Q} = (q_{ij})$ is a symmetric positively determined matrix.

We seek the control $u(t)$ that yields the minimal value to the function (2.3)

$$\min \eta = \min \int_0^T \sum_{i,j=1}^m q_{ij} u_i u_j \, dt, \qquad (2.4)$$

provided that conditions (2.1) and (2.3) hold.

1.2 Reduction of Control Problem to Equivalent Problem for Bilinear Systems

In this section we consider reduction of the nonlinear system (2.1) to a dynamically equivalent form, which allows us to simplify the solution of the corresponding optimal control problem. This simplification is achieved by the analytical solution of the optimal control problem for a dynamically equivalent bilinear system. It is shown here that the optimal control satisfies the Euler–Lagrange equation, the solutions of which are expressed analytically through the Θ-functions of Riemann surfaces. In addition, bilinear systems allow the use of known mathematical system theory results in order to investigate the system features of controlled processes (Jurdjevic, 1997; Aganović and Gajic, 1995).

Bilinearization of a nonlinear system. Write the first equation of system (2.1) as

$$\dot{x}(t) = f_0(x) + \sum_{i=1}^m u_i(t) f_i(x), \qquad x(0) = x_0, \qquad (2.5)$$

where f_i is the ith column of a matrix F. Instead of considering (2.5), we study the following bilinear system,

$$\dot{y}(t) = \left(A_0 + \sum_{i=1}^m u_i(t) A_i \right) y, \qquad y(0) = y_0, \qquad (2.6)$$

where $y = (y_1, \ldots, y_p)$ is a state vector; $A_0, \ldots, A_m$ are constant $p \times p$ matrices; $u(t) = (u_1(t), \ldots, u_m(t))$ is a restricted measurable control. To construct the system (2.6) the approach described in Krener (1973) is used.

Let $P^0(\tilde{A})$ be the set of functions $\{f_i : i = 0, \ldots, m\}$; $P^j(A) = P^{j-1}(\tilde{A}) \cup \{[f_i, c] : i = 0, \ldots, m, \ c \in P^{j-1}(\tilde{A})\}$ for $j \leq 1$. The completed system of $\tilde{A}$ is $P(\tilde{A}) = \bigcup_{j>0} P^j(\tilde{A})$, and we define $P(\tilde{A})_x = \{c(x) : c \in P(\tilde{A}) \subseteq \mathbb{R}^m\}$, $([\cdot, \cdot]$ denotes the Riemann surface Lie bracket of vector fields); and $P(B)$ is a corresponding function set for system (2.5). Consider the systems (2.5) and (2.6). Let M and N be submanifolds that carry (2.5) and (2.6) at x_0 and y_0. Then there exists such a linear map l: $\operatorname{Span} P(\tilde{A})_{x_0} \rightarrow \operatorname{Span} P(\tilde{B})_{y_0}$, such that $l(f_i(x_0)) = A_i \, (y_0)$ for $i = 0, \ldots, m$, $l([f_{i_1} \ldots , [f_{i_{h-1}}, f_{i_h}] \ldots](x_0)) = [A_{i_1}, \ldots , [A_{i_{h-1}}, A_{i_h}] \ldots](y_0)$ for $h \leq 2$, $1 \leq i_j \leq m$ if and only if there exist neighborhoods U and V and an analytical map $\lambda : U \rightarrow V$ such that λ carries (2.5) and (2.6) for the same control $u(t)$ and $x(t) \in U$ for $|t| < \varepsilon$. Furthermore l is a linear diffeomorphism if and only if λ is a local diffeomorphism.

Instead of considering (2.5), we study the following matrix bilinear system (Krener, 1975),

$$\dot{Y}(t) = \left(A_0 + \sum_{i=1}^{m} u_i(t) A_i \right) Y(t), \quad Y(0) = I, \quad u(t) \in \Omega, \qquad (2.7)$$

where $Y(t) \in Gl(p, \mathbb{R})$; $Gl(p, \mathbb{R})$ is a group of reverse $p \times p$-matrices.

Use the minimal algebra $W(M)$, containing a totality of vector fields $\{f_0, \ldots, f_k\}$, to construct the system (2.7). The Lie algebra $W(M)$ is infinitely dimensional in general, therefore instead of the algebra $W(M)$, build up the finite-dimensional algebra $\hat{\Theta}$ using the ν-order Lie bracket:

$$[f_{i_1} \cdots [f_{i_{\nu-1}}, f_{i_\nu}] \cdots], f_0, \ldots, f_h \in \hat{\Theta}. \qquad (2.8)$$

In this case, ν is stated by the accuracy of approximation of the solutions to system (2.5) by the solutions to system (2.6): the greater ν is, the higher an approximation accuracy is.

If $f_0, \ldots, f_h$ are the linearly independent elements of the algebra $\hat{\Theta}$ over the field $\mathbb{R}$, then, when the switching operation is used (up to the order ν), it is possible to obtain new elements of the algebra $\hat{\Theta}$. Assuming every Lie bracket of an order higher than ν, equal to zero, the finite-dimensional Lie algebra is derived. Let its dimensionality be equal to s. According to the Ado theorem (Ado, 1947), this algebra is isomorphic to the subalgebra $\hat{g}$ of the algebra R. Bring the $s \times s$-dimensional matrix A_i into correspondence with each f_i, and the ν-order bilinear system is obtained. The constructive building procedure, according to which the matrix bilinear system (2.7) is made up, uses the adjacent representation of the algebra $\hat{\Theta}$. If $d_1, \ldots, d_s$ is the basis of $\hat{\Theta}$, then, for each $\hat{\Theta}$, assume $ad(c)$ equal to the matrix $B = [B_{ij}]$, defined by the relation

$$[c, d_i] = \sum_{i=1}^{m} B_{ij} d_i. \qquad (2.9)$$

The adjoint representation of the $(\nu + 1)$-order canonical algebra is isomorphic with respect to the canonical algebra of the same order, and it may be used to construct an equivalent system.

Build up a mapping λ, representing a local diffeomorphism of system state spaces and converting the solutions of the system (2.5) into system (2.6) under the same controls. If the linear mapping l exists, $M = \mathbb{R}^n$, $N = \mathbb{R}^p$, $c_1(x_0), \ldots, c_n(x_0)$, then there is the basis of $\mathbb{R}^n$, and $d_1(y_0), \ldots, d_n(y_0)$ are the corresponding vectors from $p(\tilde{B})y_0$. Determine the mapping $(t, x) \rightarrow \alpha_i(t)x$ as a family of integral curves for $c_i(x)$, $i = 1, \ldots, n$, that is, such that $(d/dt)\alpha_i(t)x = c_i(\alpha_i(t)x)$, $\alpha_i(0) = x$. The mapping $(t, y) \rightarrow \beta_i(t)y$ is determined by analogy from the equation

$$(d/dt)\beta_i(t)y = d_i(\beta_i(t)y), \quad \beta_i(0)y = y, \quad i = 1, \ldots, n. \quad (2.10)$$

Introduce the variable $s = (s_1, \ldots, s_n)$ and determine the mappings $g_1 \colon s \rightarrow x$ and $g_2 \colon s \rightarrow y$, assuming $g_1(s)\alpha_n(s_n) \cdots \alpha_2(s_2)\alpha_1(s_1)x_0$ and $g_2(s) = \beta_n(s_n) \cdots \beta_2(s_2)\beta_1(s_1)y_0$. Then, $(\partial g_1/\partial s_i)(0) = c_i(x_0)$ and g_1 has the inverse mapping $g_1^{-1} \colon x \rightarrow s$ for the points $x \in U$. The mapping λ is determined by the relation $\lambda = g_2 \circ g_1^{-1}$.

Global bilinearization of nonlinear systems. Let L be the differential operator defined by

$$L(g(x)) = \sum_{i=1}^{n} f_{(i)}(x)\frac{\partial}{\partial x_i}(g(x)) = g_x(x)f(x) \quad (2.11)$$

and define the set of functions

$$S = \{h(x), Lh(x), \ldots\} \cup \left[\left\{\bigcup_{i=1}^{\hat{q}} Q_i(x), LQ_i(x), \ldots\right\}\right], \quad (2.12)$$

where Q_i is the ith column of the matrix Q; $g(x)$ is a differentiable scalar function; $f_{(i)}$ is the ith component of the vector-function $f(x)$.

If there exist integers M_i, $i = 0, \ldots, \hat{q}$, such that for $k = 1, \ldots, \hat{q}$ and all states $x(t)$, $t \in T$,

$$L^{M_0}h(x(t)) = \sum_{i=0}^{M_0-1} A_0(0, 0, i+1)L^i h(x(t))$$

$$+ \sum_{j=1}^{\hat{q}} \sum_{i=0}^{M_{\hat{q}}-1} A_0(0, j, i+1)L^i Q_i(x(t)), \quad (2.13)$$

$$\ldots \quad (2.14)$$

$$L^{M_k}Q_k(x(t)) = \sum_{i=0}^{M_0-1} A(k,0,i+1)L^i h(x(t)) \qquad (2.15)$$

$$+ \sum_{j=1}^{q} \sum_{i=0}^{M_q-1} A(k,j,i+1)L^i Q_j(x(t)), \qquad (2.16)$$

where $A_0(i,j,k)$ is a constant $(p \times p)$-matrix, and every column of $(L^i h(x))_x G(x)$ and $(L^i Q_j(x))_x G(x)$, $i = 0, \ldots, M_{j-1}$, $j = 1, \ldots, \hat{q}$ lies on S; that is, for $k = 1, \ldots, m$ and all states $x(t)$, $t \in T$, the kth column of $(L^i h(x))_x G(x)$ is equal to the expression

$$\sum_{l=1}^{M_0} A_j(k,i,0,j)L^{j-1}h(x) + \sum_{j=1}^{\hat{q}} \sum_{l=1}^{M_{\hat{q}}} A(k,i,j,l,)L^{l-1}Q_j(x), \qquad (2.17)$$

and the kth column $(L^i Q_j(x))_x G(x)$ is equal to the vector

$$\sum_{l=1}^{M_0} A_j(k,i,0,l)L^{l-1}h(x) + \sum_{r=1}^{\hat{q}} \sum_{l=1}^{M_{\hat{q}}} A_j(k,i,r,l)L^{l-1}Q_r(x), \qquad (2.18)$$

and for some positive integers M_i, $i = 0, \ldots, \hat{q}$,

$$\text{rank}\, [C, A_0'C', \ldots, (A_0')^{M_0-1}C', D_1', A_0'D_1', \ldots, (A_0')^{M_1-1}D_1', \ldots$$
$$D_{\hat{q}}'A_0'D_{\hat{q}}', \ldots, (A_0')^{M_{\hat{q}}-1}D_{\hat{q}}'] = \dim A; \qquad (2.19)$$

then (Lo, 1975) there is the bilinear system,

$$\dot{y}(t) = \left(A_0 + \sum_{i=1}^{m} A_i u_i(t) \right) y(t), \qquad (2.20)$$

$$z(t) = \left(C + \sum_{j=1}^{\hat{q}} D_j v_j(t) \right) y(t). \qquad (2.21)$$

Here A_0, A_i, C, D_j are constant matrices of appropriate sizes.

1.3   Optimal Control of Bilinear Systems

An optimal control for bilinear systems is considered here to describe the basic idea behind our new approach. It is shown that the optimal control problem can be reduced to the Euler equations. The optimal control problem solution is reduced to an integration of the Euler equations admitting a Lax representation. A soliton method (Zakharov et al., 1980) is proposed for the integration of the Euler equation.

Euler–Lagrange equation. Suppose that G is a matrix Lie group with corresponding Lie algebra g. Consider the dynamically equivalent bilinear system defined on G by

$$(d/dt)Y(t) = \left(A_0 + \sum_{i=1}^{m} u_i(t) A_i \right) Y(t), \qquad (2.22)$$

where $A_0, A_1, \ldots, A_m$ are constant $(p \times p)$-matrices; $Y(t)$ is a varying $(p \times p)$-matrix; $u(t) = (u_1(t), \ldots, u_m(t))$ is a control, that is, a measurable function belonging to an input set Ω.

Given an admissible class Ω of control functions, we wish to find $u_1(t), \ldots, u_m(t)$ in Ω that steer the state of (2.22) from $I \in G$ to $Y_1 \in G$ in T units of time in such a way as to minimize the cost functional

$$\eta = \int_0^T \sum_{i,j=1}^{m} q_{ij} u_i u_j \, dt, \qquad (2.23)$$

where $\tilde{Q} = (q_{ij})$ is a symmetric and positive-definite matrix.

We assume that the $\{A_1, \ldots, A_m\}$ in (2.22) Span g. In the light of the assumption set out in this section, this makes $\{A_1, \ldots, A_m\}$ a basis for g.

Theorem 2.1. *Let R be a nonsingular matrix either symmetric or skew-symmetric such that $R^2 = \pm I$. Suppose that*

$$g = \{C \in gl(n, \mathbb{R}) : C^t R + RC = 0\},$$
$$Y_1 \in C = \{\exp g\}_G, \quad T > 0. \qquad (2.24)$$

Then there exists an optimal control matrix

$$U^0(t) = \sum_{i=1}^{m} u_i^0(t) A_i, \qquad (2.25)$$

that steers (2.22) from I at $t = 0$ to Y_1 at $t = T$ such that (2.23) is minimized.

The optimal control matrix satisfies the differential equation (Bailleul, 1978)

$$(d/dt) \left(\sum_{i=1}^{m} u_i(t) A_i \right)$$

$$= \tilde{Q}^{-1} \left(\left[\tilde{Q} \left(\sum_{i=1}^{m} u_i(t) A_i \right), \; A_0^t + \sum_{i=1}^{m} u_i(t) A_i^t \right] \right), \qquad (2.26)$$

where $[\cdot, \cdot]$ denotes the Lie bracket of the matrices $\tilde{Q} \left(\sum_{i=1}^{m} u_i(t) A_i \right)$ and $A_0^t + \sum_{i=1}^{m} u_i(t) A_i^t$.

We can now rewrite equation (2.26) as follows,

$$(d/dt)\left[\tilde{Q}\left(\sum_{i=1}^{m} u_i(t)A_i\right)\right]$$
$$= \left[\tilde{Q}\left(\sum_{i=1}^{m} u_i(t)A_i, \ A^t + \sum_{i-1}^{m} u_i(t)A_i^t\right)\right]. \tag{2.27}$$

Using the Lie bracket, this equation can be written in the form:

$$\dot{M} = [M, \Omega], \tag{2.28}$$

where

$$\Omega = A^t + \sum_{i=1}^{m} u_i(t)A_i^t, \quad M = \tilde{Q}\left(\sum_{i=1}^{m} u_i(t)A_i\right). \tag{2.29}$$

We call equation (2.28) *the Euler–Lagrange equation* for our optimization problem.

1.4   On the Solution of the Euler–Lagrange Equation

If $A_0^t + \sum_{i=1}^{m} u_i(t)A_i^t \in \tilde{g}$, where $\tilde{g}$ is an algebra of real skew-symmetric $(n \times n)$-matrices with the ordinary operation of commutation, and S: $\tilde{g} \to \tilde{g}$ is a linear operator in g,

$$S\left(A_0^t + \sum_{i=1}^{m} u_i(t)A_i^t\right) = \tilde{Q}\left(\sum_{i=1}^{m} u_i(t)A_i\right), \tag{2.30}$$

then the Euler–Arnold equation is

$$d/dt\left(\tilde{Q}\left(\sum_{i=1}^{m} u_i(t)A_i\right)\right)$$
$$= \left[\tilde{Q}\left(\sum_{i=1}^{m} u_i(t)A_i\right), \ A_0^t + \sum_{i=1}^{m} u_i(t)A_i^t\right]. \tag{2.31}$$

It follows from (2.31) that is, the eigenvalues of the matrix $M = \tilde{Q}\left(\sum_{i=1}^{m} u_i(t)A_i\right)$ are preserved in time; that is, the traces of powers in equation (2.30), the power traces of M, are integrals of motion. In every invariant manifold, distinguished by these conditions, (2.30) is a Hamiltonian system.

An interesting example of Euler's equations in the group g is provided by the equation of free rotation of an n-dimension rigid body. In this case, $S\Omega = J \bullet \Omega + \Omega \bullet J$, where J is a symmetric positive-definite matrix (inertia tensor), which can always be regarded as diagonal, and (2.31) can be rewritten as

$$\dot{J\Omega} = \dot{\Omega}J = [J, \Omega_2]. \tag{2.32}$$

Equation (2.32) with arbitrary n was first considered by Mishchenko (1970) who discovered a series of nontrivial quadratic integrals.

By Liouville's theorem, there are sufficient Mishchenko integrals in the case $n = 4$ for proving the complete integrability of Euler's equations of a four-dimensional rigid body. For any n, equation (2.32) has $N(n)$ single-valued integrals of motion, and its general solution is expressible in terms of Θ-functions of Riemann surfaces (Manakov, 1976; Zakharov et al., 1980).

Example 2.1. Consider the bilinear system

$$(d/dt)X(t) = \begin{bmatrix} 0 & u_3(t) & -u_2(t) \\ -u_3(t) & 0 & u_1(t) \\ u_2(t) & -u_1(t) & 0 \end{bmatrix} X(t) \tag{2.33}$$

and the performance criterion $\eta = \int_0^T \left[q_1 u_1(t)^2 + q_2 u_2(t)^2 + q_3 u_3(t)^2 \right] dt$, $q > 0$.

From Theorem 2.1, we find that optimal controls steering this system between fixed endpoints satisfy

$$q_1(du_1/dt) = (q_2 - q_3)u_2 u_3, \quad q_2(du_2/dt) = (q_3 - q_1)u_1 u_3,$$
$$q_3(du_3/dt) = (q_1 - q_2)u_1 u_2. \tag{2.34}$$

Interpreting the u_i as angular velocities and q_i as moments of inertia about principal axes, the optimization problem corresponds to the problem in classical mechanics of finding the equations of motion of a rotating solid body in the absence of external torques (Baillieul, 1978). Here, η is the action, and equation (2.34) are Euler equations and Theorem 2.2 (Baillieul, 1978) show that kinetic energy and angular momentum are conserved.

Unfortunately, if we do not assume that the A_is span the Lie algebra, g is no longer a necessary condition for optimal control. Nevertheless, techniques exist that allow us to develop the requisite necessary condition, even when the A_is do not span. One approach is the maximum principle of this section. Alternatively, we can use a limiting argument coupled with (2.26).

A more direct approach to the optimization problem is to involve the high-order maximum principle developed by Krener (1977).

Theorem 2.2. *Suppose that h_0 and h_0^t have the orthogonal direct sum decomposition as*

$$h_0 = k_0 \bigoplus \cdots \bigoplus k_{r-1} \bigoplus k_{r+1} \bigoplus \cdots \bigoplus k_{s-1},$$

and let $r = 1$ in (2.31).

Suppose, moreover, that $\tilde{Q} : h_0 \to h_0$ has the form assumed above. Then, the solution to (2.26) may be written as

$$\sum_{i=1}^{m} u_i^0(t) A_i = U_0(t) + \sum_{i=1}^{s-1} U_i(t)$$

$$= \exp(A_0 t) \cdot C_1 \cdot \exp(-A_0 t)$$

$$+ \exp(A_0 t) \cdot \exp(C_1 t) \cdot \exp(C_2 - C_1) t \cdot U_0(0) \cdot \exp(-C_1 t) \cdot \exp(-A_0 t),$$

$$\sum_{i=1}^{\nu} \cdot v_i(t) A_i = U_r(t) + U_s(t)$$

$$= \exp(A_0 t) \cdot \exp(C_1 t) \cdot (U_r(0) + U_s(0)) \cdot \exp(-C_1 t) \cdot \exp(-A_0 t),$$

where

$$C_1 = \sum_{r+1}^{s-1} U_i(t),$$

$$C_2 = \sum_{r+1}^{s-1} (\lambda_{r-1} - \lambda_j)/\lambda_{r-1} U_i(0) - \lambda_{r-1}^{-1}(U_r(0) + U_s(0)).$$

Remark 2.1. In the paper by Faibusovich (1988) an application of the method of collective Hamiltonians to a class of optimal control problems can be found. A Hamiltonian system of the maximum principle is reduced to a system of differential equations on the dual to an optimal control Lie algebra of a problem (LOC) endowed with a Lie–Berezin–Kirillov Poisson structure. It enables as to construct exactly solvable cases using some techniques developed for completely integrable systems.

2.　Stability of Bilinear Systems

There has been tremendous interest in nonlinear stabilization problems in the recent years, as evidenced by several numerical research articles (Ionescu and Monopoli, 1975; Dayawansa, 1998; Hanba and Yoshihiko, 2001). The main contributing factors that have been realized are modern sensors and advanced superconducting devices, and the like, which cannot be analyzed by using linear techniques alone. More advanced techniques are necessary in order to meet the design challenges. These lead to the generalization of well-known bilinear control theories such as stabilization of active physical systems and generalization of adaptive linear systems to the bilinear settings. There exist two distinctively different approaches that are actually equivalent to feedback control: the first one is dynamic programming and the other is the regular approach (Schättler, 1998). In dynamic programming the value function is calculated as a solution of the Hamilton–Jacobi–Bellman equation. Regular synthesis is a generalization of the classical method of characteristics for first-order partial differential equations to the Hamiltonian–Jacobi–Bellman equation and hence another way to realize dynamic programming. This section describes feedback algorithms for continuous and discrete systems.

2.1   Normed Vector Space

Let $\mathbb{R}^n$ denote an n-dimensional vector space and the norm of a vector $x = (x_1, \ldots, x_n)^T$ on $\mathbb{R}^n$ be denoted by

$$\|x\| = \left(\sum_{i=1}^{n} |x_i|^2 \right)^{1/2}.$$

If A is an $n \times m$ matrix over $\mathbb{R}$, then the norm of A is defined by

$$\|A\| = \left(\sum_{i,j} |a_{ij}|^2 \right)^{1/2}.$$

If f is a linear function from $\mathbb{R}^n$ to $\mathbb{R}$ then the norm of f is defined by

$$\|f\| = \sup_{x \neq 0} \frac{\|f(x)\|}{\|x\|}.$$

Let $L^p([t_0, \infty))$ denote the set of the measurable functions g : $[t_0, \infty) \to \mathbb{R}^n$, such that

$$\|g\|_p = \left(\int_{t_0}^{\infty} \|g(t)\|^p dt \right)^{1/p} < \infty \quad \text{if } 1 \leq p < \infty$$

and

$$\|g\|_\infty = \operatorname*{ess\,sup}_{t \in (t_0, \infty)} \|g(t)\| < \infty \quad \text{if } p = \infty.$$

Let $G(t)$ be an $m \times n$ matrix. Note that $G(t)$ is bounded on $L^p([t_0, \infty))$ if

$$\|G(t)\|_p \triangleq \left(\int_{t_0}^\infty \|G(t)\|^p dt \right)^{1/p} < \infty, \quad p \neq \infty.$$

Suppose f is a continuous function (defined from $\mathbb{R}^n \to \mathbb{R}$), $f(0) = 0$ and satisfies the following hypothesis

$$\int_0^T \frac{\omega(t)}{t} dt \leq K_f T, \tag{2.35}$$

where

$$\omega(t) = \sup_{\|x\| \leq t} |f(x)|.$$

The inequality (2.35) is satisfied if either the function is linear or satisfies the Lipschitz condition and $f(0) = 0$; that is, $f(x) \leq K\|x\|$, for some constant K.

The following lemma is useful for both continuous and discrete systems.

Lemma 2.1. *Let f be a measurable function defined from $\mathbb{R}^n$ to $\mathbb{R}$ and satisfying (2.35). Suppose*

$$u(t) = f(y(t)), \quad y(t) = H(t)x(t),$$

then

$$|u(t)| \leq 4K_f \|H(t)\| \|x(t)\|.$$

Proof: Let us write

$$|u(t)| = |f(H(t)x(t))| \leq \sum_{j=0}^\infty \|Z\| \leq 2^{-j} \|H(t)\| \|x(t)\|^{\|f(Z)\|}$$

$$= \sum_{j=0}^\infty \omega(2^{-j} \|H(t)\| \|x(t)\|)$$

$$- 2 \sum_{j=0}^\infty (\omega(2^{-j} \|H(t)\| \|x(t)\|)/2^{-j+1})(2^{-j+1} - 2^{-j}). \tag{2.36}$$

It is clear that $\omega(t)$ is a bounded increasing function. Therefore, the term in the first pair of parentheses is not larger than the minimum value of the function,

$$\omega(s\|H(t)\| \|x(t)\|)/s$$

on the interval

$$s \in [2^{-j}, 2^{-j+1}].$$

So, the last sum is less than

$$2 \int_0^2 \frac{\omega(s\|H(t)\| \, \|x(t)\|)}{s} ds = 2 \int_0^{2s\|H(t)\| \, \|x(t)\|} \frac{\omega(s)}{s} ds. \qquad (2.37)$$

From (2.35), we conclude that

$$\|u(t)\| \leq 4K_f \|H(t)\| \, \|x(t)\|.$$

Remark 2.2. Here we should remark that K_f defined in (2.35) depends on the function f; for example, $K_f = \|f\|$ if f is a linear function, $K_f = K_u$ if f satisfies the Lipschitz condition: $\|f(x_2) - f(x_1)\| \leq K_u |x_2 - x_1|$, and $f(0) = 0$. Therefore, for these two kinds of special functions, linear functions, or functions in the Lipschitz class, we have a better estimation,

$$|u(t)| \leq \|f\| \, \|H(t)\| \|x(t)\|.$$

It means that we can use $\|f\|$ and K_u to substitute K_f in the following theorems when f is a linear function and K_u is a function in the Lipschitz class, respectively.

2.2 Continuous Bilinear Systems

Let us define the continuous bilinear systems (BLS) as follows,

$$\dot{x}(t) = A(t)x(t) + B(t)x(t)u(t) + c(t)u(t),$$
$$y(t) = H(t)x(t), \quad u(t) = f(y(t)), \quad x_0 = x(t_0), \qquad (2.38)$$

where $x(t)$, $y(t)$ are n-dimensional vectors, A, B, H are $n \times n$ matrices, c is an $n \times 1$ matrix, and the entries of A, B, H, c are continuous functions. We assume that the solutions $x(t)$ are continuously differentiable functions, and the solutions are uniquely defined for the initial state x_0.

For BS (2.38), the following Theorem 2.3 may be derived from Lemma 2.1.

Theorem 2.3. *Let $x(t)$ denote a solution of the system (2.38). Suppose there exists a matrix D defined on time t, such that $A = DD^{-1}$. Let*

$$\Lambda \overset{\Delta}{=} \int_0^\infty 4K_f \|H\| \, \|D^{-1}\| \, \|D\| \, \|c\| dt$$

and

$$G \triangleq \int_0^\infty 4K_f \|H\| \|D^{-1}\| \|D\|^2 \|B\| dt.$$

If

$$1 - \Lambda e^\Lambda > 0,$$

then there exists a $\delta > 0$, *such that*

$$\|x(t)\| \leq \frac{\|D^{-1}(t_0)X_0\| \|D(t)\|}{1 - (\Lambda + \|D^{-1}(t_0)X_0\|G)e^\Lambda} \tag{2.39}$$

for $\|X_0\| \leq \delta$.

Remark 2.3. From (2.39), it is not difficult to show that the equilibrium at the origin of the system is stable if $\|D(t)\| \in L^\infty([t_0, \infty))$, and asymptotically stable if $\lim_{t\to\infty} \|D(t)\| = 0$.

The proof is somewhat lengthy and given by Mohler (1973).

Example 2.2. Consider the following bilinear system

$$\begin{bmatrix} \dot{X}_1(t) \\ \dot{X}_2(t) \end{bmatrix} = \begin{bmatrix} -2 & -1 \\ 0 & -3 \end{bmatrix} \begin{bmatrix} x_1(t) \\ x_2(t) \end{bmatrix} + \begin{bmatrix} \exp(0.5t) & \exp(0.3t) \\ 0 & 0.5t \end{bmatrix} \begin{bmatrix} x_1(t) \\ x_2(t) \end{bmatrix} u(t)$$

$$+ \begin{bmatrix} (t+5)^{-3}\exp(-t) \\ \exp(-2t) \end{bmatrix} u(t), \tag{2.40}$$

$$y(t) = \begin{bmatrix} y_1(t) \\ y_2(t) \end{bmatrix} = \begin{bmatrix} 0.5(t+1)^{-1} & -0.6 \\ 0.7 & 0.2 \end{bmatrix} \begin{bmatrix} x_1(t) \\ x_2(t) \end{bmatrix}, \tag{2.41}$$

where

$$u(t) = f(y(t)), \quad f(x) = |x|\sin\frac{1}{|x|}. \tag{2.42}$$

For the application of Theorem 2.3, let D be a fundamental matrix of (2.40):

$$D(t) = \begin{bmatrix} \exp(-2t) & \exp(-3t) \\ 0 & \exp(-3t) \end{bmatrix}.$$

Let $t_0 = 0$. In the Theorem 2.3, $\|H\| = 1.08$, $\|D\| \leq \sqrt{3}\exp(-2t)$, $\|D^{-1}\| \leq \sqrt{3}\exp 3t$, $\|C\| \leq \sqrt{2}(t+5)^{-3}\exp(-t)$, $\|B\| \leq \sqrt{3}\exp(0.5t)$, $K_f = 1$: we get $\Lambda = 0.366$, $G = 77.76$. Let $x = [x_1, x_2]^T$. Then we have,

$$\|x(t)\| \leq \frac{\sqrt{3}\sqrt{(x_1+x_2)^2 + x_2^2\exp(2-t)}}{1 - 1.567 \cdot (0.366\sqrt{3G}\sqrt{(x_1+x_2)^2 + x_2^2}}.$$

Hence, the equilibrium at the origin for the system, (2.40)–(2.42), is uniformly stable and asymptotically stable.

2.3 Discrete Bilinear Systems

Consider the general form of a Discrete bilinear system with output feedback as follows,

$$x(t+1) = A(t)x(t) + \sum_{i=1}^{m} B_i(t)x(t)u_i(t) + C(t)u(t), \qquad (2.43)$$

$$y(t) = H(t)x(t), \qquad (2.44)$$

$$u(t) = (u_1(t), \ldots, u_m(t))^T = f(y(t)), \qquad (2.45)$$

where $x \in \mathbb{R}^n$, $y \in \mathbb{R}^p$, $p \le n$, $u \in \mathbb{R}^m$. $A(t)$, $B_i(t)$, $i = 1, \ldots, m$ are $n \times n$ matrices, $C(t)$ is an $n \times m$ matrix, and $H(t)$ is a $p \times n$ matrix, $f : \mathbb{R}^p \to \mathbb{R}^m$.

The following lemma is useful for the stability theory of discrete bilinear systems.

Lemma 2.2. *In the general bilinear system (2.43)–(2.45), assume that there exists $\alpha > 0$, a polynomial $h(\cdot)$ that does not include terms of degree ≤ 3, and positive coefficients, such that the following inequalities hold,*

$$\|x(t+1)\| \le \alpha_1 \|x(t)\|^2 + h(\|x(t)\|) \qquad (2.46)$$

or

$$\|x(t+1)\|^2 \le \alpha_1 \|x(t)\|^2 + h(\|x(t)\|). \qquad (2.47)$$

Then the zero state for the system (2.43)–(2.45) is uniformly stable and asymptotically stable, if $\alpha_1 < 1$.

Proof: (2.47) can be rewritten as

$$\|x(t+1)\|^2 \le \alpha_1 \|x(t)\|^2 + g(\|x(t)\|)\|x(t)\|^2,$$

where polynomial $g(\cdot)$ has a degree > 1 and positive coefficients. If we take $t = 0$, $\|x(0)\| < \delta$ then

$$\|x(1)\|^2 \le [\alpha_1 + g(\delta)]\delta^2 = \delta^2 \beta,$$

where

$$\beta \stackrel{\Delta}{=} \alpha_1 + g(\delta).$$

For every $\epsilon > 0$, one can take small enough δ such that $\beta < 1$ and $0 < \delta < \epsilon$. If $\alpha_1 < 1$, then

$$\|x(t)\|^2 \le [\alpha_1 + g(\delta\beta^{1/2})]\delta^2\beta \le \delta^2\beta^2.$$

Without difficulty, by mathematical induction, one can show that

$$\|x(t)\| \leq \delta\beta^{t/2}. \tag{2.48}$$

This implies that the zero state for the system (2.43)–(2.45) is uniformly and asymptotically stable if $\beta \leq 1$ or $\beta < 1$, respectively.

Now we first consider the simple form of the bilinear system with output feedback

$$x(t+1) = A(t)x(t) + B(t)x(t)u(t), \tag{2.49}$$
$$y(t) = H(t)x(t), \tag{2.50}$$
$$u(t) = f(y(t)), \tag{2.51}$$

where $A(t)$, $B(t)$, $H(t)$ are $n \times n$ matrices, x and y are n-vectors, and $u(t)$ is a scalar input.

Let

$$\lambda_1 \triangleq \sup_{t \geq 0} \lambda_{\max}[A^T(t)A(t)], \tag{2.52}$$

$$\lambda_2 \triangleq \sup_{t \geq 0} \lambda_{\max}[B^T(t)B(t)], \tag{2.53}$$

$$\lambda_3 \triangleq \sup_{t \geq 0} \max |\lambda[B^T(t)A(t) + A^T(t)B(t)]|. \tag{2.54}$$

Theorem 2.4. *If $f : \mathbb{R}^n \to \mathbb{R}$, $H(t)$ is uniformly bounded on Z_+, $\lambda_1 < 1$, and $\lambda_2 < \infty$, $\lambda_3 < \infty$, then the zero state for the system (2.49)–(2.51) is uniformly and asymptotically stable.*

Now we consider the more general system (2.43)–(2.45) with multiple output feedback.

Let

$$\lambda_2 \triangleq \sup_{t \geq 0} \max_{1 \leq i,j \leq m} \{\max |\lambda[B_i^T(t)B_j(t)]|\}, \tag{2.55}$$

$$\lambda_3 \triangleq \sup_{t \geq 0} \max_{1 \leq i \leq m} \{\max |\lambda[B_i^T(t)A(t) + A^T(t)B_i(t)]|\}. \tag{2.56}$$

Theorem 2.5. *Let us suppose that $C(t)$ and $H(t)$ are uniformly bounded on Z^+,*

$$\sqrt{\lambda_1} + K_1 F_H F_c < 1,$$

where K_1 is a constant that may depend on $f(\cdot)$. Then the zero state of the system (2.43)–(2.45) is uniformly and asymptotically stable.

Example 2.3. Consider the dynamical system (Mohler, 1973)

$$
\begin{bmatrix} \dot{x}_1 \\ \dot{x}_2 \\ \dot{x}_3 \end{bmatrix} = \begin{bmatrix} -R_a/L_a & 0 & 0 \\ 0 & 0 & 1 \\ 0 & 0 & -D/J \end{bmatrix} \begin{bmatrix} x_1 \\ x_2 \\ x_3 \end{bmatrix}
$$

$$
+ \begin{bmatrix} 0 & 0 & -K_a^*/L_a 0 & 0 & 0 \\ K_y/J & 0 & 0 \end{bmatrix} \begin{bmatrix} x_1 \\ x_2 \\ x_3 \end{bmatrix} u_1 + \begin{bmatrix} 1/L_a \\ 0 \\ 0 \end{bmatrix} v, \tag{2.57}
$$

$$
\begin{bmatrix} y_1 \\ y_2 \end{bmatrix} = \begin{bmatrix} 1 & 0 & 0 \\ 0 & 1 & 0 \end{bmatrix} \begin{bmatrix} x_1 \\ x_2 \\ x_3 \end{bmatrix}, \tag{2.58}
$$

$$
\dot{x} = Ax + Bxu_1 + cv, \tag{2.59}
$$

$$
y = Hx, \tag{2.60}
$$

where $x_1 = i_a$, $x_2 = \theta$, $x_3 = \omega$, $u_1 = i_a$, $v = v_a$, J is the moment of intertia, D is the viscous damping ratio, R_a is the armature resistance, L_a is the applied armature inductance, K_y, K_a are motor characteristics, K_a is the motor const, i_a is the armature current, i_e is the field current, v_a is the armature voltage, ω is the angular velocity, and θ is the angular position.

The motor control problem is to choose the functions f_1, f_2 such that the obtained system with feedback is stable.

Equations (2.57) and (2.58) can be discretized by use of a first-order Euler expansion to give

$$
x(t+1) = x(t) + TAtx(t) + TBx(t)u_1(t) + Tcv(t), \tag{2.61}
$$

$$
y(t) = Hx(t), \tag{2.62}
$$

where T is the sampling interval. Equation (2.61) can be rewritten as

$$
x(t+1) = A^*x(t) + B^*x(t)u_1(t) + c^*v(t), \tag{2.63}
$$

where

$$
A^* = I + TA = \begin{bmatrix} 1 - TR_a/L_a & 0 & 0 \\ 0 & 1 & T \\ 0 & 0 & 1 - TD/J \end{bmatrix},
$$

$$
B^* = TB = \begin{bmatrix} 0 & 0 & -K_y^*T/L_a \\ 0 & 0 & 0 \\ K_yT/J & 0 & 0 \end{bmatrix},
$$

$$
c_* = Tc = \begin{bmatrix} T/L_a \\ 0 \\ 0 \end{bmatrix}. \tag{2.64}
$$

Here,

$$\lambda_1(A^*) = 1 - TR_a/L_a, \quad \lambda_2(A^*) = 1, \quad \lambda_3(A^*) = 1 - TD/J.$$

Let $u = [u_1 v]^T = Sy(t)$, where S is a constant matrix

$$S = \begin{bmatrix} s_{11} & s_{12} \\ s_{21} & s_{22} \end{bmatrix}.$$

Thus, the corresponding feedback of system (2.61), (2.62) is

$$x(t+1) = A^*x(t) + B^*x(t)u_1(t) + c^*(t), \tag{2.65}$$
$$y(t) = Hx(t), \quad u(t) = Sy(t). \tag{2.66}$$

It can be shown that

$$\|x(t+1)\|^2 \leq \lambda_1\|x(t)\|^2 + \lambda_3\|x(t)\|^2|u_1(t)| + \lambda_2\|x(t)\|^2|u_1^2(t)|$$
$$\leq \lambda_1\|x(t)\|^2 + \lambda_3\|S\|\,\|H\|\,\|x(t)\|^3 + \lambda_2\|S\|^2\|x(t)\|^4. \tag{2.67}$$

Then, by Lemma 2.2, we conclude that the zero state of (2.65), (2.66) is uniformly and asymptotically stable when $\lambda_1 < 1$.

The principle of choosing s_{ij} is to reduce the closed-loop eigenvalues of $A^{**T}A^{**}$ where $A^{**} = A^* + CSH$. For convenience, we choose $s_{22} = 0$, and choose s_{12} such that

$$|1 - R_a/L_a + c_1 s_{21}| < 1,$$
$$-1 < 1 - R_a/L_a c_1 s_{21} < 1$$

and

$$s_{21} \in \left(\frac{-2}{c_1} + \frac{R_a}{c_1 L_a}, \frac{R_a}{c_1 L_a} \right),$$

Simulations have been made for this example.

3.   Adaptive Control of Bilinear Systems
3.1   Control of Fixed Points

We apply the adaptive control to bilinear systems in $n = 2$ and 3 dimensions with a single parameter that is varied. All systems have a stable region where asymptotically the motion goes to a fixed point attractor, as well as regimes with a limit cycle (following a Hopf bifurcation) or more complicated periodic and aperiodic behavior.

In studying control we have the following situations (Isidory, 1995). Given an initial value of the parameter, the system evolves — following equation (2.1) — to its stable steady state. At time $t = t_0$, say, the system is perturbed: here this implies an instantaneous change in the parameter value. Subsequently, the system evolves under control dynamics (i.e., equations (2.1) and (2.3)), and returns to the original steady-state. Clearly, a convenient quantifier of this process is the time of recovery τ, which depends both on the perturbation as well as on the stiffness of control.

Application to single parameter perturbation. The first system we study is a textbook example with one nontrivial degree of freedom, described by the following equations,

$$\dot{r} = u - r^3, \quad \dot{\theta} = \omega, \quad u = \alpha r. \tag{2.68}$$

The sign of α determines the dynamics: when $\alpha < 0$ the system evolves to a fixed point $(r = 0)$, and when $\alpha > 0$ there is a supercritical Hopf bifurcation (or soft excitation) and the system evolves to a (circular) limit cycle of radius $r_c = \alpha^{1/2}$. The control dynamics is determined by the error signal

$$\dot{\alpha} = \epsilon(r - \langle r \rangle), \tag{2.69}$$

where $\langle r \rangle$ is the steady-state value of r. To pull the system back to the fixed point, $\langle r \rangle$ in equation (2.69) must be 0.

The above system allows analytic treatment. Initially, $\alpha = \alpha_{in} < 0$ and the dynamics is attracted to $r = 0$. When r is suddenly perturbed to a positive value (at $t = 0$), we study how the system relaxes back to the fixed point. It is convenient to rewrite equation (2.68) as a Riccati equation,

$$\frac{dr}{d\alpha} = \frac{1}{\epsilon}(\alpha - r^2). \tag{2.70}$$

Further transformation to an equation of second order,

$$\frac{d^2u(\alpha)}{d\alpha^2} + \frac{\alpha u}{\epsilon^2} = 0, \tag{2.71}$$

allows $r(\alpha)$ to be written in terms of the two linearly independent solutions u_1 and u_2 of equation (2.71),

$$\frac{r}{\alpha}(Cu_1 + u_2) + C\frac{du_1}{d\alpha} + \frac{du_2}{d\alpha} = 0. \tag{2.72}$$

For positive α, and using the variable $z = \alpha/\epsilon^{2/3}$, equation (2.71) can be expressed as

$$\frac{d^2u(z)}{dz^2} + zu = 0, \tag{2.73}$$

which has the solutions $u_1 = Ai(z)$, $u_2 = Bi(z)$. Substituting this in equation (2.72) we get

$$r = -\epsilon^{1/3} \frac{C Ai(z) + Bi(z)}{C Ai(z) + Bi(z)}. \tag{2.74}$$

For small ϵ (i.e., large z), we can approximate the Airy functions to obtain (in terms of $\zeta = \frac{2}{3} z^{3/2}$)

$$r(\alpha) = -\epsilon^{1/3} \frac{C\left(-\frac{1}{2} \pi^{-1/2} z^{1/4} e^{-\zeta}\right) + \pi^{-1/2} z^{1/4} e^{\zeta}}{C\left(\frac{1}{2} \pi^{-1/2} z^{-1/4} e^{-\zeta}\right) + \pi^{-1/2} z^{-1/4} e^{\zeta}} \approx -\alpha^{1/2}. \tag{2.75}$$

Then solving

$$\dot{\alpha} = \epsilon r = -\epsilon \alpha^{1/2}, \tag{2.76}$$

we get

$$\alpha^{1/2} = -\frac{1}{2} \epsilon t + \alpha_0^{1/2}, \tag{2.77}$$

where α_0 is the value of α at $t = 0$. Thus

$$r(t) = -\frac{1}{2} \epsilon t + \alpha_0^{1/2}, \tag{2.78}$$

and the recovery time τ is then given approximately by

$$\tau = 2\alpha_0^{1/2}/\epsilon. \tag{2.79}$$

Although the above analysis is valid for small ϵ, in practice we find that equation (2.78) describes the recovery behavior over a wider range. This system (unlike the logistic or other unimodal maps examined by HL) has globally attracting steady states and so there is no perturbation, however large, from which the system does not recover in finite time. The recovery time τ is always close to the estimate provided by equation (2.78).

Nonlinear feedback control. We next analyze the system for which the evolution equations are

$$\dot{x} = \alpha + uy - \beta x - x, \tag{2.80}$$

$$\dot{y} = \beta x - uy, \quad u = x^2. \tag{2.81}$$

The nature of the dynamics is determined by the parameter β. When the asymptotic motion is attracted to a fixed point, the steady-state values of the BS can easily be seen to be $y_s = \beta_{in}/\alpha_{in}$ and $x_s = \alpha_{in}$. When β is perturbed, the equation for control dynamics becomes

$$\dot{\beta} = -\epsilon(y - y_s). \tag{2.82}$$

Linear stability analysis of equation (2.82) yields the following conditions for equilibrium,

$$\beta < \alpha_{in}^2 + 1 \qquad (2.83)$$

and

$$\alpha_{in}^4 + \alpha_{in}^3(\epsilon - \beta_{in}) + \alpha_{in}^2(1 - \epsilon\beta_{in}) > 0. \qquad (2.84)$$

Without loss of generality we set $\alpha_{in} = 1$ and equation (2.83) then gives the condition for attraction to the fixed point to be $\beta_{in} < 2$; for $\beta_{in} > 2$ the system is attracted to a limit cycle. Furthermore, equation (2.84) gives a stability window determined by

$$1 + (1 - \beta_{in})(\epsilon + 1) > 0. \qquad (2.85)$$

Note that equation (2.85) sets a restriction on the value of ϵ so here the range of control stiffness is limited by stability considerations.

Examination of the dependence of τ on ϵ reveals a novel feature that is not observed in the one-dimensional case. For small ϵ, $\tau \sim 1/\epsilon$. But τ does not decrease monotonically with ϵ and beyond an optimum value of ϵ, τ actually starts increasing. A rough argument accounting for the linear relation between τ and $1/\epsilon$ in the small ϵ range makes use of the observation that $y \approx_{in} /x$ (for small ϵ). Substitution in equation (2.80) gives $x(t) = 1 - \text{const} \cdot \epsilon^{-t}$, and equation (2.81) becomes

$$\dot{\beta} \approx= -\epsilon(\beta/x - \beta_{in}). \qquad (2.86)$$

Assuming small ϵ and a small perturbation, we then have

$$\beta \approx \beta_{in} - Ce^{\epsilon t} \approx \beta_{in} - \text{const}\,(1 - \epsilon t), \qquad (2.87)$$

suggesting that recovery time $\tau \propto 1/\epsilon$.

A biochemical network. The third example we study is a complex dynamical system (Loskutov and Mikhailov, 1990) which describes various biochemical processes responsible for the coherent behavior observed in spatiotemporal organization. The equations contain positive and negative feedback loops that are typically thought to occur in a variety of processes within living cells. This system, which gives rise to a variety of behavioral patterns, is

$$\dot{X}_1 = \frac{a_1}{a_2 + X_3^n} - uX_1, \qquad (2.88)$$

$$\dot{X}_2 = a_3 X_1 - \phi(X_2, X_3), \qquad (2.89)$$

$$\dot{X}_3 = a_4 \phi(X_2, X_3) - qX_3, \qquad (2.90)$$

where

$$\phi(X_2, X_3) = \frac{T X_2 (1 + X_2)(1 + X_3)^2}{L + (1 + X_2)^2 (1 + X_3)^2}. \qquad (2.91)$$

Choosing parameters (a_1, a_2, a_3, a_4, L, T and n) suitably, we have a system whose dynamics can be varied by tuning the parameters K and q. For instance, for $q = 0.5$, we get a limit cycle when $u = 0.001$, complex oscillations when $L = 0.003$, chaos when $u = 0.004$, complex oscillations and period doublings when $0.005 < u < 0.02$, a limit cycle again when $0.03 < u < 0.5$, and a steady-state when $u = 1$. For control we let u evolve as

$$\dot{u} = -\epsilon(X_1 - \langle X_1 \rangle). \qquad (2.92)$$

Equation (2.92) is very effective in returning the system back to the original steady-state ($u = 1$) when u is perturbed into any of the other above-mentioned regimes, including the chaotic region.

A discrete dissipative system. We finally apply the control to a two-dimensional discrete system given by (Loskutov and Mikhailov, 1990)

$$X_{n+1} = 1 - \frac{\alpha X_n^2}{1 + X_n^4} - u Y_n, \qquad (2.93)$$

$$Y_{n+1} = X_n, \qquad (2.94)$$

which is similar to the Hennon map with the additional feature that the global asymptotic dynamics is on an attractor for $u < 1$. When α is varied, this map gives rise to the entire repertoire of behavior seen in unimodal chaotic maps. For regulation of the steady state the control dynamics given by the equation

$$\alpha_{n+1} = \alpha_n - \epsilon(X_n - X_s) \qquad (2.95)$$

is very effective. For small ϵ the recovery time τ is proportional to $1/\epsilon$, but beyond $\epsilon = \epsilon_{\text{opt}}$, τ increases with ϵ, similar to what is observed for the Brusselator reaction-diffusion model.

The above dissipative systems with more than one degree of freedom can show novel behavior quite distinct from the one-dimensional case. For example, there is a maximum strength of shock for every value of ϵ beyond which the system does not recover. Below $\epsilon = \epsilon_c$ recovery is possible for shocks of all strengths (provided that the shock does not throw the perturbed value of α outside the allowed range). For $\epsilon > \epsilon_c$ the system fails to recover from all shocks however small in magnitude, so that the $\delta_{\max}$ versus characteristic is a step function rather than the linearly decreasing function.

Applications to multiparameter systems. Typically, a dynamical system has several parameters that govern the overall behavior. In order to regulate such systems, the control has to be applied for each relevant parameter.

A representative one-dimensional map with two parameters (which is of interest in population dynamics) is given by

$$X_{n+1} = u(1 + X_n)^{-\beta}, \quad u = \alpha X_n. \tag{2.96}$$

For specific α and β the map has a globally stable equilibrium state: when the parameters α and β are varied the map yields a rich variety of dynamical behavior. To regulate the steady state of the system we control both parameters in an obvious manner through

$$\alpha_{n+1} = \alpha_n + \epsilon(X_n - X_s), \tag{2.97}$$
$$\beta_{n+1} = \beta_n + \epsilon(X_n - X_s), \tag{2.98}$$

where ϵ is the control stiffness.

Similarly, for a two-dimensional discrete map of two driven coupled oscillators given by

$$x_{n+1} = u_1(1 - x_n) + \beta(y_n - x_n), \quad u_1 = \alpha x_n \tag{2.99}$$
$$y_{n+1} = u_2(1 - y_n) + \beta(x_n - y_n), \quad u_2 = \alpha y_n \tag{2.100}$$

the same control dynamics is effective. In both cases the recovery time remains linearly dependent on ϵ.

In the control dynamics implemented here, we have always chosen an error indicator that utilizes a single variable, $X_i - X_s$. In higher-dimensional systems there is an ambiguity regarding the choice of the state variable X, to be used in equation (2.1). Empirically, we observe that in most systems any of the variables can effect control, because the equilibrium condition will lead all other variables to their steady-state values when any one of them is forced to reach a steady-state. However, there are exceptions: for instance, in the Brusselator, using coordinate y in equation (2.81) results in control whereas using x does not. This is because the steady-state value of x ($x_s = \alpha$) does not constrain β to be the desired value, whereas the steady state of y ($y_s = \alpha\beta$) does. On the other hand, in the three-dimensional system given by equations (2.88)–(2.90), all three variables can effect control. X_1 works most efficiently, however, as the magnitude of X_1 (and hence the error signal) is small, leading to a more stable control dynamics.

One method of removing the above-mentioned ambiguity is by employing AND logic in the control, that is, by requiring that all variables reach their steady-state values, X_i^s, $i = 1, 2, \ldots, N$. The equation for control dynamics then becomes

$$\dot{\mu} = \epsilon \sum_{i=1}^{N} (X_i - X_i^2). \qquad (2.101)$$

In the examples we have studied, either equation (2.101) works equally efficiently.

3.2   Control of Limit Cycles

For the simple oscillator described in equation (2.1), limit cycles can be adaptively controlled. In this case defining the error signal is quite unambiguous as every limit cycle is uniquely characterized by its radius r_c. The difference between the actual radius and the radius of the limit cycle to be controlled can be used effectively for regulatory feedback. This is done, for example, by setting $\langle r \rangle$ in equation (2.68) equal to r_c. When perturbed onto a different limit cycle (radius not equal to r_c) or into the fixed point region ($\alpha > 0$), the system rapidly relaxes to the original limit cycle. For small $\epsilon \tau$, T is inversely proportional to ϵ, but for large ϵ we observe a different phenomenon: τ oscillates about a saturation value that is roughly constant for all values of perturbation.

When r_c is small and the system is perturbed to a much larger radius the control dynamics for small ϵ is determined by a set of equations similar to equation (2.69) (as $r > r_c$): we can therefore expect the same linear trend. When the system is kicked to the fixed point region ($r/t_0, 0$) the control dynamics is approximated by

$$\dot{\alpha} = -\epsilon r_c, \qquad (2.102)$$

so that (for small ϵ)

$$\alpha(t) = -\epsilon r_c t + \alpha_0, \qquad (2.103)$$

from which the inverse dependence of τ on ϵ follows. For large ϵ this is not valid.

More generally, we can extend the above procedure to control cycles in discrete systems. What is required is an error indicator that encodes as much information about the cycle as is necessary for its unique characterization. An error signal depending on $X_{n+2} - X_n$ suffices in bringing the system back to some period 2-cycle, but not to a specific 2-cycle: $X_{n+2} - X_n = 0$ for all period 2-cycles, and so cannot guide the control dynamics onto any particular cycle. To regulate specific cycles we require an error that is unique for every 2-cycle: one possibility is an error proportional to $|X_{n+1} - X_n| - |X_1^c - X_2^c|$, where X_1^c and X_2^c are the values of the iterates of X in the 2-cycle we want to control.

We implement this for the logistic map

$$X_{n+1} = uX_n(1 - X_n). \qquad (2.104)$$

The control dynamics follows from

$$u_{n+2} = \alpha_n - \epsilon(|X_{n+1} - X_n| - |X_1^c - X_2^c|). \qquad (2.105)$$

It is clear that this control mechanism very effectively returns the system to the desired 2-cycle. The recovery time varies inversely as ϵ. There is also a maximum strength of shock $\delta_{\max}$ (depending on ϵ), beyond which the system fails to recover: this $\delta_{\max}$ versus ϵ curve shows a step function pattern.

Similar error indicators can, in principle, be constructed for higher-order periodic orbits although the technique diminishes in utility with increase in period. This is a problem of practicality, because higher-order cycles typically have narrow windows of stability; as a consequence the control dynamics become very unstable.

For discrete dynamical systems, however, more effective algorithms can be devised. One, which employs a logical OR structure in the error indicator, is

$$\dot{\alpha} = \epsilon \prod_{i=1}^{k}(X - X_i^c), \qquad (2.106)$$

where X_i^c, $i = 1, 2, \ldots, k$ is the stable period k-orbit to be controlled. Because it implies that the desired state is either $X = X_1^c$ or $X = X_2^c$ or $\ldots X = X_k^c$ this adaptive algorithm works at every iteration step. For higher-order periodic orbits, this latter method is far superior to that embodied in equation (2.106). For controlling the 2-cycle, for instance, the control equation analogous to equation (2.106) is

$$\alpha_{n+1} = \alpha_n - \epsilon(X_n - X_1^c)(X_n - X_2^c). \qquad (2.107)$$

Even with this latter form, the inverse dependence of τ on ϵ is unchanged.

3.3   Variations in the Control Dynamics

Apart from sudden perturbations in the system environment that lead to parameters changing value drastically (the primary case studied above) there are additional noise effects that can occur. In particular, it is interesting to consider the effect of random background noise on the control algorithm. In an effort to explore this question, we study the discrete map equations (2.104) and (2.96), with additional Gaussian noise. The control dynamics remain unchanged. The variance σ of this noise clearly determines the control behavior: for small σ, recovery times with and without noise are virtually identical, and beyond a value $\sigma = \sigma_{max}$, recovery is not possible. However, most important, this control procedure is remarkably robust for $\sigma < \sigma_{max}$, and the recovery time continues to remain inversely proportional to the stiffness.

A question of some importance is whether the control algorithm is sensitive to the specific form of the control dynamics, namely the choice for $g(X - X_s)$ in equation (2.5). In realistic systems, the control dynamics that can be incorporated may be of a arbitrary functional form, arising, for example, from physicochemical or engineering design considerations specific to the system. It is thus necessary to determine whether the linear recovery we observe in the examples above is an artifact of using a linear control function, and also whether such an adaptive control is more generally applicable with different functions $g(y)$. In order to explore the features of control with different (nonlinear) functions, we have varied the control function, using $g = y^2 \, y^{1/2} \, \sin y \, 1 - e^2$ and $y(1 - y)$. Results are shown: for all functional forms recovery times remain inversely proportional to control stiffness for small ϵ. This strongly suggests that linear recovery may be a universal feature of the adaptive control algorithm.

From our study of higher-dimensional systems of varying complexity, it appears that it is possible to provide efficient regulation of the steady state of nonlinear systems through adaptive control mechanisms. The procedure studied herein utilizes an error signal proportional to the difference between the goal output and the actual output of the state variables and should be contrasted with mechanisms (Haken, 1978) using an error signal proportional to a similar difference in the parameter value. In the latter case the control will, of course, bring the parameter back to its original value, but this does not ensure that the system will regain its specific original dynamical state. An instance where this distinction is important is in systems undergoing a subcritical Hopf bifurcation or exhibiting bistability; at a given parameter value, different initial conditions lead to different dynamics. In such a case, the present adaptive control ensures that both the original parameter value and the original dynamics will be recovered.

From numerical experiments studying the dependence of controllability on the stiffness and on the strength of perturbation, we find a number of interesting phenomena quite distinct from that seen in HL, but typical of most real systems. For multiparameter systems, a simple extension of the adaptive mechanism suffices in regulating the system; furthermore it can also be adapted to regulate periodic behavior such as limit cycles.

The HL procedure is robust both to the existence of background noise, and to the variation of the form of the control function. Most interestingly, recovery times are always inversely proportional to the stiffness of control, for a small stiffness, which may be a universal feature of such adaptive control.

Biological situations where control is believed to play a crucial role include, for instance, the maintainence of homeostasis (the relative constancy of the internal environment with respect to variables such as blood pressure, pH, blood sugar, electrolytes, and osmolarity). Clinical experiments on animals show, for example, that following a quick mild hemorrhage (a sudden perturbation in arterial pressure) the blood pressure is restored to equilibrium values within a few seconds. The control of fixed points thus has a potential utility in such physicochemical contexts. Cycles are also central to a variety of biophysical and biochemical processes. Variations in these—for example, by the replacement of periodic by aperiodic behavior, or the emergence of new periodic cycles—is often associated with disease (Haken, 1978, 1988). The control of cycles has applicability in the regulation of biologically significant oscillatory phenomena.

In summary, our study confirms that adaptive control provides a simple, powerful, and robust tool for regulating multidimensional systems capable of complicated behavior. The concepts developed through the study of model systems can serve as a paradigm for understanding more complex regulatory mechanisms widespread in nature. These may also be of use in helping formulate efficient and robust design principles.

4. Notes and Sources

Much of the material on bilinear time-optimal control problems is standard in the literature on mathematical systems theory. The present exposition essentially follows Brockett (1979) and Bailleul (1978), except for the additional focus on the soliton approach, which is not specifically recognized in this literature. The soliton approach based on the Lax form is essentially taken from the study by Yatsenko (1984). The stabilization problem illustrates the classic origin of the Lyapunov method used by Gutman (1981), Gounaridis and Kalouptsidis (1986), Longchamp (1980), Quinn (1980), Ryan and Buckingham (1983), and Slemrod (1978).

Chapter 3

BILINEAR SYSTEMS AND NONLINEAR ESTIMATION THEORY

In this chapter we present an application of the concept of an adaptive estimation using an estimation algebra to the study of dynamic processes in nonlinear lattice systems. It is assumed that nonlinear dynamical processes can be described by nonlinear or bilinear lattice models. Our research focuses on the development of an estimation algorithm for signal processing in the lattice models with background additive white noise. The proposed algorithm involves solution of stochastic differential equations under the assumption that the Lie algebra, associated with the processes in the lattice system, can be reduced to a finite-dimensional nilpotent algebra. A generalization is given for the case of the lattice models, which belong to a class of causal lattices with certain restrictions on input and output signals.

The chapter is organized in the following way. In Section 3.1 we present the application of a method of adaptive estimation using an algebra-geometric approach, to the study of dynamic processes in the brain (Hopfield and Tank, 1985, Hopfield, 1994; Pardalos et. al., 2003). It is assumed that the brain dynamic processes can be described by nonlinear or bilinear lattice models. Our research focuses on the development of an estimation algorithm for a signal process in the lattice models with background additive white noise, and with different assumptions regarding the characteristics of the signal process. We analyze the estimation algorithm and implement it as a stochastic differential equation under the assumption that the Lie algebra, associated with the signal process, can be reduced to a finite-dimensional nilpotent algebra. A generalization is given for the case of lattice models, which belong to a class of causal lattices with certain restrictions on input and output signals. The

P.M. Pardalos, V. Yatsenko, *Optimization and Control*
of Bilinear Systems, doi: 10.1007/978-0-387-73669-3,
© Springer Science+Business Media, LLC 2008

application of adaptive filters for state estimation of the CA3 region of the hippocampus (a common location of the epileptic focus) is discussed. Our areas of application involve two problems: (1) an adaptive estimation of state variables of the hippocampal network, and (2) space identification of the coupled ordinary equation lattice model for the CA3 region.

In Section 3.2 we consider the problem of optimal input signal estimation for bilinear systems under input measurements. The periodogram estimates of parameters are studied. The possibilities to construct a finite-dimensional adaptive estimator for a causal dynamical system class is shown. The robust signal estimating problem is solved, when signals are estimated via application of neural networks and when nonlinear measurements are used.

In Section 3.3 we proceed to find an estimation algorithm for signal processing in the lattice models with background additive white noise. In Section 3.4 we move to the problem of a recursive realization of Bayesian estimation for incomplete experimental data. A differential-geometric structure of nonlinear estimation is studied. It is shown that the use of a rationally chosen description of the true posterior density produces a geometrical structure defined on the family of possible posteriors. Pythagorean-like relations valid for probability distributions are presented and their importance for estimation under reduced data is indicated. A robust algorithm for estimation of unknown parameters is proposed, which is based on a quantum implementation of the Bayesian estimation procedure. Section 3.6 describes a relation between nonlinear filtering problem and quantum mechanics.

1. Nonlinear Dynamical Systems and Adaptive Filters

1.1 Filtration Problems

Extensive work has been done recently on developing new types of mathematical representation of lattice models based on laws of information transformation for physical and biological systems (Haken, 1988, 1996; Hiebeler and Tater, 1997). The models of nonlinear processes arising in such systems are characterized by nondeterministic behaviors. In many cases these processes are properly described by finite-dimensional differential equations over smooth manifolds, which can be approximated by nonlinear or bilinear systems. It enables one to use well-developed methods of the mathematical theory of systems to investigate problems of identification, simulation, prediction, and the like. As the input of such systems, one may have a signal mixed with noise, which later undergoes a nonlinear dynamical transformation. It is very important to develop efficient techniques for estimation of input and output signal parameters of a lattice model.

The existence of complex chaotic, unstable, and noisy nonlinear dynamics in the brain requires a novel methodology for a constructive understanding of complex phenomena. Chaos provides us with a universal framework for understanding the onset of epilepsy (Du, Pardalos, and Wang, 2000; Freeman, 2000; Iasemidis et. al., 2001). The adopted simple chaotic models are not directly connected with epileptic seizures. By extending this approach, one can emphasize the significance of a constructive understanding as opposed to descriptive understanding. The coupled dynamical lattice (CDL) gives an example of such a constructive model for spatiotemporally complex phenomena in the epileptic human brain (Hiebeler and Tater, 1977; Kaneko, 1993, 2001). It is possible to simulate the spatiotemporal patterns of neural networks in the brain. According to the findings by Freeman and his colleagues, such spatiotemporal patterns are chaotic (Freeman and Skarda, 1985; Freeman, 2000). It will be plausible to think that the chaos observed underlies the correlated dynamics of the brain. The following steps are useful for investigation of nonlinear dynamics in CDL.

(a) Reduction of nonlinear CDL to bilinear lattice models.

(b) System analysis of controllable bilinear CDL using geometrical methods.

(c) Observe structural changes from dynamic viewpoints in phase space.

(d) Construct a model of an epileptic human brain while observing top and bottom levels from an intermediate level which is neither macroscopic nor microscopic.

(e) Construct an adequate language system to understand the epileptic human brain based on nonlinear and bilinear control theory.

Mathematical methods of low-dimensional nonlinear systems with chaos allow us to define their geometrical structures in phase space. Similar techniques are needed for high-dimensional systems. In dynamic complex systems, however, the extraction of geometrical structures is much more difficult. The dynamics of controllable nonlinear lattice systems with a large number of degrees of freedom provides a way to understand the motion dissecting structure in phase space. We analyze controllable CDLs on the basis of two techniques: Lie group methods

(Brockett, 1973) and the theory of bilinear dynamical systems (Isidory, 1995; Mohler, 1973; Pardalos et al., 2001). After identifying the underlying structure of the lattice model as a bilinear differential equation, we apply these techniques to obtain an adaptive bilinear filter.

Signal filtration is regarded as an important part of signal processing arising in neuroscience (Pardalos et al., 2003). There are many available algorithms for signal estimation under various criteria (Childers and Durling, 1975; Anderson and Moore, 1979; Kushner, 1967; Kalman and Bucy, 1961). However, these optimal algorithms suffer from intrinsic and intensive computational complexities and require a considerable amount of computer time in the case of multidimensional systems. This drawback makes it difficult or even impossible to apply these optimal algorithms to real-life problems. Nonlinear filtration is another hot topic in this field (Diniz, 1997; Lo, 1978; Brockett, 1979; Benes, 1981; Davis and Marcus, 1981; Chen, Leung, and Yau, 1996). The problem was stated, and the basic properties of signal estimation on Lie groups were established by Willsky (1973), Brockett (1981), Chiou and Yau (1994), Chikte and Lo (1981), Hazevinkel (1986), and Wong (1998). Adaptive algorithms are very useful in the case where the transfer function of the related system through which the desired signal passes is unknown a priori (Yatsenko, 1989; Diniz, 1997).

This section considers such an algorithm for the epileptic brain processes, which are described by a lattice model (Kaneko, 1993, 2001; Pardalos et al., 2003). We construct a dynamical model of the CA3 region in the form of directional coupled cells of nonlinear neurons. It is assumed that the unknown parameters belong to a certain set, and input signals of each cell can be simulated by stochastic processes, in particular, a Gauss–Markov process with additive white noise (Ikeda, 1981; Gardiner, 1985). The problems of estimation of nonlinear signal processes in every cell of the CA3 region are very essential for epileptic diagnosis. The importance of studying these problems is due to the possibility of applying the results of this research in many practical areas of biomedical engineering.

The emphasis of this section is on developing an adaptive estimation algorithm for a signal with background additive white noise, and with different assumptions regarding matrices of signal processes. We formulate conditions for the existence of the optimal filter and implement it as a stochastic differential equation under the assumption that the Lie algebra, associated with the signal process, can be reduced to a finite-dimensional nilpotent algebra. We show that in this case the filter also has the bilinear structure, but a small adjustment of its parameters with an identification algorithm is needed. We propose an algorithm for parametric identification of each cell, which is based on an expansion of the input and output signals into a series over an orthogonal basis. A generalization is given for the case where the CA3 region belongs to a class of causal systems with certain restrictions on input and output signals.

1.2 Problem Statement

In this section we discuss the use of a geometric approach to adaptive signal filtration under the assumption that the diagnosis information is given in the form of a nonlinear time series. It is assumed that the CA3 region of the brain can be considered as a lattice system.

Lattice systems (LS) are complex dynamical systems characterized by two special features: the nodes (component schemes) are all identical copies of a scheme, and they are arranged in a regular spatial lattice (Figure 3.1).

Definition 3.1. *By a neural dynamical system (neural lattice model or LM), we mean a complex dynamical system in which the nodes are all identical copies of a single controlled dynamical scheme, the standard neural cell (Hiebeler and Tater, 1997).*

We are interested in estimating a signal process $\{x^k(t)\}_{t\geq 0}$, $x^k(t) \in \mathbb{R}^q$, satisfying the following system of nonlinear differential equations,

$$\dot{x}^k(t) = f^k(x^k(t)) + G^k(x^k(t))x^k(x_j)(t), \quad t \geq 0, \tag{3.1}$$

$$\psi^k(t) = h^k(x^k(t)) + \hat{Q}^k(x^k(t))\gamma^k(t), \quad t \geq 0, \tag{3.2}$$

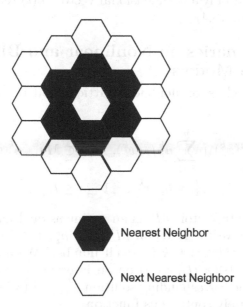

Nearest Neighbor

Next Nearest Neighbor

Figure 3.1. Hexagonal lattice of cells.

where x^k denotes state; ψ^k denotes an output; x^k and γ^k denote inputs. They are vector-valued functions of time of dimension q, l, n, and $\tilde{p}$, respectively; $x^j \in \chi$, $j = 1, \ldots, N$; χ is a neighborhood of cell k; f^k, h^k are q- and p-dimensional vector-valued functions of $x^k(t)$; G^k, Q^k are matrix-valued functions of $x^k(t)$ of appropriate sizes; $x^k(0)$ is independent of $\xi^k(\cdot)$ and γ^k processes. Moreover, it is required to find a finite-dimensional stochastic differential equation for determining the signal of each cell under several assumptions on the Lie algebra of the system (3.1), which are described in the next section.

We assume that interactions of cells can be described as the following linear Ito models (Gardiner, 1985; Pugachev and Sinitsin, 1987) for the signal and the observed process, respectively,

$$d\xi^k(t) = F^k(t)\xi^k(t)dt + (Q^k)^{1/2}(t)d\omega(t), \quad t \geq 0, \qquad (3.3)$$

$$dz^k(t) = H^k(t)\xi^k(t)dt + (R^k)^{1/2}(t)dv(t), \quad t \geq 0, \qquad (3.4)$$

where $\omega(\cdot)$ and $v(\cdot)$ are standard n- and $\tilde{p}$-dimensional independent Wiener processes, respectively; $\xi^k(t) \in \mathbb{R}^n$; $z^k(t) \in \mathbb{R}^p$ is the system output; $\xi^k(0)$ is a random vector with Gaussian distribution and zero mean, which is independent of $\omega(\cdot)$ and $v(\cdot)$; $F^k(\cdot)$, $(Q^k)^{1/2}(\cdot)$, $H^k(\cdot)$, $(R^k)^{1/2}(\cdot)$ are time-dependent matrices of appropriate dimensions; $Q^k(t)$, $R^k(t)$ are positive definite and continuously differentiable matrices for all t; $x^j \in \chi$, $j = 1, \ldots, N$; χ is a neighborhood of cell k. In many real applications, one does not actually observe the n-dimensional state vector ξ^k of the cell interaction, but only a p-dimensional vector $^zk(t)$ related to $\xi^k(t)$ by the second equation (3.4).

1.3   Preliminaries on Nonlinear and Bilinear Lattice Models

We consider the class of nonlinear lattice model described by the equations

$$\dot{x}^k = f^k(x^k) + \sum_{i=1}^{n} u_i^k f_i^k(x^k), \quad x^k \in M^k, \quad u^k \in \mathbb{R}^n, \qquad (3.5)$$

$$\psi_i^k = h_i^k(x^k), \quad 1 \leq i \leq l, \qquad (3.6)$$

where x^k is the state vector; M^k is an r-dimensional manifold; f_i^k, $0 \leq i \leq n$ are different vector fields on M^k, and h_i^k, $1 \leq i \leq l$ are different functions on M^k; $\gamma^k(t) = 0$ k is a cell number. We assume that each control function $u^k(\cdot)(x^k(t), x^j(t))$ on an interval $[0, T]$ is bounded and measurable and the corresponding solution $x_u^k(\cdot)$ of (3.5), (3.6) exists on $[0, T]$ as an absolutely continuous function.

Theorem 3.1. *Let $gl(n, \mathbb{R})$ denote the Lie algebra of $n \times n$ matrices (Brockett, 1973; Sagle, 1972). Then for any $\hat{k} > 0$ there exists a system*

$$\dot{s}^k = A_0^k s^k + \sum_{i=1}^{n} u_i^k B_i^k s^k, \quad s^k \in M^k, \quad s^k(0) = s_0^k, \quad (3.7)$$

$$\psi_i^k = p_i^k(x^k), \quad 1 \leq i \leq l, \quad (3.8)$$

where B_i^k, $0 \leq i \leq n$ are elements of $gl(M, \mathbb{R})$, k is the number of cells, and p_i^k are polynomials, such that its input-output map is the truncation of the input-output map of the initialized system (3.5), (3.6), expressed as a Chen series (Krener, 1975; Crouch, 1984), obtained by deleting all terms with $|\mu| > \hat{k}$.

The proof of the theorem follows from the condition of equivalence of nonlinear control systems (Krener, 1975).

We mention some concepts from the theory of Lie algebras (Brockett, 1973; Sagle and Walde, 1972; Lo, 1973), which are used to solve the problem of the optimal filtering of nonlinear processes in a lattice model. Consider a single cell of a bilinear matrix lattice system,

$$\hat{X}(t) = \left(A_0 + \sum_{i=1}^{n} \xi_i(t) B_i \right) X(t), \quad X(0) = I, \quad (3.9)$$

where X is a $M \times M$ matrix.

We associate with (3.9) the Lie algebra

$$L \triangleq \{A_0, B_1, \dots, B_n\}_{LA},$$

the smallest Lie algebra containing $A_0, B_1, \dots, B_n$, and the ideal L^0 in L generated by the matrices $\{B_1, \dots, B_m\}$.

We call a Lie algebra L nilpotent if the series of ideals L^n, determined by the sequence $L^0 = L, \dots, L^{l+1} = [L, L^l]$, is the trivial ideal $\{0\}$ for some l; L is abelian if $L^1 = \{0\}$.

Let L be the Lie algebra formed by the set of matrices $\{A, B_1, B_2, \dots, B_n\}$, and let L_0 be the ideal in L formed by the set

$$\{\text{Ad}_A^r(B_i), \ i = 1, 2, \dots, n; \ r = 0, 1, 2, \dots\},$$

where

$$\text{Ad}_A^0(B_i) \triangleq B_i;$$

$$\text{Ad}_A^r(B_i) \triangleq A \cdot \text{Ad}_A^{r-1}(B_i) - \text{Ad}_A^{r-1}(B_i)A, \quad \text{for } r = 1, 2, \dots.$$

Then L_0 is nilpotent (with dimension n^* and order of nilpotency n_*).

1.4   Adaptive Filter for Lattice Systems

We begin presenting the results with a proof of the following preposition.

Proposition 3.1. *Consider the interaction model (3.3), (3.4). Define a vector process $\{y^k(t)\}_{t\geq 0}$ by the equation*

$$dy^k(t) = \Theta^k y^k(t)dt + \sum_{i=1}^{n} E_i^k \xi_i^k(t) y^k(t)dt, \tag{3.10}$$

where Θ^k, $\{E_i^k\}_{i=1}^{n}$ are matrices of corresponding dimensions; k is the number of cells; $y^k(0)$ is independent of $\xi^k(0)$, $\omega(\cdot)$, and $v(\cdot)$. Then $\hat{y}^k(t/t) \triangleq E^k[y^k(t)/(z^k)^t]$ satisfies the following system of equations,

$$d\hat{y}^k(t/t) = \Theta^k \hat{y}^k(t/t)dt + \sum_{i=1}^{n}(E_i^k)^t(E^k)^t[\xi_i^k(t)y^k(t)]dt$$

$$+ \{(E^k)^t[y^k(t)(\xi^k)^T(t)] - \hat{y}^k(t/t)(\hat{\xi}^k)^T(t/t)\}(H^k)^T(R^k)^{-1}(t)dv(t),$$

$$\hat{y}^k(0/0) = E^k[y^k(0)], \tag{3.11}$$

where

$$(E^k(t))^t[\cdot] \triangleq E^k[\cdot \mid (z^k)^t] \triangleq E^k[\cdot (z^k(\tau) \mid 0 \leq \tau \leq t)],$$

$$dv(t) = dz^k(t) - H^k(t)\hat{\xi}^k(t/t)dt.$$

The proof of the proposition is based on utilization of the Kushner nonlinear filtering equation (Kushner, 1967; Lo, 1975) to the signal process $(y^k)^T(\cdot)$, $(\xi^k)^T(\cdot)$ with $z^k(\cdot)$ as the observation process.

Proposition 3.2. *Let $x = (x_0, x_1, \ldots, x_n)^T$ be a Gaussian random vector with mean vector $m = (m_0, m_1, \ldots, m_n)^T$ and covariance matrix $P = [P_{ij}]_{ij=0}^{n}$. Then the following relations are valid:*

$$E\left[e^{x_0} \prod_{i=1}^{n} x_i \right]$$

$$= \begin{cases} m_1 + P_{01}E\left[e^{x_0} \prod_{i=2}^{n} x_i \right] E\left[e^{x_0} \sum_{j=2}^{n} P_{ij} \prod_{\substack{i=2 \\ i\neq j}}^{n} x_i \right], & n > 1, \\ (m_1 + P_{01})E[e^{x_0}], & n = 1. \end{cases} \tag{3.12}$$

Proof: First note that Proposition 3.1 yields the following (Lo, 1975),

$$E\left[e^{x_0} \prod_{i=1}^{n} x_i \right] = e^{(m_0+(1/2)P_{00})} \cdot E\left[\prod_{i=1}^{n} Y_i \right], \tag{3.13}$$

where $Y = (Y_1, \ldots, Y_n)^T$ is a Gaussian random vector $(m_1 + P_{01}, \ldots, m_n + P_{0n})^T$, and $P_* = [P_{ij}]_{i,j=1}^n$ as the covariance matrix. Furthermore,

$$E\left[\prod_{i=1}^n Y_i\right] = (m_1 + P_{01})E\left[\prod_{i=2}^n Y_i\right] + \sum_{j=2}^n P_{1j}E\left[\prod_{k=2,k\neq j}^n Y_k\right]. \quad (3.14)$$

Combining (3.13) and (3.14) and using repeatedly the identity similar to (3.13) for the vector x with reduced dimensionality gives (3.12), which completes the proof.

Based on the above propositions, we can formulate the following theorem.

Theorem 3.2. *Consider the dynamical process* $\{s^k(t)\}_{t\geq 0}$, *of a single cell described by (3.1), (3.2). Let* L_0^k *be nilpotent; then the optimal estimate can be obtained from the following finite-dimensional bilinear stochastic equation,*

$$d(\hat{x}^k)^*(t/t) = \left[A^*(t)dt + \sum_{i=1}^n B_i^*(t)\hat{\xi}^k(t/t) + \sum_{i=1}^n C_i^* d\mu_i^k(t)\right](\hat{x}^k)^*(t/t),$$

$$(\hat{x}^k)^*(\cdot/\cdot) \in \mathbb{R}^{M^*}, \quad M^* \leq M\left(\frac{(n^2 n^*)^M - 1}{n^2 n^* - 1}\right),$$

$$(\hat{x}_i^k)^*(0/0) = E^k[x_i^k(0)], \quad i \leq M,$$

$$(\hat{x}_i^k)^*(0/0) = 0, \quad i \geq M,$$

$$(\hat{x}_i^k)^*(t/t) = L(t)(\hat{x}^k)^*(t/t), \quad (3.15)$$

where

$$\mu^k(t) \triangleq \int_0^t (H^k)^t(\tau)(R^k)^{-1}(\tau)d\nu(\tau)$$

is a modified innovational process; $\hat{\xi}^k(\cdot/\cdot)$ *is obtained from the standard Kalman–Bucy filter;* $L^k(\cdot)$ *is* $M \times M^*$, *and* $A^*(\cdot)$, $\{B_i^*(\cdot)\}_{i=1}^n\}$, *and* $\{C_i^*(\cdot)\}_{i=1}^n$ *are standard matrix functions of time, such that the Lie algebra generated by matrices of the form*

$$\begin{vmatrix} Ad^k & & & (B_j^*(t)) \\ & \begin{bmatrix} Ad^l & (A^*(t)) \\ C_i^*(t) & \end{bmatrix} & \end{vmatrix},$$

$$i = 1, 2, \ldots, n; \quad j = 1, 2, \ldots, n; \quad k, l = 0, 1, \ldots, \quad t \geq 0$$

is nilpotent with n_* *as the order of nilpotency.*

Proof: Equations (3.1), (3.2) can be reduced to the system (3.7) or to the bilinear equations (Lo, 1975; Yatsenko, 1984):

$$\dot{x}^k(t) = \left(A + \sum_{i=1}^{n} B_i \xi_i^k(t) \right) x^k(t),$$

$$\psi^k(t) = \left(C + \sum_{i=1}^{q} D_i \gamma_i(t) \right) x^k(t), \qquad (3.16)$$

where A, B_i, C, D_i, are constant matrices of appropriate sizes, and for some positive integers M_i, $i = 0, 1, \ldots, q$ we have

$$\text{rank}\,[C', A'C', \ldots, (A')^{M_0-1}C', D_1', A'D_1', \ldots, (A')^{M_1-1}D_1', \ldots,$$

$$D_q', A'D_q'(A')^{M_q-1}D_q'] = \dim A. \qquad (3.17)$$

To determine $\hat{x}^k$, we use the following bilinear equation,

$$dx^k(t) = Ax^k(t)dt + \sum_{i=1}^{n} B_i \xi_i^k(t) x^k(t)dt, \qquad (3.18)$$

where A, B_i are matrices of corresponding dimensions; $x^k(0)$ is independent of $\xi^k(\cdot)$, and $\omega(\cdot)$ and $v(\cdot)$ are processes.

The proof of the theorem follows from Proposition 3.2, the results in Marcus (1973), and the nilpotency assumption (Chikte and Lo, 1981; Willsky, 1973).

Define the process $\{y^k(t)\}_{t \geq 0}$

$$dy^k(t) = \left[\sum_{i=1}^{n^*} H_i^*(\xi_i^k)^*(t) \right] y^k(t)dt, \qquad (3.19)$$

where $y^k(t) = Se^{-At}x^k(t)$; S is nonsingular; $(\xi^k)^*(t) = \hat{D}(t)\xi^k(t)$; $\hat{D}(0)$ is a deterministic $n^* \times n$ matrix, and

$$H_k^* = \begin{bmatrix} h_k & h_{12}^k & h_{13}^k & \cdots & h_{1M}^k \\ 0 & h_k & h_{23}^k & \cdots & h_{2M}^k \\ 0 & 0 & h_k & \cdots & h_{3M}^k \\ \vdots & \vdots & \vdots & \cdots & \vdots \\ 0 & 0 & 0 & \cdots & h_k \end{bmatrix}, \quad k = 1, 2, \ldots, n^*. \qquad (3.20)$$

The dimension of $(x^k)^*(t) \triangleq [(y^k)^T(t), (y^{k(1)})^T, \ldots, (y^{k(M-1)})^T(t)]^T$ is defined by a differential equation (Lo, 1981). We use the following differential equation,

$$dy^{k(l)}(t) = \left[\alpha^{k(l)}(t) + \sum_{l=1}^{n} \gamma_l^{k(l)}(t)\xi_l^k(t) \right] y^{k(l)}(t)dt + \beta^{k(l)}(t)y^{k(l)}(t)dt,$$

$$y^{k(l)}(0) = 0, \quad l = 1, 2, \ldots, n, \qquad (3.21)$$

where the $M \times M$ blocks of matrices $\alpha^{k(l)}(\cdot)$, $\beta^{k(l)}(\cdot)$, k is a cell number, and $\{\gamma_i^{k(l)}(\cdot)\}_{i=1}^n$ belong to the linear manifold generated by the set $\{H_j^*\}_{j=1}^{n^*}$, and can be computed from knowledge of the covariance matrix $P(\cdot)$ of the $\xi^k(\cdot)$ process and the matrix $\hat{D}(\cdot) \stackrel{\triangle}{=} [\hat{D}_{ij}(\cdot)]$.

The filter of the form (3.15) is the result of rewriting the innovation term in the standard bilinear form.

The proof of the nilpotency of the Lie algebra $(L_0^k)^*$ is based on properties of the bilinear system (3.15). Applying the Lie bracket operations blockwise, we obtain that the matrices $\tilde{A}_{li}(t) \stackrel{\triangle}{=} \text{Ad}_{C_i^*(t)}^l(A^*(t))$, $i = 1, \ldots, n$; $l = 0, 1, \ldots$ inherit all the properties of $A^*(t)$, so that all the $M \times M$ blocks of matrices $\tilde{B}_{k,j,l,i}(t) = \text{Ad}_{\tilde{A}_{li}}^k(t)B_j^*(t)$, $k = 1, 2, \ldots$; $j = 1, \ldots, n$ are strictly triangular. Because $\{B_i^*(t)\}_{i=1}^n$ are themselves in nilpotent canonical form, the desired conclusion can be verified simply by carrying out blockwise the Lie bracket operations required in the definition of nilpotency.

1.5 Identification of Bilinear Lattice Models

In this section we describe an identification method based on the expansion of signal processes over an orthogonal basis. Using this methodology we can obtain a system of linear algebraic equations, which is used to determine the coefficients of the bilinear lattice model. By means of the least squares method we obtain estimates of the unknown parameters of the lattice model. The computational algorithm obtained has quite good accuracy.

Properties of generalized orthogonal polynomials (GOP). An orthogonal polynomial $\phi_i(t)$ can be represented by a power series

$$\Pi_i(t) = \sum_{j=0}^{i} f_{ij} t^j, \tag{3.22}$$

where f_{ij} are the coefficients of the expansion with respect to t^j. Similarly, t^i is expressed by the series

$$t^i = \sum_{j=0}^{i} \chi_{ij} \Pi_j(t). \tag{3.23}$$

Equations (3.22) and (3.23) can be written in vector form; that is,

$$\Pi = F\Theta, \quad \Theta = G\Pi = GF\Theta, \quad GF = I, \quad x_{ij} = \frac{1}{f_{ij}}, \tag{3.24}$$

where Π and Θ are the vectors of the orthogonal polynomials and the power series. F and G represent the triangular matrices of the expansion coefficients; that is,

$$\Pi(t) = [\Pi_0(t)\,\Pi_1(t)\,\Pi_2(t)\cdots\Pi_{m-1}(t)]^T,$$
$$\Theta(t) = [1\ t\ t^2\cdots t^{m-1}]^T, \tag{3.25}$$

$$F = \begin{bmatrix} f_{00} & 0 & 0 & 0 & \cdots & 0 & 0 \\ f_{10} & f_{11} & 0 & 0 & \cdots & 0 & 0 \\ f_{20} & f_{21} & f_{22} & 0 & \cdots & 0 & 0 \\ \vdots & \vdots & \vdots & \vdots & \cdots & \vdots & \vdots \\ f_{m-1,0} & f_{m-1,1} & f_{m-1,2} & f_{m-1,3} & \cdots & f_{m-1,m-2} & f_{m-1,m-2} \end{bmatrix}_{m\times m},$$

$$G = \begin{bmatrix} \chi_{00} & 0 & 0 & 0 & \cdots & 0 & 0 \\ \chi_{10} & \chi_{11} & 0 & 0 & \cdots & 0 & 0 \\ \chi_{20} & \chi_{21} & \chi_{22} & 0 & \cdots & 0 & 0 \\ \vdots & \vdots & \vdots & \vdots & \cdots & \vdots & \vdots \\ \chi_{m-1,0} & \chi_{m-1,1} & \chi_{m-1,2} & \chi_{m-1,3} & \cdots & \chi_{m-1,m-2} & \chi_{m-1,m-1} \end{bmatrix}_{m\times m}.$$

The general recursive formula of orthogonal polynomials can be represented by the expression

$$\Pi_{i+1}(t) = (a_i t + b_i)\Pi_i(t) - c_i\Pi_{i-1}(t). \tag{3.26}$$

The expansion coefficients are calculated with the above recursive formula. By substituting (3.22) into (3.26) and equating the coefficients at each power of t, we obtain the following recursive formula,

$$\left.\begin{array}{c} f_{i+1,j} = a_i f_{i,j-1} + b_i f_{ij} - c_i F_{i-1,j}, \\ f_{00} = 1, \quad f_{i,i-i} = 0, \\ f_{ij} = 0 \quad \text{for} \quad j > i. \end{array}\right\}, \tag{3.27}$$

where f_{i0} and f_{ij} are given. By integrating (3.27) we obtain

$$\int_0^t \Pi(t)dt = H\Pi(t), \tag{3.28}$$

where H is the operational matrix of integration, a generalized orthogonal polynomial $\Pi(t)$, introduced by M. Wang (Andreev, 1982). Using a similar approach, we integrate Θ:

$$\int_0^t \Theta(t) = E\Theta(t), \tag{3.29}$$

where E is the operational matrix of the GOP for the forward integration. The operational matrix H is calculated by the formula

$$H = FEG. \tag{3.30}$$

Any integrable function can be represented by a power or orthogonal series

$$x(t) = \sum_{i=0}^{m-1} x_i t^i = \sum_{i=0}^{m-1} \alpha_i \Pi_i. \tag{3.31}$$

If we estimate the value of x_i using the power series and the least squares method, essential error can occur.

Thus we propose to use the least squares method combined with the orthogonal polynomials in order to estimate the sensor parameters. By substituting (3.22) into (3.31) and comparing the expansion coefficients, we obtain

$$x_i = \sum_{j=1}^{m-1} \alpha_j f_{ji}, \tag{3.32}$$

where the values of α_j are determined from the condition of orthogonality; that is,

$$\alpha_j = \int_f^b W(t) x(t) \Pi_j(t) dt / \int_a^b W(t) \Pi_j^2(t) dt, \tag{3.33}$$

and integration (3.33) is approximated with Simpson's method. If $i \geq m$ then (3.23) is approximated by the series

$$t^i = \sum_{j=0}^{m-1} \chi_{ij} \Pi_j(t). \tag{3.34}$$

Applying (3.33) we obtain

$$t^i = \sum_{l=0}^{m-1} \left[\sum_{j=1}^{m-1} \chi_{ij} f_{jl}(t) \right] t^l. \tag{3.35}$$

From (3.35) it follows

$$t^{m+k} = \sum_{l=0}^{m-1} r_{kl} t^l, \tag{3.36}$$

where

$$r_{kl} = \sum_{i=1}^{m-1} \chi_{m+k,j} f_{jl}. \tag{3.37}$$

Multiplying each component Θ by t^i gives the expression

$$t^i \Theta = R_i \Theta(t), \tag{3.38}$$

where

$$R_i = \left[\begin{array}{ccccc|cccccc}
0 & 0 & 0 & \cdots & 0 & 1 & 0 & 0 & \cdots & 0 \\
0 & 0 & 0 & \cdots & 0 & 0 & 1 & 0 & \cdots & 0 \\
0 & 0 & 0 & \cdots & 0 & 0 & 0 & 1 & \cdots & 0 \\
\vdots & \vdots & \vdots & \cdots & \vdots & \vdots & \vdots & \vdots & \cdots & \vdots \\
0 & 0 & 0 & \cdots & 0 & 0 & 0 & 0 & \cdots & 1 \\
\hline
r_{00} & r_{0i} & r_{02} & & \cdots & & & & & r_{0,m-1} \\
r_{i0} & r_{ii} & r_{i2} & & \cdots & & & & & r_{i,m-1} \\
r_{20} & r_{2i} & r_{22} & & \cdots & & & & & r_{2,m-1} \\
\vdots & \vdots & \vdots & & \cdots & & & & & \vdots \\
r_{i-i,0} & r_{i_i,i} & r_{i-1,2} & & \cdots & & & & & r_{i-1,m-1}
\end{array}\right]_{m\times m}. \tag{3.39}$$

Define the matrix Λ_i as

$$\Lambda_i = [I_i 0]_{i\times m}.$$

Then (3.39) can be written as

$$R_i = \left[\begin{array}{c}
O_{(i)} \ \vdots \ I_{(m-1)} \\
\hdotsfor{1} \\
\Lambda_i R_m
\end{array}\right]_{m\times m}.$$

Analysis of the bilinear lattice model. Consider a general bilinear model (BM)

$$\dot{x}^k(t) = A(t)x^k(t) + L(t)u^k(t) + \sum_{j=1}^{n} B_j(t)x^k(t)u_j^k(t),$$

$$x^k(0) = x_0^k, \tag{3.40}$$

where $x^k(t)$ is an M-dimensional state vector; k is a cell number; $u^k(t)$ is the n-dimensional input vector of the kth cell; the matrices $A(t)$, $L(t)$, and $B_j(t)$ have the appropriate dimensions.

If the input signal is defined, then (3.40) can be written in the form

$$\dot{x}^k(t) = V^k(t)x^k(t) + L(t)u^k(t), \tag{3.41}$$

where

$$V^k(t) = A(t) + \sum_{j=1}^{n} B_j(t)u_j^k(t).$$

By expanding $\dot{x}^k(t)$, $x^k(t)$, $u^k(t)$, $V^k(t)$, and $L(t)$ into generalized orthogonal polynomials series, we obtain

$$\dot{x}^k(t) = \sum_{i=0}^{m-1} d_i^k t^i = \hat{D}^k \Theta^k(t), \tag{3.42}$$

$$x^k(t) = \sum_{i=0}^{m-1} x_i^k t^i = X^k \Theta^k(t), \tag{3.43}$$

$$u^k(t) = \sum_{i=0}^{m-1} u_i^k t^i = U^k \Theta^k(t), \tag{3.44}$$

$$V^k(t) = \sum_{i=0}^{m-1} V_i^k t^i, \tag{3.45}$$

$$L(t) = \sum_{i=0}^{m-1} L_i t^i. \tag{3.46}$$

From (3.22) we have $X^k = X^k(0) + D^k E^k$, where
$$X^k(0) = [x^k(0), 0, 0, \ldots, 0].$$
The product of matrices $V^k(t)x^k(t)$ and $L(t)u^k(t)$ can be expressed with the GOP-series by using (3.38), (3.43), (3.45), and (3.46):

$$V^k(t)x^k(t) = \sum_{i=0}^{m-1} V_i^k X^k R_i^k \Theta^k(t),$$

$$L(t)u^k(t) = \sum_{i=0}^{m-1} L_i^k U^k R_i^k \Theta^k(t). \tag{3.47}$$

Substituting the equations (3.29), (3.42), (3.44), and (3.47) into (3.41) we obtain

$$\hat{D}^k = \hat{C}^k + \sum_{i=0}^{m-1} V_i^k D^k E^k R_i^k, \tag{3.48}$$

where $\hat{C}^k$ is given by the expression

$$\hat{C}^k = \sum_{i=0}^{m-1} (V_i^k X^k(0) + L_i^k U^k) R_i^k = [c_0^k, c_1^k, \ldots, c_{m-1}^k]_{n \times m}.$$

Define matrices $\hat{D}^k$ and $\hat{C}^k$ as

$$
\hat{D}^k = \begin{bmatrix} d_0^k \\ d_1^k \\ d_2^k \\ \vdots \\ d_{m-1^k} \end{bmatrix}_{n \times m \times 1}, \qquad \hat{C}^k = \begin{bmatrix} c_0^k \\ c_1^k \\ c_2^k \\ \vdots \\ c_{m-1^k} \end{bmatrix}_{nm \times 1},
$$

and $\quad V_i^k = V_i^k \otimes (E^k R_i^k)^T, \quad i = 0, 1, 2, \ldots, m-1.$

Thus (3.48) can be rewritten as

$$
\hat{D}^k = \left(I - \sum_{i=0}^{m-1} V_i^k \right)^{-1} \hat{C}^k, \tag{3.49}
$$

where $\otimes$ is the Kronecker matrix multiplication. From (3.49) it can be seen that the coefficients of expansion $\hat{D}^k$ of the variable $x^k(t)$ into generalized orthogonal series are computed directly.

Estimation of the lattice parameters. In this section we show an application of the GOP-approximation for estimating the parameters of the bilinear lattice model, when the input signal and the measured state parameters are known. Consider the bilinear lattice model

$$
\hat{x}^k(t) = A x^k(t) + L u^k(t) + \sum_{j=1}^{n} B_j x^k(t) u_j^k(t), \tag{3.50}
$$

where A, L, and B_j are unknown parameters to be estimated; k is a cell number. By the generalized product of orthogonal series we mean

$$
u_j^k(t) = \sum_{t=0}^{m-1} u_{jl}^k t^l,
$$

$$
x^k(t) u_j^k(t) = \sum_{l=0}^{m-1} u_{jl}^k X^k t^l \Pi(t) = \sum_{l=0}^{m-1} u_{jl}^k X^k R_l^k \Pi^k(t).
$$

The integration of (3.50) gives

$$
x^k(t) - x^k(0) = A \int_0^t x^k(t') dt' + L \int_0^t u^k(t') dt'
$$

$$
+ \sum_{j=1}^{n} L_j \int_0^t x^k(t') u_j^k(t') u_j^k(t') dt'. \tag{3.51}
$$

Substituting (3.43), (3.44), (3.46) into (3.51) and using (3.29), (3.38) we obtain

$$X^k\Pi^k - X^k(0)\Pi^k = AX^k E^k \Pi^k LU^k E^k \Pi^k$$

$$+ \sum_{j=1}^{n} B_j X^k \left[\sum_{t=0}^{m-1} u_{jl} R_l^k\right] E^k \Pi^k. \tag{3.52}$$

Substituting the expression for Θ into (3.51) gives

$$X^k G^k \Pi^k - \sum_{j=1}^{n} X^k(0) G^k \Pi^k = AX^k E^k G\Pi^k + LU^k E^k G^k \Pi^k$$

$$+ \sum_{j=1}^{n} B_j X^k \left[\sum_{j=1}^{m-1} u_j^k R_j^k\right] E^k G^k \Pi^k(t)$$

or

$$X^k G^k - X^k(0) G^k = AX^k E^k G^k + LU^k E^k G^k$$

$$+ \sum_{j=1}^{2} B_j X^k \left[\sum_{l=0}^{m-2} u_{jl}^k R_l^k\right] E^k G^k,$$

$$Z^k S^k = (X^k - X^k(0)) G^k, \tag{3.53}$$

where Z^k is the parameter vector; that is,

$$Z^k = [AL\, B_1\, B_2\, \cdots\, B_n].$$

1.6 A Generalization for Nonlinear Lattice Models

Consider the lattice model, which is described by the following equations

$$d\xi^k(t) = F(t)\xi^k(t)dt + G(t)dw(t), \tag{3.54}$$

$$dx^k(t) = f_0(x^k(t))dt + \sum_{i=1}^{n} f_i(x^k(t))\xi_i^k(t)dt, \tag{3.55}$$

$$dz^k(t) = H(t)\xi^k(t)dt + R^{1/2}(t)dv(t), \tag{3.56}$$

where $\xi^k(t)$, $x^k(t)$, $z^k(t)$ are n, q, p-dimensional vectors, respectively; w and v are independent standard processes of the Brownian type; $R > 0$, $\xi^k(0)$ is a Gaussian random variable independent of w and v; $x^k(0)$ is a variable independent of $\xi^k(0)$, w, and v; $\{f_i, i = 0, \ldots, n\}$ are analytical functions of x^k. Let $[F(t), G(t), H(t)]$ be completely controllable and observable matrices (Kalman, Falb, and Arbib, 1969), and define

$$Q(t) \stackrel{\triangle}{=} G(t)G'(t). \tag{3.57}$$

It is required to estimate the conditional mean $\hat{x}^k(t/t) = E^t[x^k(t)/(z^k)^t]$ and $\hat{\xi}^k(t/t)$ by the observation $(z^k)^t \stackrel{\triangle}{=} \{z^k(s), 0 < s < t\}$. The Volterra series for the ith component of x^k is given by the expression

$$x_i^k(t) = \omega_{0i}(t) + \sum_{j=1}^{\infty} \int_0^t \cdots \int_0^t \sum_{k_1,\ldots,k_j=1}^{\check{n}} \omega_{ji}^{(k_1,\ldots,k_j)}(t, \sigma_1, \ldots, \sigma_j)$$
$$\times \xi_{k_1}^k(\sigma_1) \cdots \xi_{k_j}^k(\sigma_j) d\sigma_1 \ldots d\sigma_j, \tag{3.58}$$

where the jth order kernel $w_{ji}^{(k_1,\ldots,k_j)}$ is a locally bounded and piecewise continuous function. We consider triangular kernels that satisfy the condition $w_{ji}^{(k_1,\ldots,k_j)}(t, \sigma_1, \ldots, \sigma_j) = 0$, if $\sigma_{l+r} > \sigma_r$; $l, r = 1, 2, 3, \ldots$. We say that the kernel $w(t, \sigma_1, \ldots, \sigma_j)$ is separable if it can be expressed by a finite sum

$$w(t, \delta_1, \ldots, \delta_j) = \sum_{i=1}^{r} \gamma_0^i(t)\gamma_1^i(\delta_1)\gamma_2^i(\delta_2) \cdots \gamma_j^i(\delta_j). \tag{3.59}$$

Consider a linear system (3.54), (3.56) and define the scalar-valued process

$$\eta^k(t) = \int_0^t \int_0^{\delta_1} \cdots \int_0^{\delta_{j-1}} \xi_{k_1}^k(\delta_{r_1}) \cdots \xi_{k_i}^k(\delta_{r_1}) \gamma_1(\delta_1) \cdots \gamma_j(\delta_j) d\delta_1 \cdots d\delta_j,$$
$$x^k(t) = e^{\xi_l^k(t)} \eta^k(t), \tag{3.60}$$

where $\{\gamma_i^k\}$ are deterministic functions of t and $i > j$. Then the conditional means $\hat{\eta}^k(t/t)$ and $\hat{x}^k(t/t)$ can be defined by a finite-dimensional system of nonlinear stochastic equations that are supplied by the input process of the form $d\nu^k(t) \stackrel{\triangle}{=} dz^k(t) - H(t)\hat{x}^k(t/t)dt$. The following theorem is valid (Marcus, 1973; Marcus and Willsky, 1976).

Theorem 3.3. *Suppose we have a scalar process*

$$x^k(t) = e^{\xi_l^k(t)} \eta^k(t), \tag{3.61}$$

where η^k is a finite Volterra series in ξ^k (i.e., the expansion (3.58) has a finite number of terms with separable kernels). Then $\hat{\eta}^k(t/t)$ and $\hat{x}^k(t/t)$

can be computed using the finite-dimensional system of nonlinear stochastic differential equations driven by the innovations

$$dv^k(t) \stackrel{\triangle}{=} dz^k(t) - H(t)\hat{x}^k(t/t)dt. \tag{3.62}$$

Using the results of Brockett (1975) on finite Volterra series it is easy
to show that each term in (3.61) can be represented by a bilinear system
of the form

$$\dot{x}^k(t) = \xi_j^k(t)x^k(t) + \sum_{j=1}^{n} B_j(t)\xi_j^k(t)x^j(t), \tag{3.63}$$

where x^k is a vector of state; k is a vector; B_j are strictly upper-triangular
matrices (zeroes on and below the main diagonal). For such a system the
Lie algebra L_0 is nilpotent. Conversely, if the Lie algebra L_0^k corresponding to the bilinear system

$$\dot{X}^k(t) = \left(A_0 + \sum_{i=1}^{n} \xi_i^k(t)B_i\right) X^k(t), \quad X^k(0) = I, \tag{3.64}$$

where X^k is a $\hat{k}$-vector, is nilpotent, then each component of the solution
(3.64) can be written as a finite sum of terms (3.61). This leads to the
following result.

Consider the lattice model (3.54), (3.56), and (3.64), and assume that
L_0^k is a nilpotent Lie algebra. Then the conditional expectation $\hat{x}^k(t/t)$
can be calculated by means of a finite-dimensional system of nonlinear
differential equations driven by the innovations.

LM (3.55) with bounded interactions of cells and causal and continuous map "input-output" of the LS (Susmann, 1983; Fliess, 1975) can be
uniquely approximated by a bilinear system of the form (3.64), in which
$A_0, B_1, \ldots, B_n$ are all strictly upper-triangular. For such a LS both L_0
and L are also nilpotent Lie algebras.

1.7   Estimation of the State Vector of CA3 Region

The bilinear lattice model based on the two-variable reduction of the
Hodgkin–Hukley model (Hodgkin and Hukley, 1952; Traub, Miles and
Jeffreys, 1993; Gröbler, 1998) initially was proposed by Morris and Lecar
(1981) as a model for barnacle muscle fiber, but is also useful in modeling
the pyramidal cells in the network of the CA3 region of the hippocampus
(Figure 3.2).

The system of equations for our proposed network model is given by

$$\dot{x}_1^k = a_1 + b_{11}x_1^k + c_{12}x_2^k u_1^k + u_2^k + b_{13}x_3^k, \tag{3.65}$$

$$\dot{x}_2^k = a_2 + b_{22}x_2^k, \tag{3.66}$$

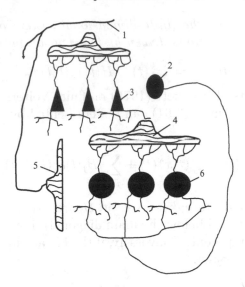

Figure 3.2. A network showing interconnections between an excitatory pathway, a population of pyramidal cells, and a population of inhibitory interneurons (1 is excitatory pathway, 2 is population of excitatory synapses, 3 is pyramidal neurons, 4, 5 are populations of inhibitory synapses, and 6 depicts inhibitory neurons).

$$\dot{x}_3^k = b_{31} x_1^k + d_3 u_2^k, \tag{3.67}$$

$$\psi^k(t) = h^k(x^k(t)) + Q^k(x^k(t))\gamma^k(t). \tag{3.68}$$

This system can by represented by a pair of equations of the form

$$\dot{x}^k(t) = f^k(x^k(t)) + G^k(x^k(t))\xi^k(t), \quad t \ge 0, \tag{3.69}$$

$$\psi^k(t) = h^k(x^k(t)) + Q^k(x^k(t))\gamma^k(t), \quad t \ge 0, \tag{3.70}$$

where x^k (state), ψ^k (output), ξ^k, and γ^k (inputs) are 3-, 3-, 2-, and 3-dimensional vector-valued functions of time, respectively; $f^k \doteq \{a_1 + b_{11}x_1^k + c_{12}x_2^k x_1^k + b_{13}x_3^k, a_2 b_{22}x_2^k, b_{31}x_1\}$, h^k are 3-dimensional vector-valued functions of $x^k(t)$; G^k, Q^k are matrix-valued of $x^k(t)$ of appropriate sizes;

$$G^k = \begin{bmatrix} c_{12}x_2^k & 1 \\ 0 & 0 \\ 0 & d_3 \end{bmatrix},$$

$x^k(0)$ is independent of ξ^k, $\omega(\cdot)$, and $v(\cdot)$ processes; $a_1 = g_{c_a}m_\infty$; $a_2 = (\phi\omega_\infty)/\tau_\omega$; $b_{11} = g_{c_a}m_\infty - g_L V^L$; $b_{13} = -\alpha_{inh}$; $b_{22} = -\phi/\tau_\omega$; $b_{31} = \alpha_{inh}$; $c_{12} = -g_k$; $d_3 = bc$; $m_\infty = f_1(x_1^i, y_1)$; $w_\infty = f_2(x_1^i, y_3, y_4)$; $\alpha_{exc} = f_3(x_1^i, y_5, y_6)$; $\alpha_{inh} = f_4(x_1^i, y_6, y_7)$; $\tau_\omega = f_5(x_1^i, y_3, y_4)$; x_1^k and x_3^k are the membrane potentials of the pyramidal and inhibitory cells, respectively; x_2^k is the relaxation factor which is essentially the fraction of open potassium channels in the population pyramidal cells; all three variables apply to node i in the lattice. The parameters g_{c_a}, g_k, and g_L are the total conductances for the populations of Ca, K, and leakage channels, respectively. V_i^K is the Nerst potential for potassium in the node; V_L is a leak potential, τ_ω is a voltage-dependent time constant for W_i, I is the applied current, and ϕ and b are temperature scaling factors. The parameter c differentially modifies the current input to the inhibitory interneuron; $u_1^k = V_i^k$; $u_2^k = I^k$.

The equations (3.69), (3.70) can be reduced to the following system of bilinear equations,

$$\dot{Y}^k(t) = \left(A + \sum_{i=1}^{2} B_i \xi_i^k(t) \right) Y^k(t),$$

$$\psi^k(t) = \left(C + \sum_{i=1}^{3} D_i \gamma_i^k(t) \right) Y^k(t), \qquad (3.71)$$

where A, B_i, C, and D_i are constant matrices of appropriate sizes.

In the deterministic case, the model of the CA3 region with bounded interactions of cells can be uniformly approximated by a bilinear system of the form (3.9) in which $A_0, B_1, \ldots, B_n$ are all strictly upper-triangular. For such a bilinear lattice both L_0^k and L^k are nilpotent Lie algebras.

The Hodgkin–Hukley model can be represented in the form

$$\dot{x}^k(t) = A + \left(B + u_1^k(t)C \right) x^k(t) + D_2 u_2^k, \qquad (3.72)$$

with the A, B, C, D matrices and the u_j^k is a scalar function of time and x_j^k. Here

$$A = \begin{bmatrix} a_1 \\ a_2 \\ 0 \end{bmatrix}, \quad B = \begin{bmatrix} b_{11} & 0 & b_{13} \\ 0 & b_{22} & 0 \\ b_{31} & 0 & 0 \end{bmatrix}, \quad C = \begin{bmatrix} 0 & c_{12} & 0 \\ 0 & 0 & 0 \\ 0 & 0 & 0 \end{bmatrix}, \quad D = \begin{bmatrix} 1 \\ 0 \\ d_3 \end{bmatrix}.$$

Let $A \approx 0$, $B_1 = C$, $D \approx B_2 x^k$; that is, the system (3.72) is already in the form given by (3.19), (3.20) and hence we may take, following the notation of the proof, $\hat{D}(\cdot) \approx I_2$, $(\xi^k)^*(\cdot)$, and $y^k(\cdot) \approx x^k(\cdot)$. Because $n = n^* = 2$, $M = 3$ we have $M^* = 219$. In this case $\hat{D}_{ij}(\cdot) \approx I$ so that the number of resulting augmenting states can be reduced by a factor of $2^2 = 4$, by combining them as follows. Define

$$y^{k(l)}(t) = \sum_{k,\, i=j=1}^{2} y^{k(k,i,j)}(t), \quad l = 1, 2 \qquad (3.73)$$

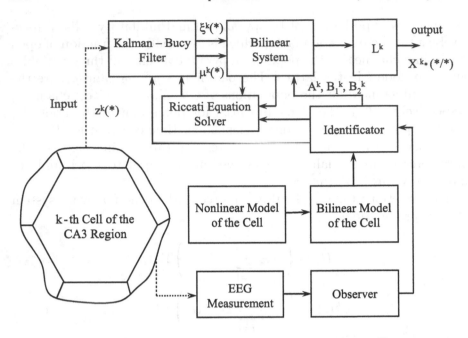

Figure 3.3.   Block schematic of the adaptive nonlinear filter

and

$$y^{k(l,k')}(t) = \sum_{i=j=1}^{2} \sum_{i'=j'=1}^{2} y_{k',i',j'}^{k(l,i,j)}(t), \quad l, k' = 1, 2. \qquad (3.74)$$

This yields an $M^* = 21$-dimensional adaptive filter. The structure of the adaptive filter for a single cell is shown in Figure 3.3.

This filter can be used for analysis and has a rich variety of different dynamical behaviors of the CA3 region. The bilinear lattice model can reproduce single-action potentials as well as sustained limit cycle oscillations for different values of parameters. On the basis of the proposed adaptive filter we can find simple limit cycles as well as periodic state and aperiodic behavior. The periodic behavior corresponds to a phase-locked mixed-mode state on a torus attractor. We can observe a large variety of a mixed-mode state interspersed with regions of apparent chaotic behavior. Using the proposed algorithms we can analyze the complexity for different values of c, that is, when the current to the inhibitory cells is relatively low compared to the current input to the pyramidal cells. A low degree of inhibition results in a system that is more likely to be periodic, suggesting the type of spatiotemporal coherence that could exist with a seizure. When inhibition is completely absent, the dynamics of a system goes to a fixed point corresponding to a state of total depolarization of the network. An intersection of the transition regions between mixed-mode states undergoes a periodic-doubling sequence as the underlying torus attractor breaks up into a fractal object. When the parameters b and y_6 are controllable, a similar behavior is observed in a series of system investigations.

1.8 Detection and Prediction of Epileptic Seizures

In this section we discuss possible developments of a decision support system (DS) for analyzing the spatiotemporal dynamical changes in the EEG. We introduce the definitions of T-index (Iasemidis et al., 2001) as a measure of distance between the mean values of pairs of STL_{max} (Short Time Lyapunov exponent) profiles over time.

Definition 3.2. *By the T-index (or T-signal) at time t between electrode sites i and j, we mean the variable*

$$T_{ij} = \sqrt{N} \times E(STL_{max,i} - STL_{max,j}/\sigma_{ij})(t), \qquad (3.75)$$

where $E(\cdot)$ is the sample average difference for the $(STL_{max,i} - STL_{max,j})$ estimated over the moving time window $\omega_t(\lambda)$ defined as,

$$\theta_t(\lambda) = \begin{cases} 1 & if \ \lambda \in [t - N - 1, t], \\ 0 & if \ \lambda \notin [t - N - 1, t], \end{cases} \qquad (3.76)$$

where N is the length of the moving window; $\sigma_{ij}(t)$ is the simple standard deviation of the STL_{max} differences between electrode site i and j within the moving window $\theta_t(\lambda)$; STL (Short Time Lyapunov) is the operator of numerical estimation of L_{max}; L_{max} is the maximum Lyapunov exponent (the Kolmogorov–Sinai entropy).

Let us consider a decision support system registering the T-signal. To each T-signal (TS) corresponds a real sensor signal of certain form, which is transmitted to the sensors and appears in its input possibly distorted, and corrupted by random noise. The proposition that the TS (epileptic signal) was transmitted is equivalent to the hypothesis about the composition of the input $T(t)$ to the DS during a certain interval of time; we denote these two hypotheses by H_j, $0 \leq j \leq 1$. The DS must choose one of these hypotheses on the basis of input $T(t)$ during the *observation interval*, say $0 \leq t \leq t_f$.

The input $T(t)$ is a stochastic process, described in terms of probability density functions. For simplicity we suppose that the input during $(0, t_f)$ has been appropriately sampled and can be represented by n

samples $T_1, T_2, \ldots, T_n$. We designate these data collectively by a vector $T_1, T_2, \ldots, T_n$, and we represent T as a point in an n-dimensional Cartesian space $\mathbb{R}_n$.

Under hypothesis H_j, that is, when the jth of the $M = 2$ signals has been transmitted, the n samples T are random variables having a joint probability density function such as $p_j(T)$ is a nonnegative function whose integral over the entire space $\mathbb{R}_n$ equals 1:

$$\int_{-\infty}^{\infty} \cdots \int_{-\infty}^{\infty} p_j(T_1, T_2, \ldots, T_n) dT_1) dT_2, \ldots, dT_n = \int_{\mathbb{R}_\times} p_j(T) d^n T = 1;$$

$dT_1, dT_2, \ldots, dT_n$ is the volume element in the data space. The probability under hypothesis H_j that the point T representing a particular set of samples lies in a arbitrary region ∇ of that space is

$$Pr(T \in \nabla | H_j) = \int_{\nabla} p_j(T) d^n T = 1, \quad j = 1, 2.$$

On the basis of the observed values of $(T_1, T_2, \ldots, T_n)$ the DS is to decide between the two hypotheses H_0, H_1. That is, DS must choose which of the two probability density functions $p_j(T)$ it believes actually characterizes the input $T(t)$ during $(0, t_f)$. The scheme by which the DS makes these choices is called a *strategy*. It must assign a definite selection between H_0, H_1 to each possible datum $(T_1, T_2, \ldots, T_n)$ into two disjoint regions R_1, R_2. When the point T falls into region R_j, the DS chooses hypothesis H_j, deciding that the jth epileptic signal was transmitted.

Let

$$\left\{ \Lambda(T) = \frac{p_1(T)}{p_0(T)} \geq \lambda \right\} \Longrightarrow H_1 \qquad (3.77)$$

be the likelihood ratio (Helstrom, 1995). The likelihood ratio $\Lambda(T)$ is a function of the random variables $(T_1, T_2, \ldots, T_n)$ and is itself a random variable. The rule (3.77) is optimum (Helstrom, 1995). The decision level λ is determined by the preassigned value of the false-forecast (FF) probability Q_0,

$$Q_0 = Pr(\Lambda(T) \geq \lambda | H_0) = \int_{\lambda}^{\infty} P_0(\Lambda) d\Lambda, \qquad (3.78)$$

where $P_0(\Lambda)$ is the probability density function of the random variable $\Lambda(T)$ under hypothesis H_0.

Let us suppose that the regions R_0 and R_1 are separated by the decision surface D given by $\Lambda(T) = \lambda$, with λ chosen to satisfy (3.78). Then the probability of epileptic detection is given by

$$Q_d = \int_{R_1} p_1(T)d^nT = \int_{R_1} \Lambda(T)p_0(T)d^nT, \qquad (3.79)$$

by virtue of (3.77).

In this section we described three types of lattice models with deterministic and stochastic interactions between neurons. Based on the nonlinear lattice model we have shown that there exists a bilinear lattice model, such that its input-output map is the truncation of the initialized nonlinear lattice as a Chen series. Because of the inherent nonlinearity, the solution can provide a rich repertoire of spatiotemporal patterns. We considered the analysis of EEG patterns according to basic concepts of system analysis and synergetics as an example. We have shown that the mathematical model of the CA3 region with bounded interaction of cells can be uniformly approximated by a bilinear lattice in which A_0^k, B_i^k are all strictly upper-triangular.

Based on the bilinear lattice models, we have developed an effective adaptive filtration algorithm to deliver better performance than linear models. Then we have proposed a general strategy for estimation of stochastic nonlinear lattices with different restrictions on the nonlinear interaction between neurons. This will allow us to solve the problem of suboptimal estimation for quite a wide class of stochastic conditions.

The behavior of this adaptive estimate has been illustrated by means of simulations of the CA3 region of the hippocampus (a common location of the epileptic focus) using a bilinear representation. This region is the self-organized information flow network of the human brain. This high coordination also becomes macroscopically visible through EEG measurements of brain activity under different circumstances.

This model consists of hexagonal LS of nodes, each describing a controlled neural network consisting of a group of prototypical excitatory pyramidals cells and a group of prototypical inhibitory interneurons connected via excitatory and inhibitory synapses.

We have shown that the prediction of epileptic seizures can be obtained by consideration of the spatiotemporal dynamical changes in the EEG using the Neyman–Pearson criterion for the T-index. Clearly, further work is needed to carefully probe experimentally observed dynamics in the epileptic brain and to clarify the bifurcation and self-organization structure that display complex probability distribution functions.

A coupled stochastic model for the CA3 region of the hippocampus is developed. This model consists of a lattice of cells, each describing a subnetwork consisting of a group of prototypical excitatory pyramidal cells and a group of prototypical inhibitory interneurons connected via on/off excitatory and inhibitory synapses. We simulate the weak interaction between cells using nonlinear mechanisms such as diffusion and thermal noise.

2.   Optimal Estimation of Signal Parameters Using Bilinear Observations

In this section we consider the problem of optimal input signal estimation for bilinear systems under input measurements. The periodogram estimates of parameters are studied. The possibilities to construct a finite-dimensional adaptive estimator for a causal dynamical system class are shown. The robust signal estimating problem is solved, when signals are estimated via application of neural networks and when nonlinear measurements are used.

2.1   Estimation Problem

A controllable highly sensitive sensor was developed on the basis of a new physical phenomenon (Yatsenko, 1989). To use all the potential sensor ability, it is necessary to estimate an input signal against the background of a random noise. The known approximate solution to the problem is based on a bilinearized observation model and on application of a linear filter. If the estimation is performed by a linear observation model, the problem estimation and solution evidently make it possible to yield more accurate estimates. Introduce an extra control into a multisensor system (MS), and the latter is highly sensitive, robust, and controllable with respect to a useful signal.

Assume that an MS is described (Yatsenko, 1989; Andreev, 1982; Butkovskiy and Samoilenko, 1990) by the equations

$$\dot{x}(t) = \widehat{A}x(t) + u_1(t)\widehat{B}_1 x(t) + u_2(t)\widehat{B}_2 x(t),$$
$$y(t) = \widehat{c}x(t), \quad x(0) = x_0, \tag{3.80}$$

where $x(t)$ is a two-dimensional vector of MS state; $u_1(t)$ is a piecewise smooth scalar control defined on $(0, \infty)$; $u_2(t) = r(t) + s(t)$; $r(t) = A_0\varphi(\omega_0 t)$ is a useful signal; A_0 and ω_0 are constant values larger than 1; φ is almost periodic function of the form

$$\varphi(t) = \sum_{k=-\infty}^{\infty} c_k e^{i\lambda_k t}, \tag{3.81}$$

$$\sum_{k=-\infty}^{\infty} |c_k| < \infty, \quad \lambda_k \geq 0, \quad k \geq 0, \tag{3.82}$$

$c_k = \overline{c}_{-k}$, $\lambda_k = -\lambda_{-k}$, $|\lambda_l - \lambda_k| \geq \Delta > 0$, under $l \neq k$; $\{s(t), \ t \in \mathbb{R}^1\}$ is a noise that is a real stochastic process which is stationary in a narrow sense, $Ms(t) = 0$, $(Ms(t_1)s(t + t_1)) = s(t)$, and that satisfies the strong mixing condition

$$\sup_{\tilde{A} \in F_{-\infty}^{l}, \tilde{B} \in F_{t+\tau}^{\infty}} |P(\tilde{A}\tilde{B}) - P(\tilde{A})P(\tilde{B})| = \alpha(\tau) \leq C/\tau^{1+\varepsilon} \tag{3.83}$$

with some fixed positive numbers $\tau > 0$, $C > 0$, and $\varepsilon > 0$; $F_a^b = \sigma\{s(t), \ t \in [a, b]\}$ is the smallest σ-algebra generated by a stochastic process $s(t)$, $t \in [a, b]$; and for some $\delta > 4/\varepsilon$, $\exists \varepsilon > 0$,

$$M|s(t)|^{4+\delta} < \infty. \tag{3.84}$$

The spectral density $f(\lambda)$, which is a continuous function and bounded on $\mathbb{R}^1$, is associated with $s(t)$ by the relation

$$s(t) = \int_{-\infty}^{\infty} e^{i\lambda t} f(\lambda) d\lambda.$$

It is necessary to estimate the unknown parameters A_0 and ω_0 by means of the observation $y(t)$ on $t \in [0, T]$. To solve this problem, we introduce some definitions that are concerned with the notion of MS invertibility.

2.2 Invertibility of Continuous MS and Estimation of Signal Parameters

Let F be a finite-dimensional Lie group (cf. Dubrovin, Novikov, and Fomenko, 1984). The right multiplication mapping $L_x : y \to yx$ from $F \to F$ has differential dL_x (cf. Butkovskiy and Samoilenko, 1990; Susmann and Jurdjevic, 1972).

Definition 3.3. *A vector field X on F is called right-invariant, if it satisfies the condition*

$$dL_x X(y) = X(yx) \quad \text{for all} \ \ y \in F. \tag{3.85}$$

Definition 3.4. *A single–input–single–output bilinear model of MS is a control system of the form*

$$\dot{x}(t) = \widehat{A}x(t) + u_2(t)\widehat{B}_2 x(t), \quad x(0) = x_0,$$
$$y(t) = \widehat{c}x(t), \tag{3.86}$$

where the state $x \in \mathbb{R}^2$; $\widehat{A}$ and $\widehat{B}$ are 2×2 matrices over $\mathbb{R}$; $\widehat{c}$ is a 1×2 matrix over $\mathbb{R}$, and $u_2(t) \in \mathcal{U}$, the class of piecewise real analytic functions on $(0, \infty)$.

It is often convenient to express the solution of system (3.80) as $x(t) = X(t)x_0$, where $X(t)$ is a 2×2 matrix-valued function of t which is the trajectory of the corresponding matrix bilinear system.

A single-input matrix bilinear system is a system of the form

$$\dot{X}(t) = \widehat{A}X(t) + u_2(t)\widehat{B}_2 X(t), \quad X(0) = X_0,$$
$$Y(t) = \widehat{C}X(t), \tag{3.87}$$

where A, B, and C are 2×2 matrices over $\mathbb{R}$; $u_2 \in \mathcal{U}$; and $\widehat{C}$ is a 2×2 matrix over $\mathbb{R}$. Let X_0 be invertible; then $X(t) \in GL(2, \mathbb{R})$, the Lie group of invertible 2×2 in real matrices (cf. Brockett, 1972; Susmann and Jurdjevic, 1972).

The matrix system (3.87) is a special case of the more general class of the right-invariant systems studied in Susmann and Jurdjevic (1972).

Definition 3.5. *A right-invariant model of MS (RIMMS) is a system of the form*

$$\dot{x}(t) = \widehat{A}(x(t)) + u_1(t)\widehat{B}_1(x(t)) + u_2(t)\widehat{B}_2(x(t)),$$
$$y(t) = Px(t), \quad x(0) = x_0 \in F, \tag{3.88}$$

where $u_1, u_2 \in \mathcal{U}$; F is a Lie group; P is a Lie subgroup of F with Lie algebra $\mathcal{P}$; and $\widehat{A}, \widehat{B}_1, \widehat{B}_2 \in \mathcal{F}$, the Lie algebra of right-invariant vector fields on F. The examples for the sets P and F are presented in Brockett (1972) and Susmann and Jurdjevic (1972).

Definition 3.6. *A single-input right-invariant system is a system of the form:*

$$\dot{x}(t) = \widehat{A}(x(t)) + u_2(t)\widehat{B}_2(x(t)),$$
$$y(t) = Px(t), \quad x(0) = x_0 \in F. \tag{3.89}$$

The properties of a right-invariant MS are related to the structure of the Lie algebra $\mathcal{F}$. This algebra is a vector space with a nonassociative "multiplication" defined as follows,

$$[X, Y](m) = X(m)Y - Y(m)X,$$

where $X, Y \in \mathcal{F}$ (Dubrovin, Novikov, and Fomenko, 1984). We define $\mathrm{ad}_X^m Y$ inductively as follows,

$$\mathrm{ad}_X^0 Y = Y, \ldots, \ \mathrm{ad}_X^k Y = [X, \mathrm{ad}_X^{k-1} Y].$$

For matrix bilinear systems with $X, Y \in \mathcal{F}$ right-invariance means that $X(M) = XM$ and $[X, Y](M) = (YX - XY)M$.

Let $\widehat{\mathcal{F}}$ be a subset of the Lie algebra $\mathcal{F}$. We define $\{\widehat{\mathcal{F}}\}_{LA}$ to be the Lie algebra generated in $\widehat{\mathcal{F}}$ in $\mathcal{F}$. Thus $\{\widehat{\mathcal{F}}\}_{LA}$ is the smallest Lie subalgebra of $\mathcal{F}$ containing $\widehat{\mathcal{F}}$. For each $x \in F$ let $\widehat{\mathcal{F}}(x) = \{N(x) : N \in \widehat{\mathcal{F}}\}$. It is known that the structure of the reachable set for (3.88) is related to the structure of the Lie algebras:

$$\mathcal{N} = \left\{ \widehat{A}, \widehat{B}_1, \widehat{B}_2 \right\}_{LA},$$
$$\mathcal{N}_0 = \left\{ \operatorname{ad}_{\widehat{A}}^k \widehat{B}_i : k = 0, 1, \ldots, i = 1, 2 \right\}_{LA},$$
$$\mathcal{B} = \left\{ \widehat{B}_1, \widehat{B}_2 \right\}_{LA}.$$

Thus each right-invariant system has associated with it the chain of Lie algebras:

$$\mathcal{F} \supset \mathcal{N} \supset \mathcal{N}_0 \supset \widehat{\mathcal{B}}.$$

If $\exp : \mathcal{F} \to F$ is the standard exponential mapping in Lie theory then $\exp \mathcal{N} = \{\exp N : N \in \mathcal{N}\} \subset F$ and the group generated by $\exp \mathcal{N}$, $\{\exp \widehat{\mathcal{F}}\}_G$ is a Lie subgroup of F (Dubrovin, Novikov, and Fomenko, 1984). Thus each right-invariant MS gives rise to the chain of Lie groups $F \supset \Psi \supset \Psi_0 \supset \Theta$, where $\Psi = \{\exp \mathcal{N}\}_G$, $\Psi_0 = \{\exp \mathcal{N}_0\}_G$ $\Theta = \{\exp \widehat{\mathcal{B}}\}_G$.

Because $\mathcal{N}_0$ is an ideal in $\mathcal{N}$ (i.e., for each $N_0 \in \mathcal{N}_0$, $N \in \mathcal{N}$ and $[N_0, N] \in \mathcal{N}_0$), we know that Ψ_0 is a normal subgroup of Ψ. The following theorem relates the structure of the trajectories of a bilinear MS to the above group decomposition (Dubrovin, Novikov, and Fomenko, 1984).

Theorem 3.4. *Let the system (3.88) have the corresponding solution* $t \to x(t)$ *for arbitrary* $u_1, u_2 \in \mathcal{U}$. *Then* $x(t) \in (\exp t \widehat{A}) \Psi_0 x_0$ *for all* $t \geq 0$, *where* $(\exp t \widehat{A}) \Psi_0 x_0 = \{\exp t \widehat{A} \cdot \psi \cdot x_0 : \psi \in \Psi_0\}$.

To determine $u_2(t)$ on the basis of the observation $y(t)$, it is necessary to introduce basic results.

Definition 3.7. *A right-invariant mathematical model of MS (MMMS) is said to be invertible, if the observation* $\nu \to y(\nu)$ *on any interval* $0 \leq \nu \leq t$ *uniquely determines the input* $\nu \to (u_1, u_2)$ *for* $0 \leq \nu < t$.

The following theorem can be formulated (cf. Hirschorn, 1977).

Theorem 3.5. *The right-invariant MMMS is invertible, if and only if*

$$\operatorname{ad}_{\widehat{A}}^k \widehat{B}_2 \notin \mathcal{P},$$

for some positive integer $k = \{0, 1, \ldots, n-1\}$, *where* n *is the dimension of* $\mathcal{N}$ *and* $\mathcal{P}$ *is the Lie algebra of* P.

The invertibility criterion of MMMS follows from Theorem 3.6.

Theorem 3.6. *Matrix MMMS (3.87) is invertible if and only if*

$$\widehat{C} \operatorname{ad}_{\widehat{A}}^k \widehat{B}_2 \neq 0$$

for some positive integer $k = \{0, 1, \ldots, n^2 - 1\}$.

Definition 3.8. *The relative order* k_1 *of the MMMS is the least positive integer* k *such that*

$$\widehat{C} \operatorname{ad}_{\widehat{A}}^{k-1} \ddot{B}_2 \neq 0, \tag{3.90}$$

or $k_1 = \infty$ *if*

$$\widehat{C} \operatorname{ad}_{\widehat{A}}^k \widehat{B}_2 = 0,$$

for all $k > 0$.

Definition 3.9. *A left-inverse for the MMMS is called the system of the form*

$$\hat{y}(t) = d(\hat{x}(t)) + \hat{u}_2(t)\mu(\hat{x}(t)), \quad \hat{x}(0), \hat{x}(t) \in M, \tag{3.91}$$

such that

$$\hat{y}(t) = u_2(t) \quad under \quad \hat{u}_2(t) = y^{(k_1)}(t). \tag{3.92}$$

Here $\dot{\hat{x}}(t) = a(\hat{x}(t)) + \hat{u}_2 b(\hat{x}(t))$, M *is a differentiable manifold,* $a(\cdot)$ *and* $b(\cdot)$ *are smooth vector fields on* M, $d(\cdot)$, $\mu(\cdot)$ *are smooth vector functions on* M, *and* (k_1) *is the* k_1*th derivative of the observation* $y(t)$.

Theorem 3.7. *Assume that system (3.86) is invertible,* $k_1 < \infty$, *and*

$$\widehat{c} \operatorname{ad}_{\widehat{A}}^{k_1-1} \widehat{B}_2 \neq 0.$$

Then the left-inverse (3.91) exists, and

$$a(\hat{x}) = \widehat{A}\hat{x} - (\hat{c}\widehat{A}^{k_1}\hat{x}/\hat{c}\widehat{A}^{k_1-1}\widehat{B}_2\hat{x})\widehat{B}_2\hat{x},$$

$$b(\hat{x}) = (1/\hat{c}\widehat{A}^{k_1-1}\widehat{B}_2\hat{x})\widehat{B}_2\hat{x},$$

$$d(\hat{x}) = -(\hat{c}\widehat{A}^{k_1}\hat{x}/\hat{c}\widehat{A}^{k_1-1}\widehat{B}_2\hat{x}),$$

$$\mu(\hat{x}) = (1/\hat{c}\widehat{A}^{k_1-1}\hat{x}\widehat{B}_2\hat{x}). \tag{3.93}$$

If $\tilde{u}(t) = y^{(k_1)}(t)$, *then* $\hat{y}(t) = u_2(t)$.

Consider the functional

$$Q_T(\omega) = \left| \frac{2}{T} \int_0^T \hat{y}(t) e^{i\omega t} dt \right|^2. \tag{3.94}$$

Let ω_T be the value of $\omega \geq 0$, for which $Q_T(\omega)$ achieves its maximum value. Because $Q_T(\omega)$ with probability 1 is a continuous function of ω, and $Q_T(\omega) \to 0$ as $\omega \to \infty$, the value ω_T is determined with probability 1.

Theorem 3.8. *Let conditions (3.81)–(3.84) hold, $|c_{i_0}| > |c_i|$, $i \neq \pm i_0$, $i_0 > 0$, and $f(\lambda_{i_0}\omega_0) > 0$. Then:*

(a) $\lim Q_T(\omega_T) = \lim Q_T(\lambda_{i_0}\omega_0) = 4A_0^2|c_{i_0}|^2$ *for $T \to \infty$.*

(b) $T(\omega_T/\lambda_{i_0} - \omega_0) \to 0$ *with probability 1 for $T \to \infty$.*

(c) $A_T = 1/2|c_{i_0}|^{-1}Q_T(\omega_T)$ *is a strongly consistent estimate of A_0.*

(d) $T^{3/2}(\omega_T - \lambda_{i_0}\omega_0)$ *is an asymptotically normal random variable with mean zero and variance*

$$\sigma^2 = 12\pi A_0^{-2}|c_{i_0}|^{-2}f(\lambda_{i_0}\omega_0).$$

(e) $\xi_T = \sqrt{T}(A_T - A_0)$ *is an asymptotically normal random variable with the parameters $(0, \pi|c_{i_0}|^{-2}f(\lambda_{i_0}\omega_0))$.*

Theorem 3.8 is proved on the basis of the results presented in Knopov (1984).

A flowchart of the signal estimation algorithm is depicted in Figure 3.4.

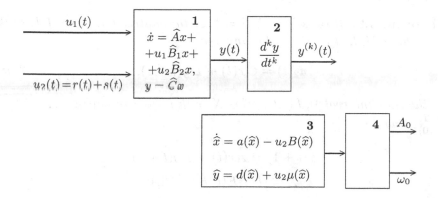

Figure 3.4. Schematic diagram of proposed observer: 1, bilinear MS; 2, differentiator; 3, inverse dynamic system; 4, computation of extremum of $Q_T(\omega)$.

2.3   Estimation of Parameters of an Almost Periodic Signal Under Discrete Measurements

Consider a bilinear model of MS

$$x(t+1) = Ax(t) + B_1 x(t) u_1(t) + B_2 x(t) u_2(t), \qquad (3.95)$$
$$y(t) = Cx(t), \qquad (3.96)$$

where $x(t)$ is a two-dimensional state vector, $u_1(t)$ is a scalar control, $u_2(t) = r(t) + s(t)$, $y(t)$ is a two-dimensional vector of the MS output, and A, B_i, and C are real constant matrices. The assumptions concerned with the functions $r(t)$ and $s(t)$ are the same as in Section 2.

It is necessary to estimate the unknown parameters A_0 and ω_0 using observations of $y(t)$ on $[0, T]$.

The matrix B_i can be expressed as $B_i = e_i v_i'$, $i = 1, 2$, where e_i and v_i are two-dimensional vectors and the dash is the symbol of transposition.

Let us introduce notations

$$\widehat{X} = R^2 - \bigcup_{i=1}^{2} V_i,$$
$$\Psi(x) = \operatorname{diag}\{\gamma_1(x)\gamma_2(x)\},$$
$$v_i'(x)x(t) = \gamma_i(x),$$
$$V_i = \{x \mid v_i'x = 0\},$$
$$E = [e_1, e_2], \quad u = (u_1, u_2), \qquad (3.97)$$

and $\mathcal{U}_{x(0)}^{0,T}$ is a set of the sequences $\{u(0), u(1), \ldots, u(T)\}$ such that $x(0) \in \widehat{X}$, $x(t) \in \widehat{X}$, $t = 0, \ldots, T$.

Theorem 3.9. *If* $\operatorname{rank} B_i = 1$, $i = 1, 2$, *the matrix* CE *has a full rank, then the MMMS (3.95) has the inverse*

$$u_2(t) = \tilde{C}x(t) + \tilde{D}y(t+1), \qquad (3.98)$$

on the time interval $[0, T]$ *at* $x(0) \in \widehat{X}$ *with respect to* $\{u(0), \ldots, u(T)\} \in \mathcal{U}_{x(0)}^{0,T}$, *and*

$$x(t+1) = \tilde{A}x(t) + \tilde{B}y(t+1);$$
$$\tilde{A} = [I - E(CE)^{-1}C]A;$$
$$\tilde{B} = E(CE)^{-1}; \quad \tilde{C} = -\Psi^{-1}(x)(CE)^{-1}CA;$$
$$\tilde{D} = \Psi^{-1}(x)(CE)^{-1}.$$

Proof: If a unit shift operator is applied to equation (3.96) (see Kotta, 1983) and equation (3.95) is substituted into it, the result is

$$y(t+1) = CAx(t) + CB_1x(t)u_1(t) + CB_2x(t)u_2(t). \tag{3.99}$$

Equations (3.95) and (3.99) taken together give algebraic equations (cf. Kotta, 1983)

$$RZ = Z_0, \tag{3.100}$$

where

$$R = \begin{bmatrix} I_2 & -B_1x(t) & -B_2x(t) \\ 0 & CB_1x(t) & CB_2x(t) \end{bmatrix};$$

$$Z = \begin{bmatrix} x(t+1) \\ u(t) \end{bmatrix}; \quad Z_0 = \begin{bmatrix} Ax(t) \\ y(t+1) - CAx(t) \end{bmatrix}.$$

The equation (3.100) can be solved for $x(t+1)$, $u(t)$ uniquely if and only if R has an inverse.

The matrix R

$$R = \begin{bmatrix} I_2 & -E\Psi(x) \\ 0 & CE\Psi(x) \end{bmatrix}$$

has the inverse, if the matrix CE has a full rank and $\gamma_i(x) = v_i^T x(t) \neq 0$, $i = 1, 2$, $t = 0, 1, \ldots$.

If the matrix R is invertible, then

$$R^{-1} = \begin{bmatrix} I_2 & E(CE)^{-1} \\ 0 & \Psi^{-1}(x)(CE)^{-1} \end{bmatrix},$$

and the solution of equation (3.100) is

$$x(t+1) = [I - E(CE)^{-1}C]Ax(t) + E(CE)^{-1}y(t+1),$$
$$\hat{u}(t) = -\Psi^{-1}(x)(CE)^{-1}CAx(t) + \Psi^{-1}(x)(CE)^{-1}y(t+1). \tag{3.101}$$

Write the system (3.101) in the following way,

$$x(t+1) = \tilde{A}x(t) + \tilde{B}y(t+1),$$
$$u(t) = \tilde{C}x(t) + \tilde{D}y(t+1).$$

Consider the functional

$$Q_T(\omega) = \left| \frac{2}{T} \sum_{i=0}^{T} u_2(t)e^{i\omega t}dt \right|^2, \quad i = 0, \Delta T, 2\Delta T, \ldots, T.$$

Let ω_T be the value of $\omega \geq 0$, under which $Q_T(\omega)$ achieves its maximum value. With probability 1, $Q_T(\omega)$ is the continuous function of ω, and $Q_T(\omega) \to 0$ as $\omega \to \omega_T$, therefore ω_T is determined with probability 1.

The main result in this section is the following theorem.

Theorem 3.10. *Let the conditions (3.81)–(3.83) be satisfied and*

$$|c_{i_0}| > |c_i|, \quad i \neq \pm i_0, \quad i_0 > 0,$$
$$f(\lambda_{i_0}\omega_0) > 0. \tag{3.102}$$

Then, with probability 1

$$T(\omega_T/\lambda_{i_0} - \omega_0) \to 0 \quad for \ T \to \infty,$$
$$\lim_{T\to\infty} Q_T(\omega_T) = \lim Q_T(\lambda_{i_0}\omega_0) = 4A_0^2|c_{i_0}|^2,$$

and $A_T = 1/2|c_{i_0}|^{-1}Q_T(\omega_T)$ is the strongly consistent estimate of A_0.

To obtain an asymptotic distribution for ω_T and A_T we use the following formulation of the central limit theorem.

Theorem 3.11. *Let the function $a_T(t)$ satisfy the following conditions.*

(a) *$a_T(t)$ is a real function, defined for $t \geq 0$, and such that for each $T \geq 0$,*

$$W^2(T) = \sum_{t=0}^{T} a_T^2(t) < \infty.$$

(b) *For some constant $0 < C < \infty$*

$$W^{-1}(T) \operatorname*{Sup}_{\substack{0 \leq t \leq T \\ T \to \infty}} |a_T(t)| \leq \frac{C}{\sqrt{T}}, \quad W(T) \to \infty.$$

(c) *For any real v the limit*

$$\lim_{T\to\infty} \frac{1}{W^2(T)} \sum_{t=0}^{T} a_T(t + |v|)a_T = \rho(v),$$

exists and the function $\rho(v)$ is continuous.

(d) *The stochastic process $s(t)$ satisfies (3.81)–(3.83) and (3.102).*

Then $1/W(T) \sum_{t=0}^{T} a_T(t)s(t)$ *is asymptotically normal as $T \to \infty$ with parameters 0 and $\sigma^2 = 2\pi \int_{-\infty}^{\infty} f(\lambda)d\mu(\lambda)$, where $\mu(\lambda)$ is a monotone nondecreasing function bounded on R^1; the value $T^{3/2}(\omega_T - \lambda_{i_0}\omega_0)$ is asymptotically normal with mean zero and the variance $\sigma^2 = 12\pi A_0^{-2} \times |c_{i_0}|^{-2}f(\lambda_{i_0}\omega_0)$; and the value $\xi_T = \sqrt{T}(A_T - A_0)$ is asymptotically normal with the parameters $(0, \pi|c_{i_0}|^{-2}f(\lambda_{i_0}\omega_0))$.*

The proof for discrete time follows from the theorems presented in Knopov (1984).

2.4 Neural Network Estimation of Signal Parameters

The present section considers one more general signal estimation case, when a signal influences a nonlinear dynamic system input. The problem is solved by a neural network algorithm, allowing us to determine an unknown input influence as for a signal, experimentally measured at a nonlinear system output. A neural network processes an output signal as an inverse dynamic system. It is possible to obtain an optimal estimate of a useful signal by means of the above-mentioned approach or when an optimal filter (see Kalman, Falb, and Arbib, 1969) is used to process neural network output. As opposed to the previously published results, the proposed approach does not require a number of the "strict" conditions to be met for bilinear dynamic system invertibility. The sensor equation is assumed known with some uncertainty.

Consider the following dynamic system

$$\dot{x}(t) = f_0(x,t) + \sum_{i=1}^{m} f_i(x,t)u_i(t) = f_0(x,t) + F(x,t)u(t), \quad (3.103)$$

where $x(t)$ is an l-dimensional state vector; $f_0(x,t)$ is a vector-function, nonlinearly dependent on x; $u(t)$ is an m-dimensional input influence, which belongs to the class of continuous functions; and $u_s(t) = r(t) + s(t)$ is an additive mixture of a useful signal with noise.

The present section solves the problem, concerned with estimation of parameters of an input influence $u(t)$, performed with respect to the observation

$$y(t) = h(x,t) + \omega(t), \quad (3.104)$$

where $h(x,t)$ is a nonlinear C^∞-class function; and $\omega(t)$ is an observation noise.

To estimate the unknown parameter $u(t)$, find the estimate of the state vector derivative. Assume this estimate is numerically determined by the nth degree polynomial

$$\widehat{x}_i(t) = \sum_{j=0}^{n} c_{ji}t^j, \quad i = 1,\ldots,l. \quad (3.105)$$

The polynomial coefficients c_{ji} are determined by one of the existing methods. Given the derivative for the ith component of the state vector, then the estimate of this derivative is obtained by the expression

$$\frac{d\widehat{x}_j(t)}{dt} = \sum_{j=1}^{n} jc_{ji}t^{(j-1)}, \quad i = 1,\ldots,l. \quad (3.106)$$

Substituting $\widehat{x}(t)$ and $d\widehat{x}/d(t)$ in equation (3.103) we obtain

$$\left[\frac{d\widehat{x}(t)}{dt} - f_0(x,t)\right] = F(\widehat{x},t)u(t) + \omega(t), \qquad (3.107)$$

where $\omega(t)$ is the one-dimensional noise that reflects our uncertainty of the model. Let us define the new noisy preprocessed observation $z(t)$, as follows,

$$z(t) = [d\widehat{x}(t) - f_0(\widehat{x},t)] = F(\widehat{x},t)u(t) + \omega(t). \qquad (3.108)$$

These observations represent new input to the neural network. During the training weights are set, which minimize an error between the true known signals and the signals, generated by means of the neural network. In each experiment a set of influences $u(k)$ is generated, by which vector $z(t)$ is calculated. The training sequence

$$n^T(k) = [z_1(k)\ldots z_l(k), z_1(k-1)\ldots z_l(k-1)z_1(k-I)$$
$$\ldots z_l(k-I)] = [n_1(k)n_2(k)\ldots n_{l\times I}(k)] \qquad (3.109)$$

is delivered to the first neural network layer. At the time moment k, the output $\gamma_i(k)$ of the second layer of the ith neuron is determined by the expression

$$\gamma_i(k) = q\left[\sum_{j=1}^{l}\alpha_{ij}n_j(k)\right], \quad i = 1,\ldots,N, \qquad (3.110)$$

where α_{ij} are unknown weights; N is a number of neurons; and the nonlinearity q is defined as

$$q(x) = \frac{1}{1 + \exp(-x)}. \qquad (3.111)$$

The output of the ith neuron of the last layer yields the following estimate of an unknown signal

$$\widehat{u}_i(k) = \sum_{j=1}^{N}\beta_{ij}\sigma_j(k), \quad i = 1,\ldots,N, \qquad (3.112)$$

where β_{ij} are unknown weights. The derived estimate is used for signal estimation.

On-line training. The preprocessed observation $z(k)$ is the desired output and it drives a neural network; it is also compared to a feedback

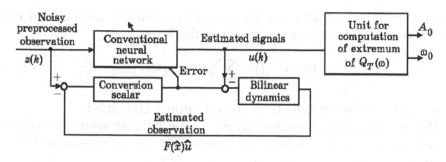

Figure 3.5. Neural network estimator of signals.

quantity $F(\widehat{x}, t)\widehat{u}$, which is the actual output. A conventional network is used to estimate signal $\widehat{u}(k)$. The error signal is used to update the estimates of the α_{ij} and β_{ij} weights of the neural network, using the backpropagation algorithm.

A flowchart of the signal estimation system is depicted in Figure 3.5.

2.5 Finite-Dimensional Bilinear Adaptive Estimation

Consider the linear Gauss–Markov system

$$da(t) = F(t)\alpha(t)dt + G(t)d\omega(t), \qquad (3.113)$$

$$dz(t) = H(t)\alpha(t)dt + R^{1/2}(t)dv(t), \qquad (3.114)$$

where $\alpha(t)$ is an n-vector, $z(t)$ is a p-vector, ω and v are independent standard Brownian motion processes, $R > 0$, and $\alpha(0)$ is a Gaussian random variable independent of ω and v. It is assumed that $[F(t), G(t), H(t)]$ is completely controllable and observable. Also, we define $Q(t) \triangleq G(t)G'(t)$.

Consider the system described by (3.113), (3.114) and the bilinear system

$$\dot{X}(t) = \left(A_0 + \sum_{i=1}^{n} \alpha_i(t)A_i \right) X(t), \quad X(0) = I, \qquad (3.115)$$

where X is a $k \times k$ matrix. It is easy to show (see Brockett, 1975) that each term has the form

$$x(t) = e^{\alpha_j(t)}\eta(t), \qquad (3.116)$$

where η is a finite Volterra series in α with separable kernels, and can be implemented by a bilinear system of the form

$$\dot{x}(t) = \alpha_j x(t) + \sum_{k=1}^{n} A_k(t)\alpha_k(t)x(t), \qquad (3.117)$$

where x is a k-vector and A_j are strictly upper triangular.

If the Lie algebra L corresponding to the bilinear system (3.115) is nilpotent, then the conditional expectation $\hat{x}(t|t)$ can be computed with a finite-dimensional system of nonlinear differential equations driven by the innovation (see Marcus and Willsky, 1976)

$$d\nu(t) \overset{\triangle}{=} dz(t) - H(t)\hat{x}(t|t)dt. \qquad (3.118)$$

The parameters of matrices A_0, A_i can be estimated via an identification algorithm.

2.6   Example

Assume that the MS is described by the equation

$$\dot{X}(t) = \widehat{A}X(t) + u_2(t)\widehat{B}_2 X(t), \quad X(0) = I,$$
$$Y(t) = \widehat{C}X(t), \qquad (3.119)$$

where

$$\widehat{A} = \begin{bmatrix} 0 & 0 \\ 0 & 1 \end{bmatrix}, \quad \widehat{B} = \begin{bmatrix} 1 & 0 \\ -1 & 0 \end{bmatrix}, \quad \widehat{C} = \begin{bmatrix} 1 & 1 \end{bmatrix},$$
$$u_2(t) = r(t) + s(t).$$

Then,

$$\mathrm{ad}_{\widehat{A}}\widehat{B}_2 = \begin{bmatrix} 0 & 0 \\ 1 & 0 \end{bmatrix}, \quad [\widehat{B}_2, \mathrm{ad}_{\widehat{A}}\widehat{B}_2] = \begin{bmatrix} 0 & 0 \\ 1 & 0 \end{bmatrix},$$
$$\mathrm{ad}_{\widehat{A}}^2\widehat{B}_2 = -\mathrm{ad}_{\widehat{A}}\widehat{B}_2, \quad \mathrm{ad}_{\widehat{A}}[\widehat{B}_2, \mathrm{ad}_{\widehat{A}}\widehat{B}_2] = -[\widehat{B}_{21}\,\mathrm{ad}_{\widehat{A}}\widehat{B}_2],$$
$$\mathrm{ad}_{\widehat{B}_2}^2\,\mathrm{ad}_{\widehat{A}}\widehat{B}_2 = I\widehat{B}_2, \quad \mathrm{ad}_{\widehat{A}}\widehat{B}_2.$$

Because $\widehat{C}\widehat{B}_2 = 0$ and $\widehat{C}\,\mathrm{ad}_{\widehat{A}}\widehat{B}_2 \neq 0$, then $k = 2$, and, therefore, according to Theorem 3.6, system (3.119) is invertible, and has left-inverse (3.91) (satisfying condition (3.92)). Using conditions (3.91) and (3.93), we obtain

$$\dot{\hat{x}} = a(\hat{x}) + \hat{u}_2 b(\hat{x}), \quad \hat{x}_0 = x_0, \quad \hat{y} = d(\hat{x}) + \hat{u}_2 e(\hat{x}),$$

where

$$a(\hat{x}) = \begin{bmatrix} \hat{x}_2 \\ 0 \end{bmatrix}, \quad b(\hat{x}) = \begin{bmatrix} -1 \\ 1 \end{bmatrix}, \quad d(\hat{x}) = \hat{x}_2/x_1, \quad e(\hat{x}) = -1/\hat{x}_1.$$

The function $\hat{y}(t)$ is used to determine the extremum of the functional (3.94). This problem can be solved with the numerical algorithm described in Danilin and Piyavsky (1967).

The section proposes new mathematical models for estimating almost periodic signal parameters on the basis of discrete and continuous bilinear observations. The proposed algorithms are more efficient as compared with the known ones. Also, the considered algorithms make it possible to find a value of a minimally detectable signal. However, efficient global optimization procedures are needed. The global optimization algorithms described in Horst and Pardalos (1995) seem to be useful, when this problem is solved. The proposed algorithms are directly used to estimate a value of a gravity signal, present in controlled cryogenic sensors (Yatsenko, 1989, 2003).

To detect signals against the background of a noise influencing a multisensor system input, the neural network approach is proposed. This neural network functions as an inverse dynamic system. Numerical simulation proves that such a network is able to distinguish relay, steady, and harmonic signals. It is also stated that the three-layer neural network is efficiently trained in the strongly invertible nonlinear and bilinear dynamic controlled system class. As a whole, along with the adaptive filtering algorithm, the neural network may be considered as a robust signal parameter observer.

The geometric structure of certain bilinear systems with variable parameters is examined, and the optimal estimators for these systems are thus shown to be finite-dimensional. The results described here may be used to construct suboptimal estimators for a broad class of nonlinear stochastic systems.

3.   Bilinear Lattices and Nonlinear Estimation Theory

In this section we present an application of the concept of an adaptive estimation using an estimation algebra to the study of dynamic processes in nonlinear lattice systems. It is assumed that nonlinear dynamic processes can be described by nonlinear or bilinear lattice models.

3.1   Lattice Systems and DMZ Equations

Weakly nonlinear coupled systems (discrete self-trapping model, discrete nonlinear Schrödinger equation, etc.) belong to the universal, widely applicable, highly illustrative, and thoroughly studied models of nonlinear physics. These models have applications to molecular crystals, molecular dynamics, nonlinear optics, and biomolecular dynamics. It has been shown that weakly nonlinear coupled systems exhibit unusual dynamical phenomena pertaining only to nonlinear discrete systems, such as existence of intrinsic localized modes. Here we present an application of the concept of a nonlinear estimation using an estimation algebra to the study of dynamic process in nonlinear lattice systems. It is assumed that a nonlinear dynamic process can be described by nonlinear or bilinear lattice models (LM). Our research focuses on the development of an estimation algorithm for a signal process in the lattice models with background additive white noise, and with different assumptions regarding the characteristic of the signal process.

A cell model of LM is based on the following signal observation model,

$$dx(t) = f(x(t))dt + g(x(t))dv(t), \quad x(0) = x_0,$$
$$dy(t) = h(x(t))dt + dw(t), \quad y(0) = 0 \qquad (3.120)$$

in which x, v, y, and w, are, respectively, $\mathbb{R}^n$-, $\mathbb{R}^n$-, $\mathbb{R}^m$-, and $\mathbb{R}^m$-valued processes. The values v and w are independent standard Brownian processes. We assume that f and h are C^∞ smooth. We refer to $x(t)$ as the state vector of the system at time t and $y(t)$ as the observation at time t.

Let $\rho(t, x)$ denote the conditional probability density of the state $x(t)$ given the observation $\{y(s) : 0 \leq s \leq t\}$. It is governed by the well-known (Davis and Marcus, 1981) *Duncan–Mortensen–Zakai* (DMZ) equation which is a stochastic partial differential equation in terms of an unnormalized version of $\rho(t, x), \sigma(t, x)$. The DMZ can be written as

$$d\sigma(t, x) = L_0\sigma(t, x)dt + \sum_{i=1}^{m} L_i\sigma(t, x)dy_i(t), \quad \sigma(0, x) = \sigma_0, \qquad (3.121)$$

where

$$L_0 = \frac{1}{2}\sum_{i=1}^{n}\frac{\partial^2}{\partial x_i^2} - \sum_{i=1}^{n} f_i\frac{\partial}{\partial x_i} - \sum_{i=1}^{n}\frac{\partial f_i}{\partial x_i} - \frac{1}{2}\sum_{i=1}^{m} h_i^2 \qquad (3.122)$$

and for $i = 1, \ldots, m$, L_i is the zero-degree differential operator of multiplication by h_i. The value σ_0 is the initial probability density. Equation (3.121) is a stochastic partial differential equation. If we define an unnormalized density

$$\xi(t, x) = e^{-\sum_{i=1}^{m} h_i(x)y_i(t)}\sigma(t, x),$$

then $\xi(t, x)$ satisfies the following time-varying partial differential equation, which is called the robust DMZ equation:

$$\frac{\partial \xi}{\partial t}(t, x) = L_0 \xi(t, x) + \sum_{i=1}^{m} y_i(t)[L_0, L_i]\xi(t, x)$$

$$+ \frac{1}{2} \sum_{i,j=1}^{m} y_i(t)y_j(t) \left[[L_0, L_i], L_j\right] \xi(t, x), \quad \xi(0, x) = \sigma_0, \qquad (3.123)$$

where $[\cdot, \cdot]$ is the Lie bracket as described by the following definition.

Definition 3.10. *If X and Y are differential operators, then the Lie bracket of X and Y, $[X, Y]$, is defined by*

$$[X, Y]\phi = X(Y\phi) - Y(X\phi)$$

for any C^∞ function ϕ.

The objective of constructing a robust finite-dimensional filter to (3.120) is to find a smooth manifold M with complete C^∞ vector fields μ_i on M and C^∞ functions ν on $M \times \mathbb{R} \times \mathbb{R}^n$ and ω_i on $\mathbb{R}^m$, such that the solution to $\xi(t, x)$ can be represented in the form

$$\frac{dz}{dt}(t) = \sum_{i=1}^{k} \mu_i(z(t))\omega_i(y(t)), \quad z(0) \in M,$$

$$\xi(t, x) = \nu(z(t), t, x). \qquad (3.124)$$

The concept of an estimation algebra provides a systematic tool to deal with questions concerning finite-dimensional filters. We say that system (3.120) has a robust universal finite-dimensional filter if for each initial probability density ξ_0 there exists a z_0 such that (3.123) holds if $z(0) = z_0$, and μ_i, ω_i are independent of σ_0.

The Wei–Norman approach (Wei and Norman, 1963, 1964) is based on the observation that Lie algebraic ideas can be used to solve time-varying linear differential equations. Based on this idea, Wei and Norman considered the differential equation

$$\frac{d}{dt}X(t) = A(t)X(t) \equiv \sum_{i=1}^{} ma_i(t)A_iX(t), \quad X(0) = I, \qquad (3.125)$$

where X and A_i are $n \times n$ matrices and a_i are scalar-valued functions. They showed that if $B_1, \ldots, B_l$ is a base of the Lie algebra generated by $A_1, \ldots, A_m$, then the Wei–Norman theorem states that locally in t, $X(t)$ has a representation of the form

$$X(t) = e^{b_1(t)B_1} \cdots e^{b_l(t)B_l}X_0, \qquad (3.126)$$

where the b_i satisfy an ordinary differential equation of the form

$$\frac{db_i}{dt} = c_i(b_1, \ldots, b_l), \quad b_i(0) = 0$$

for all i. The functions c_i are determined by the structure constants of the Lie algebra (generated by the A_i relative to the basis $\{B_1, \ldots, B_l\}$).

By considering the DMZ equation as an equation of the form (3.125), the Wei–Norman approach suggests constructing a finite-dimensional filter if the estimation algebra is finite-dimensional. For this purpose, we introduce the concept of the estimation algebra of (3.120) and examine its algebraic structure.

Definition 3.11. *The estimation algebra E for the system defined by (3.120) is the Lie algebra generated by $\{L_0, L_1, \ldots, L_m\}$.*

The following theorem was announced in Tam, Wong, and Yau (1990) and proved in detail in Chiou and Yau (1994), which includes the Kalman–Bucy filtering system as a special case.

Theorem 3.12. (Chiou and Yau, 1994). *Let E be an estimation algebra of (3.120) satisfying $\partial f_j/\partial x_i - \partial f_i/\partial x_j = c_{ij}$, where the c_{ij} are constants for all $1 \leq i, j \leq n$. Suppose that E is a finite-dimensional estimation algebra of maximal rank. Then E has a basis of the form $1, x_1, \ldots, x_n, D_1, \ldots, D_n$, and L_0, where $D_i = \partial/\partial x_i - f_i$, and $\eta := \sum_{i=1}^{n} \partial f_i/\partial x_i + \sum_{i=1}^{n} f_i^2 + \sum_{i=1}^{m} h_i^2$ is a degree two polynomial of the form $\sum_{i,j=1}^{n} a_{ij} x_i x_j + \sum_{i=1}^{n} b_i x_i + d$. The robust DMZ equation (3.122) has a solution for all $t \geq 0$ of the form*

$$\xi(t, x) = e^{T(t)} e^{r_n(t)x_n} \cdots e^{r_1(t)x_1} e^{s_n(t)D_n} \cdots e^{s_1(t)D_1} e^{tL_0} \sigma_0, \quad (3.127)$$

where $T(t), r_1(t), \ldots, r_n(t), s_1(t), \ldots, s_n(t)$ satisfies the following ordinary differential equations. For $1 \leq i \leq n$,

$$\frac{dr_i(t)}{dt} = \frac{1}{2} \sum_{j=1}^{n} s_j(t)(a_{ij} + a_{ji}), \quad (3.128)$$

$$\frac{ds_i}{dt}(t) = r_i(t) + \sum_{j=1}^{n} s_j(t)c_{ji} + \sum_{k=1}^{m} h_{ki} y_k(t), \quad (3.129)$$

where $h_k(x) = \sum_{j=1}^{n} h_{kj} x_j + e_k$, for $1 \leq k \leq m$, h_{kj} and e_k are constant; and

$$\frac{dT}{dt}(t) = -\frac{1}{2}\sum_{i=1}^{n} r_i^2(t) - \frac{1}{2}\sum_{i=1}^{n} s_i^2(t)\left(\sum_{j=1}^{n} c_{ij}^2 - a_{ii}\right)$$

$$+ \sum_{i=1}^{n} r_i(t) - \sum_{j=2}^{n}\sum_{i=1}^{j} s_j(t)c_{ij} - \sum_{i,j=1}^{n} s_i(t)r_j(t)c_{ij}$$

$$+ \sum_{1 \le i < k \le n} s_i(t)s_k(t)\left(\sum_{j=1}^{n} c_{ij}c_{jk} + \frac{1}{2}(a_{ik} + a_{ki})\right)$$

$$+ \frac{1}{2}\sum_{i=1}^{n} s_i(t)b_i + \frac{1}{2}\sum_{i,j=1}^{m} y_i(t)y_j(t)\left(\sum_{k=1}^{n} h_{ik}h_{jk}\right). \tag{3.130}$$

3.2 Structure of Estimation Algebra

Recently Chen, Leung, and Yau (1996) have made important progress in the program of classification of finite-dimensional estimation algebras of maximal rank. They study the quadratic forms in E and show that the Ω-matrix is linear in the sense that all ω_{ij} are degree-one polynomials.

Definition 3.12. *Let Q be the space of quadratic forms in n variables, namely, the real vector space spanned by $x_i x_j$, $1 \le i \le j \le n$. Let $X = (x_1, \ldots, x_n)^T$. For any quadratic form $p \in Q$, there exists a symmetric matrix A such that $p(x) = X^T A X$. The rank of the quadratic form is denoted $r(p)$ and is defined to be the rank of the matrix A. A fundamental quadratic form of the estimation algebra E is an element $p_0 \in E \cap Q$ with the greatest positive rank; that is, $r(p_0) \ge r(p)$ for any $p \in E \cap Q$. The maximal rank of quadratic forms in the estimation algebra E is defined to be $k = r(p_0)$ and is called the quadratic rank of E.*

After an orthogonal transformation, p_0 can be written as

$$p_0(x) = c_1 x_1^2 + c_2 x_2^2 + \cdots + c_k x_k^2, \quad c_i \neq 0, \quad 0 \le k \le n.$$

From $p_0(x)$, we can construct a sequence of quadratic forms in $E \cap Q$ as follows,

$$q_0(x) = p_0(x),$$

$$q_j(x) = [[L_0, q_{j-1}], q_0] = \sum_{i=1}^{k} 4^j c_i^{j+1} x_i^2.$$

In view of the invertibility of the Vandermonde matrix, we can assume that

$$p_0(x) = x_1^2 + x_2^2 + \cdots x_k^2 \in E. \tag{3.131}$$

Lemma 3.1. (Chen, Leung, and Yau, 1996). *If p is a quadratic form in the estimation algebra E of (3.120), then it is independent of x_j for $j \geq k$, where $k = r(p_0)$ is the quadratic rank of E. In other words, $\partial p / \partial x_j = 0$ for $k + 1 \leq j \leq n$.*

Let $p_1 \in E \cap Q$ be an element with least positive rank; that is, $0 < r(p_1) \leq r(q)$ for any nonzero $q \in E \cap Q$. After an orthogonal transformation that fixes $x_{k+1}, \ldots, x_n$ variables (i.e., an orthogonal transformation on $x_1, x_2, \ldots, x_k$), and the Vandermonde matrix procedure as above, we can assume

$$p_1 = \sum_{i=1}^{k_1} x_i^2 \in E, \quad 1 \leq k_k \leq k. \tag{3.132}$$

Notice that the orthogonal transformation on $x_1, \ldots, x_k$ leaves p_0 invariant. In summary, we deduce that $p_0 = \sum_{i=1}^{k} x_i^2$ has the greatest positive rank and $p_1 = \sum_{i=1}^{k_1} x_i^2$ has the least positive rank.
   Define

$$S_1 = \{1, 2, \ldots, k_1\} \subseteq S = \{1, 2, \ldots, k\} \tag{3.133}$$

and $Q_1 = $ real vector space spanned by $\{x_i x_j : k_1 + 1 \leq i \leq j \leq k\} \subseteq Q$. If $k_1 < k$, then $Q_1 \cap E$ and is a nontrivial space, because $p = p_0 \in E \cap Q$. In a similar procedure as above, there exist $k_2 > k_1$ and

$$p_2 = \sum_{i=k_1+1}^{k_2} x_i^2 \in E \cap Q \tag{3.134}$$

with the least positive rank in $E \cap Q$. By induction, we can construct a series of S_i, Q_i, and p_i such that

$$S_i = \{k_{i-1} + 1, \ldots, k_i\}, \quad 0 = k_0 < k_1 < \cdots < k_i < \cdots \leq k.$$

$Q_i = $ real vector space spanned by $\{x_i x_j : k_i + 1 \leq l \leq j \leq k\}$,

$$p_i = \sum_{j=k_{i-1}+1}^{k_i} x_j^2 = \sum_{j \in S_i} x_j^2, \quad i > 0 \tag{3.135}$$

and p_i has the least positive rank in $E \cap Q_{i-1}$, for $i > 0$.

Lemma 3.2. (Chen and Yau, 1996). *If $p \in E \cap Q$, then there exists a constant λ such that*

$$p(0, \ldots, 0, x_{k_{i-1}+1}, \ldots, x_{k_i}, 0, \ldots, 0) = \lambda p_i \quad for \ i > 0.$$

Lemma 3.3. (Chen and Yau, 1996). *If $p \in E \cap Q$, then for $i > 0$*

$$p(x_1, \ldots, x_{k_{i-1}+1}, 0, \ldots, 0, x_{k_{i+1}}, \ldots, x_n) \in E.$$

Lemma 3.4. (Chen and Yau, 1996). *Let $p = \sum_{i \in S_{l_1}} \sum_{i \in S_{l_2}} 2a_{ij} x_i x_j \in E$, where $a_{ij} \in \mathbb{R}$ and $l_1 < l_2$. Then $|S_{l_1}| = |S_{l_2}|$ and $A = (a_{ij}) = bT$ where b is a constant and T is an orthogonal matrix.*

The following theorem is the main result of Chen, Leung, and Yau (1996).

Theorem 3.13. *If E is a finite-dimensional estimation algebra of maximal rank, then all the entries $\omega_{ij} = \partial f_i/\partial x_i - \partial f_i/\partial x_j$ of Ω are degree-one polynomials. Let k be the quadratic rank of E. Then there exists an orthogonal change of coordinates such that ω_{ij} are constants for $1 \leq i, j \leq k$; ω_{ij} are degree-one polynomials in $x_1, \ldots, x_k$ for $1 \leq i \leq k$ or $1 \leq j \leq k$; and ω_{ij} are degree-one polynomials in $x_{k+1}, \ldots, x_n$ for $k_1 \leq i, j \leq n$. Let $\Omega = (\omega_{ij})$, where $\omega_{ij} = \partial f_j/\partial x_i - \partial f_i/\partial x_j$, is the matrix introduced by Wong (1997, 1998). For the finite-dimensional estimation algebra of maximal rank, it is easy to see that ω_{ij} is in E. In view of Ocone's theorem, ω_{ij} is a polynomial of degree 2. Let $\omega_{ij}^{(2)}$, $\omega_{ij}^{(1)}$, and $\omega_{ij}^{(0)}$ be the homogeneous part of degree 2, 1, and 0 respectively.*

Lemma 3.5. *Suppose that E is a finite-dimensional estimation algebra of maximal rank. Then*

(i) *$\omega_{ji}^{(2)}$ depends only on $x_1, \ldots, x_k$ for $i \leq k$ or $j \leq k$.*

(ii) *$\omega_{ij}^{(2)} = 0$, $\forall\, k+1 \leq i, j \leq n$.*

(iii) *$\dfrac{\partial \omega_{ij}^{(2)}}{\partial x_l} + \dfrac{\omega_{il}^{(2)}}{\partial x_i} + \dfrac{\omega_{li}^{(2)}}{\partial z_j} = 0 \;\forall\, 1 \leq i, j, l \leq n.$*

(iv) *$\dfrac{\partial \omega_{ij}^{(1)}}{\partial x_l} + \dfrac{\omega_{il}^{(1)}}{\partial x_i} + \dfrac{\omega_{li}^{(1)}}{\partial z_j} = 0 \;\forall\, 1 \leq i, j, l \leq n.$*

Lemma 3.6. *Let E be a finite-dimensional estimation algebra with maximal rank. Then $\langle 1, x_1, \ldots, x_n, D_1, \ldots, D_n, L_0 \rangle \subseteq E$.*

Theorem 3.14. *Let E be a finite-dimensional estimation algebra of maximal rank k and k be the quadratic rank of E. Then ω_{ij} are constants for $1 \leq i, j \leq k$ or $k+1 \leq i, j \leq n$; ω_{ij} are degree-one polynomials in $x_1, \ldots, x_k$ for $1 \leq i \leq k$ or $1 \leq j \leq k$. Moreover $\alpha_j = \sum_{l=1}^{k} x_l \omega_{jl}$, for $k+1 \leq j \leq n$, are in E.*

The following theorem can be shown using the similar method to that discussed in Chen, Leungand, and Yau (1996).

Theorem 3.15. *Suppose that the state-space dimension is six. If E is the finite-dimensional estimation algebra with maximal rank corresponding to the filtering system (3.120), then E is a real Lie algebra of dimension 14 with basis given by $\{1, x_1, \ldots, x_6, D_1, \ldots, D_6, L_0\}$. Furthermore, η is a quadratic function and $\partial f_j/\partial x_i - \partial f_i/\partial x_j$, $1 \leq i, j \leq n$, are constant functions.*

4.   Notes and Sources

The material contained in Chapter 3 is a synthesis of early papers of Kalman and Bucy (1961), Hazewinkel (1982, 1986, 1995), Hazewinkel and Marcus (1982), and Brockett (1981). The problem was stated and the basic properties of signal estimation on Lie groups were established by Willsky (1973), Chiou and Yau (1994), and Chitke and Lo (1981). A dynamical model of the CA3 region of the brain in the state space form and an adaptive filter are essentially taken from Chitke and Lo (1981) and Pardalos et al. (2001). The concept of a robust recursive Bayesian estimation is due to Kulhavý (1996). The quantum interpretation of nonlinear filtering is fundamentally inspired by the papers of Mitter (1979, 1980).

Chapter 4

CONTROL OF DYNAMICAL PROCESSES AND GEOMETRICAL STRUCTURES

Modern control theory basically deals with dynamical systems on smooth manifolds. However, many practical systems such as multiple agents do not have such structures. The axiomatic control theories should adequately reflect in terms of their internal language of notions and control problems. In terms of these theories, the control structures can make up various hierarchies. According to Kalman, for example, the most general structure is represented by a controllability–reachability structure over which the optimal control structure is built. This approach regarding the structure of optimal control and Yang–Mills fields was discussed in Yatsenko (1985) and Butkovskiy and Samoilenko (1990).

In this chapter, the geometrical description problem of multiple agents is studied. We discuss mathematical aspects of the "unified game theory (UGT)" and "theory of control structure (TCS)". We consider a game as a hierarchical structure. It is assumed that each agent can be described by a fiber bundle. A joint maneuver has to be chosen to guide each agent from its starting position to its target position while avoiding conflicts. Among all the conflict-free joint maneuvers, we aim to determine the one with the least overall cost. The cost of an agent's maneuver is its energy, and the overall cost is a weighted sum of the maneuver energies of all individual agents, where the weights represent priorities of the agents.

As an example, we consider the hierarchical structure of such a multiagent system in Figure 4.1. Each agent of the system can be described by a stochastic or deterministic differential equation with control. In this chapter we first reduce the model to a hierarchical geometric representation using fiber bundles. Then we consider an integrated geometrical model where the separated model of agents is integrated into a single model. For example, the interaction between six robots in Figure 4.2 can

P.M. Pardalos, V. Yatsenko, *Optimization and Control of Bilinear Systems*, doi: 10.1007/978-0-387-73669-3,
© Springer Science+Business Media, LLC 2008

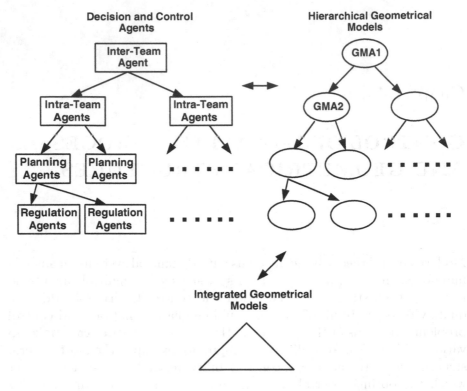

Figure 4.1.   Hierarchical structure of multiple agents.

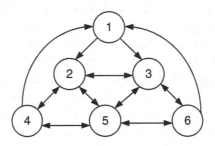

Figure 4.2.   Hierarchical structure of multiple robot.

be described by a hierarchical structure. The integrated model allows solving controllability, observability, and cooperative control problems.

In Section 2, we consider geometric aspects of the nonlinear control systems. The section constructs a formal model, where the optimal control structure appears independently of the controllability/reachability structure and that of the space of local system states. The efficiency of this axiomatic approach is illustrated using structural analysis of a general problem of optimal control. In Section 3, we analyze in detail the relationship among gauge fields, identification problems, and control systems. The result of the analysis is an estimation algebra of a nonlinear estimation problem. The estimation algebra turns out to be a useful concept to explore finite-dimensional nonlinear filters. In Section 4, we consider a Lie group related to Yang–Mills gauge groups. We show that the estimation algebra of the identification problem is a subalgebra of the current algebra. Section 5 focuses on nonlinear control systems and Yang–Mills fields. Section 6 is devoted to geometric models of multiagent systems as controlled dynamic information objects. It is shown that these systems can be described by commutative diagrams that allow us to analyze a symmetry.

1. Geometric Structures

We briefly describe the role of topological, metric, and orderness structures. Note that each standard ordinary differential control system or inclusion $x' \in I(x)$, $x \in X$, generates two independent topological structures on X. One of them is generated by a family of inclusions of $x \in X$, that is, the family of reachable sets $O(x_0, \varepsilon)$ for x_0 at time $\varepsilon \geq 0$, and another one by a family of a controllability set $O(x_0, \varepsilon)$ for x_0 at time $\varepsilon \geq 0$ (observability topology).

Let (X, τ) be a topological space, where X is an abstract nonempty set and τ is a topology on X.

Definition 4.1. *Control (or admissible control) $\gamma(a, b)$ in (X, τ) is an image of the continuous (in the sense of topology τ) map $\varphi \colon [0, 1] \to X$,*

$$x = \varphi(t), \quad 0 \leq t \leq T, \quad x \in X, \tag{4.1}$$

$$a = \varphi(0), \quad b = \varphi(T), \tag{4.2}$$

where $a \in X$ is an initial point and $b \in X$ is a final point of the control $\gamma(a, b)$.

Thus, the control $\gamma(a, b)$ is a pathwise connected and linearly ordered subset (sequence) of X where $a \in \gamma(a, b)$ and $b \in \gamma(a, b)$ are the smallest and the largest of its elements, respectively. As results, maps (4.1) and (4.2) are admissible parameterizations of the control $\gamma(a, b)$.

The verification of Definition 4.1 consists of a validation of the controllability and finding optimal control of systems without using any differential or difference structure. Furthermore, we consider that there is a metric in topological spaces which allows us to analyze control problems at various levels of generality. We look for the "minimal" but not trivial structures, which can be responsible for controls.

1.1   Metric Spaces

The concepts of a metric and a metric space are introduced by the following definitions.

Definition 4.2. *A metric space* (X, ρ) *is a pair* (X, ρ) *where* X *is an arbitrary non-empty set and* p *is a metric structure of* X, *that is,* ρ *is a real-valued function* $\rho = \rho(x, y)$, $(x, y) \in X^2 = X \times X$, *or map*

$$\rho : X^2 \to \mathbb{R} \tag{4.3}$$

with the metric axioms:

$$\rho(a, b) \geq 0 \quad for \ \forall \, (a, b) \in X^2, \tag{4.4}$$

$$\rho(a, a) = 0 \quad for \ \forall \, a \in X, \tag{4.5}$$

$$\rho(a, b) < \rho(a, c) + \rho(c, b) \tag{4.6}$$

for any $a \in X$, $b \in X$, $c \in X$, *and (4.6) is called a 'triangle inequality'. Sometimes* ρ *is also called a global metric on* X *or distance in* X.

The metric introduced by Definition 4.2 differs from the usual concept of a metric: there is neither the symmetry axiom ($\rho(a, b) = \rho(b, a)$ for $\forall \, a \in X, \forall \, b \in X$) nor the requirement $\rho(a, b) > 0$ if $a \neq b$. So, the given concept of a metric is more adequate to the situation in typical control problems. As is known, the metric space (X, ρ) can also be considered as a topological space (X, τ), where topology T is induced by metric ρ.

But the metric can measure control $\gamma(a, b)$ introduced by Definition 4.1. This can be done by the following definition.

Definition 4.3. *The length* $l[\gamma(a, b)]$ *of the control* $\gamma(a, b)$ *is a real-valued function*

$$l\,[\gamma(a, b)] = \lim_{N \to \infty} l_N\,[\gamma(a, b)]\,, \tag{4.7}$$

where

$$l_N\,[\gamma(a, b)] = \sum_{i=0}^{N} \rho(x_i, x_{i+1}), \tag{4.8}$$

where $a = x_0 < x_1 < \cdots < x_N < x_{N+1} = b$ *is the Nth partition* T_N *of* $\gamma(a, b)$, *and the partition* T_N *becomes finer with* $N \to \infty$. *Of course, it is necessary to prove or admit the existence and uniqueness of (4.7). If so,* $\gamma(a, b)$ *is called measurable (in metric* ρ). *The set* $\gamma(a, b)$ *of all measurable* $\gamma(a, b)$ *is denoted* $\Gamma(a, b)$:

$$\Gamma(a, b) = \{\gamma(a, b)\}. \tag{4.9}$$

So, in the metric space (X, ρ) the *admissible control* $\gamma(a, b)$ is just a measurable (in sense of metric ρ) sequence and vice versa.

If we have several sequences in X:

$$\gamma_1(x_0, x_1), \gamma_2(x_1, x_2), \ldots, \gamma_n(x_n, x_{n+1}) \tag{4.10}$$

then we can define their sum

$$\gamma(x_0, x_{n+1}) = \sum_{i=1}^{n} \gamma_i(x_{i-1}, x_i) \tag{4.11}$$

which is also a sequence.

Inversely, if $\gamma(a, b)$ is a sequence and $x_i \in \gamma(a, b)$, $i = 1, \ldots, n$, $x_1 < \cdots < x_n$, then $\gamma(a, b)$ can be represented as the sum of sequences:

$$\gamma(a, b) = \gamma_1(a, x_1) + \gamma_2(x_1, x_2) + \cdots + \gamma_n(x_n, b). \tag{4.12}$$

Definition 4.4. *The sequence $\gamma_i(x_{i-1}, x_i)$ in (4.11) is called a piece of the sequence $\gamma(a, b)$. We accept that functional (4.7) is additive one:*

$$\rho(a, b) \geq 0 \quad for \ \forall \, (a, b) \in X^2, \tag{4.13}$$
$$\rho(a, a) = 0 \quad for \ \forall \, a \in X. \tag{4.14}$$

1.2   Optimal Control

Consider the following problem of optimal control in (X, ρ).

1. Determine

$$\bar{l}(a, b) = \{\inf l[\gamma(a, b)] : \ \gamma(a, b) \in \Gamma(a, b)\}. \tag{4.15}$$

2. Determine $\bar{\gamma} = \bar{\gamma}(a, b)$, if it exists, such that

$$l[\bar{\gamma}(a, b)] = \bar{l}(a, b). \tag{4.16}$$

   This admissible $\bar{\gamma}(a, b)$ is called the minimal of the optimal control problem.

3. Describe all sets $\{\bar{\gamma}(a, b)\}$ for fixed $(a, b) \in X^2$ and for all $(a, b) \in X^2$.

A simple but important property of the minimal $\bar{\gamma}(a, b)$ is given by the following theorem.

Theorem 4.1. *If the admissible $\gamma(c, d)$ is the minimal of the optimal control problem, then the sequence $\bar{\gamma}(a, b)$ is also minimal.*

This is a consequence of the additivity property of (4.12). If any admissible sequence $\gamma(a, b)$ is minimal, it does not mean that $\gamma(a, b)$ is also minimal.

It is easy to prove the inequality

$$\rho(a, b) \leq \bar{l}(a, b). \tag{4.17}$$

Definition 4.5. *The metric space (X, ρ) is an obstacleless metric space if there exists at least one point $(a, b) \in X^2$ such that*

$$\rho(a, b) \leq \bar{l}(a, b).$$

The metric space (X, ρ) is a generalized metric space if

$$\rho(a, b) = \bar{l}(a, b) \quad for \; \forall \, (a, b) \in X^2.$$

An example of a generalized space is the Euclidean space $\mathbb{R}^n$.
The following theorem is valid.

Theorem 4.2. $\bar{l} = \bar{l}(a, b)$ *is also metric on X, and $(X, \bar{l})$ is also a metric space.*

Definition 4.6. *Metric $\bar{l} = \bar{l}(a, b)$ is called a secondary metric.*

Generally, $\bar{l}(a, b)$ is distinguished from the initial metric $\rho(a, b)$ on X.

Definition 4.7. *If the secondary metric $\bar{l}$ coincides with the metric ρ then ρ is called a self-secondary metric.*

The following theorems are valid.

Theorem 4.3. *The secondary metric is a self-secondary metric.*

This is similar to the property of projection operator $P : P^2 = P$.

Theorem 4.4. *The metric space (X, ρ) is a generalized space if the metric ρ is the self-secondary metric.*

We illustrate the application of the above-introduced concepts by the following.

Theorem 4.5. *(Sufficient condition for minimal.) The sequence $\gamma(a, b)$ in (X, ρ) is minimal if for any of its admissible $\gamma(c, d)$ the next relation is true:*

$$\bar{l}(c, d) = \bar{l}(c, x) + \bar{l}(x, d) \quad for \; \forall \, x \in \gamma(c, d), \tag{4.18}$$

where $\bar{l}$ is a secondary metric of ρ.

It might seem that for $\gamma(a, b)$ to be minimal just one identity is sufficient:

$$\bar{l}(a, b) = \bar{l}(a, x) + \bar{l}(x, b) \quad for \; \forall \, x \in \gamma(a, b). \tag{4.19}$$

But it is not true; there exists a contrary example.

From a topology standpoint the secondary metric $\bar{l}$ generally is weaker (rougher) than the "initial" or "first" metric ρ. In other words, topology (X, ρ) is stronger (thinner) than secondary topology $(X, \bar{l})$.

1.3 Identification of Nonlinear Agents and Yang–Mills Fields

In this section we consider models where each agent of the hierarchical system is described by a stochastic differential equation.

Consider the stochastic differential system:

$$d\theta = 0, \tag{4.20}$$

$$dx_t = A(\theta)x_t dt + b(\theta)dw_t, \tag{4.21}$$

$$dy_t = \langle (\theta), x_t \rangle dt + dv_t. \tag{4.22}$$

Here $\{w_t\}$ and $\{v_t\}$ are independent, scalar, and standard Wiener processes, and $\{x_t\}$ is an $\mathbb{R}^n$-valued process. Assume that θ takes values in a smooth manifold $\Theta \to \mathbb{R}^N$, and the map $\theta \to \Sigma(\theta) := (A(\theta), b(\theta), c(\theta))$ in a smooth map taking values in minimal triples. By the identification problem we mean the nonlinear filtering problem associated with equation (4.21), that is, the problem of recursively computing conditional expectations of the form $\pi_t(\phi) \triangleq E[\phi(x_t, \theta)|Y_t]$, where Y_t is the σ-algebra generated by the observations $\{y_s : 0 \le s \le t\}$ and ϕ belongs to a suitable class of functions on $\mathbb{R}^n \times \Theta$.

For a given y_t, the joint unnormalized conditional density $\rho \triangleq \rho(t, x, \theta)$ of x_t and θ satisfy the stochastic partial differential Stratonovitch equation

$$d\rho = A_0\rho dt + B_0\rho dy_t, \tag{4.23}$$

where the operators A_0 and B_0 are given by

$$A_0 := \frac{1}{2}\left\langle b(\theta), \frac{\partial}{\partial x}^2 \right\rangle - \left\langle \frac{\partial}{\partial x}, A(\theta)x \right\rangle - \langle c(\theta), x \rangle^2 / 2, \tag{4.24}$$

$$B_0 := \langle c(\theta), x \rangle. \tag{4.25}$$

From the Bayes formula it follows that

$$\pi_t(\phi) = \sigma_t(\phi)/\sigma_t(l), \tag{4.26}$$

where

$$\sigma_t(\phi) = \int_\Theta \int_{\mathbb{R}^n} \phi(x, \theta)\rho(t, x, \theta)|dx||d\theta|, \tag{4.27}$$

where $|dx|$ and $|d\theta|$ are fixed volume elements on $\mathbb{R}^n$ and Θ, respectively. Furthermore, if $Q(t, \theta)$ denotes the unnormalized posterior density of θ given t, then it satisfies the equation:

$$dQ = E[\langle c(\theta), x_t|\theta \rangle, Y_t]Q(t, \theta)dy_t. \tag{4.28}$$

The papers on nonlinear filtering theory by Hazewinkel (1982, 1986) show that it is natural to look at equation (4.23) formally as a deterministic partial differential equation,

$$\frac{\partial \rho}{\partial t} = A_0 \rho + \dot{y} B_0 \rho. \tag{4.29}$$

By the Lie algebra of the identification problem, we mean the operator Lie algebra $\bar{G}$ generated by A_0 and B_0. For more general nonlinear filtering problems, estimation algebras analogous to $\bar{G}$ have been emphasized by Brockett (Mitter, 1979, 1980) and others as being objects of central interest. In the paper by Krishnaprasad and Marcus (1982) the Lie algebra $\bar{G}$ is used to classify identification problems and to understand the role of certain sufficient statistics.

1.4   The Estimation Algebra of Nonlinear Filtering Systems

To understand the structure of the estimation algebra it is well worth considering an example.

Example 4.1. Let $dx_t = \theta dw_t$; $d\theta = 0$; $dy_t = x_t dt + dv_t$. Then $A_0 = (\theta^2/2)(\partial^2/\partial t^2) - (x^2/2)$ and $B_0 = x$, and $\bar{G} = \{A_0, B_0\}_{I..A}$ is spanned by the set of operators $(\theta^2/2 - x^2/2)$, $(\theta^{2n}x)_{n=0}^{\infty}$, $(\theta^{2n}(\partial/\partial x))_{n=1}^{\infty}$ and $\{\theta^{2n}1\}_{n=1}^{\infty}$. We then notice that,

$$\tilde{G} \subseteq \mathbb{R}\left[\theta^2\right] \otimes \left\{\frac{\partial^2}{\partial x^2}, x\frac{\partial}{\partial x}, \frac{\partial}{\partial x}, x^2, x, 1\right\} L.A.$$

is a subalgebra of the Lie algebra obtained by tensoring the polynomial ring $\mathbb{R}\left[\theta^2\right]$ with a six-dimensional Lie algebra. Here, L.A. stands for the Lie algebra generated by the elements in the brackets.

The general situation is very much as in this example. Consider the vector space (over the reals) of operators spanned by the set

$$S := \left\{\frac{\partial^2}{\partial x_i \partial x_j}, x_i\frac{\partial}{\partial x_j}, \frac{\partial}{\partial x_i}, x_i x_i, x_j, 1\right\},$$

$$i = 1, 2, \ldots, n, \quad j = 1, 2, \ldots, n. \tag{4.30}$$

This space of operators has the structure of a Lie algebra henceforth denoted $\tilde{G}_0$ (of dimension $3n^2 + 2n + l$) under operator commutation (the commutation rules being $(\partial^2/\partial x_i \partial x_j)$, $x_k = \delta jk(\partial/\partial x_i) + \delta ik(\partial/\partial x_j)$ etc., where δjk denotes the Kronecker symbol. For each choice Θ, A_0 and B_0 take values in $\tilde{G}_0$. It follows that in general A_0 and B_0 are smooth maps from Θ into $\tilde{G}_0$. Thus, let us consider the space of smooth maps $C^{\infty}(\Theta; \tilde{G}_0)$. This space can be given by the structure of a Lie algebra (over the reals) in the following way, given φ, $\phi \varepsilon C^\infty(\Theta; \tilde{G}_0)$,

define the Lie bracket $[\cdot, \cdot]_C$ on $C^\infty(\Theta; \tilde{G}_0)$ by

$$[\phi, \psi]_C(P) = [\phi(P), \psi(P)] \quad \text{for every } P \in \Theta. \tag{4.31}$$

Here the bracket on the right-hand side of equation (4.31) is in $\tilde{G}_0$. We denote as $\tilde{G}_0$ the Lie algebra $(C^\infty(\Theta; \tilde{G}_0); [., .]_C)$. Whenever the dimension of Θ is greater than zero, $\tilde{G}_0$ is infinite-dimensional and is an example of a *current algebra*. Current algebras play a fundamental role in the physics of Yang–Mills fields where they occur as Lie algebras of gauge transformations. Elsewhere in mathematics they are studied under the guise of local Lie algebras. The following is immediate.

Proposition 4.1. *The Lie algebra $\tilde{G}$ of operators generated by*

$$A_0 := \frac{1}{2} \left\langle b(\theta), \frac{\partial}{\partial x} \right\rangle^2 - \left\langle \frac{\partial}{\partial x}, A(\theta)x \right\rangle - \langle c(\theta), x \rangle^2 / 2 \tag{4.32}$$

and $B_0 := \langle c(\theta), x \rangle$ is a subalgebra of the current algebra $C^\infty(\Theta; \tilde{G}_0)$.

1.5 Estimation Algebra and Identification Problems

It is known (Marcus and Willsky, 1976) that $\tilde{G}$ admits a faithful representation as a Lie algebra of vector fields on a finite-dimensional manifold. Specifically, consider the system of equations,

$$d\theta = 0,$$

$$dz = \left[A(\theta) - Pc(\theta)c^T(\theta)\right] zdt + Pc(\theta)dy_t,$$

$$\frac{dP}{dt} = A(\theta)P + PA^T(\theta) + b(\theta)b^T(\theta) - Pc(\theta)c^T(\theta)P,$$

$$ds = \frac{1}{2}\langle c(\theta), z \rangle^2 dt - \langle c(\theta), z \rangle dy_t. \tag{4.33}$$

The system of equations (4.33) evolves on the product manifold $\Theta \times \mathbb{R}^{n(n+3)/2+1}$. Associated with equations (4.33) are the pair of vector fields (first-order differential operators),

$$a_0^* = \langle (A(\theta) - Pc(\theta)c^T(\theta))z, \partial/\partial z \rangle$$
$$+ \operatorname{tr}\left((A(\theta)P + PA^T(\theta) + b(\theta)b^T(\theta) - Pc(\theta)c^T(\theta)P), \partial/\partial P\right)$$
$$+ 1/2\langle c(\theta), z \rangle^2 \partial/\partial s$$

and

$$b_0^* = \langle P(\theta), \partial/\partial z \rangle - \langle c(\theta), z \rangle \partial/\partial z.$$

Here $\partial/\partial P = [\partial/\partial P_{ij}] = (\partial/\partial P)^T = n \times n$ symmetric matrix of differential operators. Consider the Lie algebra of vector fields generated by a_0^* and b_0^*. Because a_0^* and b_0^* are vertical vector fields with respect to the fibering $\Theta \times \mathbb{R}^{n(n+3)/2+1} \to \Theta$, then every vector field is in this Lie algebra. One of the main results is the following (Marcus and Willsky, 1976).

Theorem 4.6. *The map*

$$\Phi_k : \tilde{G}_0 \to \bigcup \Theta \times \mathbb{R}^{n(n+3)/2+1}$$

defined by

$$b_0^* = \langle P(\theta), \partial/\partial z \rangle 1/2 \langle c(\theta), z \rangle \partial/\partial s$$

is a faithful representation of the Lie algebra of the identification problem as a Lie algebra of (vertical) vector fields on a finite-dimensional manifold fibered over Θ.

Example 4.2. To illustrate Theorem 4.5, consider the Lie algebra of Example 4.1. The embedding equations (4.33) take the form

$$d\theta = 0,$$
$$dp = \left(\theta^2 - p^2\right) dt,$$
$$dz = -pz dt + p dy_t,$$
$$ds = z^2/2 dt - z dy_t.$$

Then

$$\Phi_k(B_0) = \Phi_k(x) = b_0^* = p\frac{\partial}{\partial z} + (-z)\frac{\partial}{\partial s}.$$

The induced maps on Lie brackets are given by

$$\Phi_k(\theta^{2k}\partial/\partial z) = \theta^{2k}\partial/\partial z, \quad k = 0, 1, 2, \ldots,$$
$$\Phi_k(\theta^{2k}x) = \theta^{2k}(p\partial/\partial z - z\partial/\partial s), \quad k = 1, 2, \ldots,$$
$$\Phi_k(\theta^{2k}l) = \theta^{2k}\partial/\partial s, \quad k = 1, 2, \ldots.$$

The embedding equations have the following statistical interpretation. Assume that the initial condition for (4.12) is of the form

$$\rho_0(x, \theta) = \left(2\pi \det \sum(\theta)\right)^{-n/2}$$

$$\times \exp\left(-\left\langle x - \mu(\theta), \sum(\theta)^{-1}(x - \mu(\theta))\right\rangle\right) \cdot Q_\theta,$$

where $\theta \rightarrow (\mu(\theta), \Sigma(\theta), Q_0(\theta))$ is a smooth map, $\sum(\theta) > 0$, $\theta \in \Theta$, and $Q_0 > 0$ for $\theta \in \Theta$. Suppose equation (4.33) is initialized at

$$(\theta_0, z_0, P_0, s_0) = \left(\theta_0, \mu(\theta_0), \sum(\theta_0), -\log(Q_0(\theta))\right). \tag{4.34}$$

Append to the system (4.11) an output equation,

$$\bar{Q}_t = e^{-s_t}. \tag{4.35}$$

Now if (4.33) is solved with initial condition (4.34) one can show by differentiating $\bar{Q}_t$ that $\bar{Q}_t$ satisfies the nonlinear equation. In other words, the system (4.33) with initial condition (4.34) is a finite-dimensional recursive estimation for the posterior density $Q(t, \theta_0)$. We have thus verified the homomorphism principle of Brockett (1979), that finite-dimensional recursive estimators must involve Lie algebras of vector fields which are homomorphic images of the Lie algebra of operators associated with the unnormalized conditional density equation.

2.　Lie Groups and Yang–Mills Fields

It has been remarked elsewhere that the Cauchy problem associated with (4.8) may be viewed as a problem of integrating a Lie algebra representation. In this connection one should be interested in whether there is an appropriate topological group associated with $\tilde{G}$. We have the following general procedure.

Let M be a compact Riemannian manifold of dimension d. Let L be a Lie algebra of dimension $n < \infty$. We can always view L as a subalgebra of the general linear Lie algebra $g\ell(m; \mathbb{R})$, $m > n$ (Ado's theorem).

Assumption 4.1. *Let $G = \{\exp(L)\}_G \subset g\ell(m; \mathbb{R})$ be the smallest Lie group containing the exponentials of elements of L. We assume that G is a closed subset of $g\ell(m; \mathbb{R})$.*

Define

$$\mathcal{R} = C^\infty(M; g\ell(m; \mathbb{R})),$$
$$\mathcal{L} = C^\infty(M; L),$$
$$\mathcal{D} = C^\infty(M; G).$$

Clearly $\mathcal{R}$ is an algebra under pointwise multiplication and

$$\mathcal{L} \subset \mathcal{R}, \quad \mathcal{D} \subset \mathcal{R}.$$

Let $(U\alpha, \varphi_\alpha)$ be a C^∞ atlas for M. Then for an $f_1, f_2 \in \mathcal{R}$, define

$$\| f_1 - f_2 \| = \left[\int_{\varphi_\alpha(U_\alpha)} d\mathrm{vol} \sum_{\ell=0}^{k} \left| D^\ell(f_1 - f_2)\varphi_\alpha^{-1} \right|^2 \right]^{1/2}, \tag{4.36}$$

where

$$|f|^2 = \text{tr}\,(f'f). \tag{4.37}$$

(Here $k = d/2 + s$, $s > 0$.) Let $\mathcal{R}_k$ be the completion of $\mathcal{R}$ and $\mathcal{D}_k$, the completion of $\mathcal{D}$ in the norm $\|\cdots\|_k$ ($\mathcal{D}_k$ is closed in $\mathcal{R}_k$). By the Sobolev theorem, $\mathcal{R}_k$ is a Banach algebra and the group operation

$$\mathcal{D}_k \times \mathcal{D}_k \dashrightarrow \mathcal{D}_k,$$
$$(f_1, f_2) \to f_1 f_2 \tag{4.38}$$

when $(f_1 f_2)(m) = f_1(m) f_2(m)$ is continuous. Thus $\mathcal{D}_k$ is a topological group.

By proceeding as before, one can give a Sobolev completion of $\mathcal{L}$ to obtain $\mathcal{L}_k$, an infinite-dimensional Lie algebra, where once again by the Sobolev theorem the bracket operation

$$[.,.]\mathcal{L}_k \times C_k \to \mathcal{L}_k,$$
$$(f_1, f_2) \to [f_1, f_2]$$

with $[f_1, f_2](m) = [f_1(m), f_2(m)]$ is continuous. Now, for a small enough neighborhood $V(0)$ of $0 \in \mathcal{L}$, one can define

$$\exp : V(0) \to \mathcal{D}_k,$$
$$\xi \to \exp(\xi)$$

by pointwise exponentiation. This permits us to provide a Lie group structure on $\mathcal{D}_k$ with $\mathcal{L}_k$ canonically identified as the Lie algebra of $\mathcal{D}_k$.

The procedure outlined above appears to play a significant role in several contexts (the index theorem Yang–Mills fields (Mitter, 1979, 1980)).

For our purposes $\mathcal{L}$ is identified with a faithful matrix representation of $\tilde{G}_0$. Thus we associate with the identification problem a Sobolev Lie group, which is a subgroup of $\mathcal{D}_k$ corresponding to $\tilde{G}_0$

Remark 4.1. One of the important differences between the problem of filtering and the problems of Yang–Mills theories is that in the latter case there are natural norms for Sobolev completion. This follows from the fact that in Yang–Mills theories the algebra $\mathcal{L}$ is compact (semisimple) and one has the Killing form with which to work. In filtering problems, $\tilde{G}_0$ is never compact.

We use a representation of the form

$$\rho(t, x, \theta) = \exp(g_1(t, \theta)A^1) \ldots \exp(g_n(t, \theta)A^n)\rho_0 \tag{4.39}$$

for the solution to equation (4.8). In the case of Example (4.1), this takes
the form

$$
\rho(t, x\theta) = \exp\left(g_1(t, \theta)\left(\frac{\theta^2}{2}\frac{\theta^2}{\theta_x} - \frac{x^2}{2}\right)\right)\exp\left(g_2(t, \theta)\theta^2\frac{\partial}{\partial x}\right)
$$
$$
\times \exp\left(g_3(t, \theta)x\right)\exp\left(g_4(t, \theta)l\right)\rho_0.
$$

Differentiating and substituting in (4.29), we can obtain

$$
\frac{\partial g}{\partial t}(t, \theta) = 1,
$$

$$
\frac{\partial g_2}{\partial t}(t, \theta) = \cosh(g_1, \theta)\dot{y},
$$

$$
\frac{\partial g_3}{\partial t} = -\frac{1}{\theta}\sinh(g_1, \theta)\dot{y},
$$

$$
\frac{\partial g_4}{\partial t} = \frac{\partial g_3}{\partial t}(t, \theta)g_2(t, \theta) \tag{4.40}
$$

and $g_i(0, \theta) = 0$ for $i = 1, 2, 3, 4$, $\theta \in \Theta$. The above first-order partial
differential equations may be easily solved by quadrature and one has
the representation

$$
\rho(t, x, \theta) = \int_{-\infty}^{\infty}\sqrt{\frac{1}{2\pi \sinh(|\theta|t)}}\exp\left(-\frac{1}{2}\coth^2\left(\frac{|x|^2}{|\theta|} + z\right)t|\theta|\right)
$$
$$
\times \exp\left(\frac{xz}{\sqrt{|\theta|\sinh(|\theta|t)}}\right)\exp\left(g_4\left(t, \theta\right)\theta^2\right)
$$
$$
\times \exp\left(g_2(t, \theta)\sqrt{|\theta|z}\right)\rho_0\left(g_3(t, \theta)\theta^2\sqrt{|\theta|z}, \theta\right)dz, \tag{4.41}
$$

where $\rho_0(, \theta) \in L_2(\mathbb{R})$ for every $\theta \in \Theta$ and is smooth in θ. Furthermore
$\Theta\mathbb{R}$ is a bounded set and 0 closure Θ.

In equation (4.39), g_1 should be viewed as canonical coordinates of
the second kind on the corresponding Sobolev Lie group. Now expand
g_2 and g_3 to obtain

$$
g_2(t, \theta) = \sum_{k=0}^{\infty}\theta^2\int_0^t\frac{\sigma^{2k}}{(2k)!}\dot{y}_\sigma d\sigma, \quad k = 1, 2, \ldots,
$$

$$
g_3(t, \theta) = -\sum_{k=0}^{\infty}\theta^{2k}\int_0^t\frac{\sigma^{2k+1}}{(2k+1)!}\dot{y}_\sigma d\sigma, \quad k = 1, 2, \ldots. \tag{4.42}
$$

It follows that all the "information" contained by the observations $\{y_\sigma : 0 \leq \sigma \leq t\}$ about the joint unnormalized conditional density is contained
in the sequence

$$\mathrm{T}\Delta\left\{\int \frac{\sigma^k}{k!}\dot{y}_\sigma d\sigma; \quad k = 0, 1, 2, \dots\right\}. \tag{4.43}$$

Thus T is nothing but a joint sufficient statistic for the identification problem.

3.   Control of Multiagent Systems and Yang–Mills Representation

Consider an object, the motion equation for which can be represented as

$$\dot{x} = r(x, u), \tag{4.44}$$

where $x = (x_1, x_2, x_3) \in Q \subset \mathbb{R}^3$; a function $r(x, u)$ is derived when an equation for dynamics of a particle in a field is reduced to Cauchy form, and the field is characterized by a variable u. The equations similar to (4.44) are widely used in physics and its applications. The equations of the concrete particle dynamics are considered in Daniel and Viallet (1980) and in many other papers. At present, control dynamics equation construction problems deserve great attention. For instance, these problems include controllable models of dynamics of particles in scalar, vector, and spinor fields.

This section builds up a controllable model for dynamics of a particle in electromagnetic and charged fields. The model is based on the gauge field concept (Daniel and Viallet, 1980), which allows us to formulate different principles for automatic control of the dynamics of the particles.

Constructing a controllable model means creating a transformation from a field u to a Yang–Mills field. The essence of this transition is as follows (Mitter, 1979). Instead of u, consider an n-component vector field $\widehat{f}(\widehat{x})$, $\widehat{x} \in T^1$ in a four-dimension space-time T^1. Let $M(\widehat{x})$ be local gauge transformations such that

$$\widehat{f}(\widehat{x}) = M(\widehat{x})\widehat{f}'(\widehat{x}) \tag{4.45}$$

and, for a fixed x, $M(x)$ form a group $G_1 \in GL(n)$. Introduce an operator ∇_α; that is,

$$\nabla_\alpha \widehat{f} = \left[\partial_\alpha + K_\alpha(\widehat{x})\widehat{f}(\widehat{x})\right], \tag{4.46}$$

which satisfies the conditions

$$M(\widehat{x})\nabla'_\alpha \widehat{f}'(\widehat{x}) = \nabla_\alpha \widehat{f}(\widehat{x}), \quad \nabla'_\alpha = \partial_\alpha + K'_\alpha, \tag{4.47}$$

where $K_\alpha = -Q_b C_\alpha^b$; $\{Q_b\}$ is a basis of Lie algebra $\widehat{g}$ for a group G_1, and $[G_a, Q_b] = g_{ab}^c Q_c$; G_{ab}^c are structural constants of the Lie algebra $\widehat{g}$. The equations for the values C_α^b are derived from the Lagrangian $Y_{\alpha\beta}^a Y_a^{\alpha\beta}$, where

$$Y^a_{\alpha\beta} = \frac{\partial C^a_\beta}{\partial \widehat{x}^a} - \frac{\partial C^a_\alpha}{\partial \widehat{x}^\beta} - \frac{1}{2} g^a_{bc} \left(C^b_\alpha C^c_\beta - C^b_\beta C^c_\alpha \right), \tag{4.48}$$

and the Lagrangian has the following form,

$$\partial_\beta Y^{\alpha\beta} = Y^{\alpha\beta}_b g^b_{ac} C^c_\beta.$$

Relation (4.47) yields the law of transformation for a field of matrices K_α:

$$K'_\alpha(\widehat{x}) = M^{-1}(\widehat{x}) K_\alpha(\widehat{x}) M(\widehat{x}) + M(x)^{-1} \frac{\partial M(\widehat{x})}{\partial \widehat{x}^\alpha}.$$

Such transformation satisfies the law group g. A set of these transformations forms a gauge group, formally denoted

$$\widetilde{g} = \prod_x g.$$

It is shown in Yatsenko (1985), that the values C^b_α are Yang–Mills fields. The Yang–Mills field describes a parallel transfer in a charge field and states its curvature. Such a field can be brought in correspondence with the notion of connectedness in some main fiber bundle $(P, T^1, \widetilde{g})$, $\pi \colon P \to T^1$, where T^1 is a base and $\widetilde{g}$ is a structure group.

A control in $(P, T^1, \widetilde{g})$, $\pi \colon P \to T^1$ is understood as a connectedness C^b_α. Notice that one can consider a projection π as a control. Thus, it is possible to deal with a "controllable" fiber bundle $(P, T^1, \widetilde{g})$, $\pi \colon P \to T^1$ and a vector field $r(\widehat{x}, u(C^b_\alpha))$ on P instead of the initial object described by equation (4.44).

To solve control problems, it is necessary to construct equivalent and aggregated models. We construct an equivalent model of a controllable object $(P, T^1, \widetilde{g})$, $\pi \colon P \to T^1$ as follows. Let $\pi \colon P \to T^1$ be a main $\widetilde{g}$ fiber bundle and let $l \colon Z \to T^1$ be some m-dimensional $\widetilde{g}$-vector fiber bundle with a trivial action, exerted by $\widetilde{g}$ onto Z. Assume also that a structure of a $k > 1$-dimensional cellular set can be introduced on T^1. An equivariant embedding of π into 1 is understood as an embedding $h \colon P \to Z$, commutating with projections. If $K > m$, that is, an action, exerted by $\widetilde{h}$ onto Z is free outside a zero section for 1, then the main $\widetilde{g}$ fiber bundle $\pi \colon P \to T^1$ can be equivariantly embedded into $l \colon Z \to T^1$. An equivalent model of a controlled process is understood as a ternary $(Z, T^1, \widetilde{g})$. In its turn, an equivalent model admits an exact aggregation, performed by means of a factorization of an induced. In this case, it is possible to assume, that a vector fiber bundle is specified by an interrelation system ω on some set X^1. Introduce an equivalence relation S on X^1. This relation generates an object of the same nature, as the initial object X^1, and a factor-object (F-object) is obtained, which possesses a factorizing equivalence relation S. If (X_1, ω) generates an object of the same nature, possessed by an initial object, and this generation is carried out on a subset X_1 of X_1, then a subobject $X_1(\widetilde{\omega})$ (P-object) is derived. By using the language of mathematical structure theory, it is possible to create a general theory of aggregation of invariant models for nonlinear systems.

Consider the main automatic particle dynamics control principles, which electromechanical systems with distributed parameters as an example. It is shown in Yatsenko (1985), that a closed distributed automatic control system can be represented by two subsystems S_1 and S_2, interrelated by the electromagnetic field

$$S = S_1 \cup S_2.$$

Represent the field B of a whole control system state by a field of internal states of each subsystem B_1 and B_2, of an interaction field B_0 and by external field B_3. In addition, represent B_0 by two components X and U; that is, $B_0 = X + U$, where X is an information carrier and U is a control field. Consider U as the result of an influence exerted by X onto the control medium and simulated by an operator dependence $U = \widehat{B}(X, E)$, where $\widehat{B}$ is a control operator and E is the control medium power supply field. An external field B is also divided into two components V and N, where V is a field of control that is carried out according to a fixed space–time program, and N is a field of disturbing effects. The control object is described by a fiber bundle $(P, T^1, \widetilde{g})$ with a control $C_\alpha^b(X, Y, U, V, N)$, where Y is a field of an internal state. It is clear that a section is only one in $(P, T^1, \widetilde{g})$, if U, V, and N are physically implementable and uniquely specified. The general problem, concerning the calculation of an electromagnetic field of a control system, consists of finding such a physically implementable operator $\widehat{B}$ and programmed controlling influence V, under which the particle dynamics would meet certain previously formulated requirements.

4. Dynamic Systems, Information, and Fiber Bundles

There has been active research of controlled multiagent objects as information transforming systems during the last few years. Despite the achievements that have been made in this area, effective mathematical methods for investigating such systems have not yet been developed. One possible approach is based on the differential geometry methods of system theory (Van Der Shaft, 1982, 1987). This section is devoted to one of the problems of this area of research, that of developing a method for analyzing a class of mathematical models of symmetric controlled processes. Assuming that the process is described by a commutative diagram (Van der Shaft, 1982, 1987) which is based on the lamination concept, we propose a geometric method for "identifying" its hidden structure.

Investigation of the information-transformation laws in various systems is one of the most essential stages in the creation of new agents. The goal of experimental and theoretical research is the implementation of optimal strategy using complex structure nonequilibrium processes in such systems. To investigate these processes it is required to develop the corresponding mathematical methods. In this context we propose an approach which is based on the assumption that one can use models from mathematical system theory to adequately describe informational processes. The essence of this approach is in the following.

Some dynamic system S that implements a transformation F, or an input informational action U, into an output one X, is considered. It is assumed that one can affect the information-transforming process by a reconfiguring action that changes the dynamic behavior, structure, symmetry, and so on of the process. We refer to the objects described in the preceding S as dynamic information-transforming systems (DITS).

The connection between the input and output actions is necessary for obtaining answers to questions about the method of programming the entire system, optimizing the flow of informational signals, and the interconnections among the global system properties (stability, controllability, etc.) and the corresponding local properties of the various subsystems. One also has to answer those questions when solving pattern-recognition problems, constructing an associative memory. A generalized description of DITS that contains a large number of subsystems (e.g., a neural network) is postulated in this section: the controlled process in the DITS is described adequately by a commutative diagram that generalizes the concept of a nonlinear controlled dynamic system on a manifold. Taking into account the symmetry concept that is characteristic of classical mechanics (Arnold, 1983), one has to transfer it to the DITS, "identify" the hidden structure of the informational process, and demonstrate that the proposed model admits local and/or global decompositions into smaller-dimensionality feedback subsystems.

We note that the decomposition idea was first applied to discretely symmetric automatic control systems by V. Yatsenko (1985). Continuous symmetry group dynamic systems were considered by Van der Shaft (1987). Substantive results on the decomposability of systems with symmetries have been obtained by A.Y. Krener (1973) and others. However, this question remains open for DITSs.

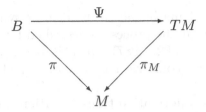

Figure 4.3.   Diagram of a nonlinear controlled DITS.

Necessary concepts and definitions. Some definitions and concepts that are necessary for describing the DITS structure and the conditions for its decomposability are presented in this section. The necessary notions about manifolds, connectivities, and distributions are given in Griffiths (1983). We introduce the definition of a nonlinear DITS.

Definition 4.8. *Consider a triple $F(B, M, \psi)$, where B is a smooth fiber over M with the projection $\pi : B \to M$; π_M is the natural projection of TM on M, and ψ is a smooth mapping such that the diagram presented in Figure 4.3 is commutative, by a 'geometrical model of the agent'.*

We interpret the manifold M as the DITS state space and the $\pi^{-1}(x) \in B$ layer as the space of input action values which depends in the general case, on the current system state. If one chooses the coordinates (x, u), which correspond to the B_x layer, then this definition of the DITS, F, corresponds locally to the nonlinear transformation $\psi : (x, u) \to (x, \psi(x, u))$ and the dynamic system

$$\dot{x}(t) = \psi(x(t), u(t)), \quad u(t) \in U, \tag{4.49}$$

where x is the DITS state vector, $u = (u^1, u^2)$ are the control actions, $u^1(\cdot, \cdot)$ is the vector of the coded input informational action that depends in general on time and on the current state, and $u^2(\cdot, \cdot)$ is the action used to reconfigure the dynamic properties of the DITS and to train it.

The control algorithm u^2 inputs to the system the capability of transforming the set of input actions into a set of output signals that allows one to identify the input images uniquely. In essence, it realizes the decoding process, which identifies the input images. In the simplest case, it can be realized on the basis of the successive input action segmentation method. Such a method facilitates a unique separation of the input images by the use of the simplest binary decoding rule.

Definition 4.9. *Let M be a smooth manifold. We say that the smooth mapping $Q : G \times M \to M$ such that:*

1. $Q(e, x) = x$ for all $x \in M$, and

2. $Q(g, Q(h, x)) = Q(gh, x)$ *for any g and $h \in G$, and all $x \in M$, is the left action (or G-action) of the G Lie group on M.*

We fix one of the variables for various time instants and examine the Q action as a function of the remaining variables. Let $Q_g : M \to M$ denote the function $x \mapsto Q(g, x)$ and $Q_x : G \to M$ the function $g| \to Q(g, x)$. We note that because $(Q_g)^{-1} = Q_g^{-1}$, Q_g is a diffeomorphism.

We introduce the definition of group action on a manifold.

Definition 4.10. *Let Q be the action of G on M. We say that the set $G \cdot x = \{Q_g(x) | g \in G\}$ is the orbit (Q-orbit) of the point $x \in M$. The action is free at x if $g| \to Q_g(x)$ is one-to-one. It is free on M if and only if it is free at all $x \in M$.*

We now introduce the concept of global symmetry of a controlled DITS.

Definition 4.11. *Let $\hat{F}(B, M, \psi)$ be a nonlinear controlled DITS, and θ and Q be actions of G on B and M, respectively. Then, F has symmetry (G, θ, Q) if the diagram presented in Figure 4.4 is commutative for all $g \in G$.*

We consider, within the framework of the presented definition, the special case in which the symmetry lies "entirely within the state space."

Definition 4.12. *Let $B = M \times U$, where U is some manifold. Then, (G, Q) is a symmetry of the state space of the system $\hat{F}(B, M, \psi)$ if (G, θ, Q) is a symmetry of $\hat{F}$ for $\theta_g = (Q_g, Id_U) : (x, u) \to (Q_g(x), u)$.*

Global state space symmetry can be defined only for DITS B_x which is a trivial lamination because otherwise the input spaces would depend on the state and the problem made substantially more complicated.

We introduce now the definition of local symmetry.

Definition 4.13. *We assume that $Q : G \times M \to M$ is an action and that $\varepsilon \in T_e G$. Then, $Q^\xi(R \times M \to M) : (t, x)| \to Q(\exp t\xi, x)$, where*

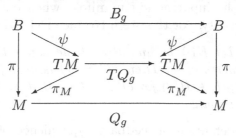

Figure 4.4. A commutative diagram of an DITS with symmetries.

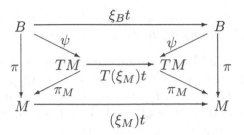

Figure 4.5.   Diagram of a symmetric DITS.

$\exp : T_e G \to G$ *is the usual exponential mapping, is the* $\mathbb{R}$*-action on* M, *and* Q^{ξ} *is the complete flow on* M. *We say that the corresponding vector field on* M, *which is defined by the expression*

$$\xi_m(x) = \frac{d}{dt}Q(\exp t\xi, x)\Big|_{t=0}, \tag{4.50}$$

is the infinitesimal action generator, which corresponds to ξ.

Let X_t denote the flow of the vector field X; that is, $X_t = F_t(X_0)$. It is obvious from the definition of the infinitesimal generator that if (G, θ, Q) is a symmetry of the $\hat{F}(B, M, \psi)$ system, then the diagram presented in Figure 4.5 is commutative for all $t \in \mathbb{R}$ and $\xi \in T_e G$.

On the basis of the local commutativity property we present the following definition of infinitesimal DITS symmetry.

Definition 4.14. *Let* $\hat{F}(B, M, \psi)$ *be a nonlinear DITS. Then,* (G, θ, Q) *is an infinitesimal symmetry of* F *if, for each* $x_0 \in M$, *there exist an open neighborhood* $\hat{O}$ *of the point* x_O *and* $\xi > 0$ *such that*

$$(\xi_M)_t * \psi(\xi) = \psi((\xi_b)_t(b)), \tag{4.51}$$

for all $b \in \pi^{-1}(\hat{O})$, $|t| < \xi$, *and* $\| \xi \| < 1$, $\xi \in T_e G$, *where* $\| \cdot \|$ *is an arbitrary fixed norm on* $T_e G$.

One can define an infinitely small state space symmetry for nontrivial laminations of the input action manifold when one can introduce integratable connectivity. For this we introduce Definition 4.15.

Definition 4.15. *Let* $H(\cdot)$ *be an integratable connectivity on* B *and* (G, θ, Q) *be a symmetry of* F. *Then,* (G, θ, Q) *is an infinitesimal state space symmetry if* $\xi_B(b) \in H(b)$ *for all* $\xi \in T_e G$; *that is, the infinitesimal generators* θ *are horizontal.*

We introduce a definition of feedback equivalence of two DITSs in analogy with Van der Shaft (1982).

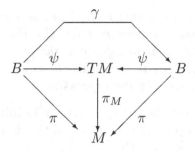

Figure 4.6. Diagram of feedback-equivalent DITSs.

Definition 4.16. *A system, $F(B, M, \psi)$, is feedback equivalent to a system $F'(B, M, \widetilde{\psi})$ if there exists an isomorphism $\gamma : B \to B$ such that the diagram presented in Figure 4.6 is commutative.*

Isomorphism means that, for $x \in M$, γ_x is a mapping from the layer over x' into the layer over x', and it is a diffeomorphism. Consequently, this corresponds to a 'control feedback'.

The local structure of DITSs with symmetries. Inasmuch as we are interested in the local structure of DITS, we have to assume that the system has an infinitesimal symmetry, which satisfies some nonsingularity condition. For this, we set the dimensionality of M to n and that of G to k, where $k < n$. We note that the action $Q : G \times M \to M$ is free at the point $m \in M$ if $Q_m : G \to M$ is one-to-one. This is equivalent to saying that the tangent mapping Q is of full rank; that is, rank $Q = \dim G$. Hence, Q is free on M if and only if it is free in some neighborhood of m. We say that an action which satisfies this condition is nonsingular at the point m.

The basic result of this section is that the existence of an infinitesimal symmetry in a neighborhood of a singular point in DITS makes it possible to decompose the system into a cascade union of simpler subsystems. The structure of these subsystems depends, in general, on the symmetry group G. If, for example, G has a nontrivial center, then one of the subsystems is in fact a quadrature subsystem.

Let, in addition, $C = h \in G|jg = gh$ for all $g \in G$ be the center of the group G to which the kernel C_+ of the Lie semialgebra T_eG, which has the same dimensionality as C, corresponds. Hence, if G has an l-dimensional center, then there exist linearly independent vectors $\xi^1, \ldots, \xi^k \in T_eG$ such that $[\xi^i, \xi^j] = 0$ for all $1 \le i \le l$ and $1 \le j \le k$.

Using the results of Van der Shaft and Markus investigations (Van der Shaft, 1982, 1987; Markus (1973); Marcus and Willsky (1976)) that deal with the properties of systems with symmetries as applied to DITSs, one can formulate the following theorems.

Theorem 4.7. *Let us assume that $\hat{F}(B, M, \xi)$ is a controlled DITS with an infinitesimal state space symmetry (G, θ, Q), that G has an l-dimensional center, and that Q is nonsingular at the point $m \in M$. Then, the B coordinates $(x_1, \ldots, x_n, u)$ in a neighborhood of m exist such that $\hat{F}$ is given in these coordinates by the expression.*

Using the obtained results for systems for infinitesimal state space symmetries, one can propose the structure of the decomposed system. It suffices to demonstrate for this that the decomposed system with infinitesimal symmetry is locally feedback-equivalent to the original system with infinitesimal state space symmetry.

Definition 4.17. *Let $\hat{F}(B, M, \psi)$ be a controlled DITS and $\hat{O}$ be an open subset of M. Then, we say that a system of the form $\hat{F}(\pi^{-1}(\hat{O}), \hat{O}, \psi)|\pi^{-1}(O)$ is $\hat{F}|\hat{O}$ ($\hat{F}$ bounded on $\hat{O}$).*

Theorem 4.8. *Let $\hat{F}(B, M, \psi)$ have an infinitesimal symmetry (G, θ, Q) and Q be nonsingular at the point m. There exist a neighborhood of m and a system F with infinitesimal symmetry (G, θ, Q) such that $\hat{F}|\hat{O}$ is feedback-equivalent to the $\hat{F}$ system.*

Let $\hat{F}(B, M, \psi)$ be a controlled DITS with symmetry (G, θ, Q) and Q be nonsingular at the point m. Then, in a neighborhood of m, $\hat{F}$ is feedback-equivalent to $\hat{F}$ with infinitesimal symmetry and has the structure shown in Figure 4.7, where γ is the feedback function, the L^i are nonlinear subsystems of dimensions $n-k$ and $k-l$, respectively, and Q is an l-dimensional "quadrature" system

$$\dot{x}_i = f_i(x_1, \ldots, x_{n-k}, u), \quad i = 1, \ldots, n-k,$$
$$\dot{x}_j = f_j(x_1, \ldots, x_{n-1}, u), \quad i = n-k+1, \ldots, k. \tag{4.52}$$

The global structure of DITS. The decomposability of DITS with global symmetries is the result of factoring the DITS state space, which follows from the properties of symmetry.

We introduce the definition of proper action.

Definition 4.18. *Let Q be a G-action on M. We say that Q acts properly if $(g, m) \rightarrow m$ is a proper mapping, that is, if the preimages of compact sets are compact.*

This definition is equivalent to the following assertion: whenever x_n converges on M and $Q_{g_n}(x_n)$ converges on M, g_n includes a subsequence that converges in G. Hence, if G is compact, this condition is satisfied automatically. Membership in the same Q-orbit is an equivalence relation on M. Let M/G be the set of equivalence classes and $p : M \rightarrow M/G$ be

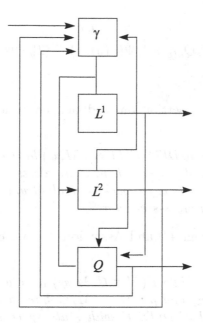

Figure 4.7.   Local structure of DITS with infinitesimal symmetries.

specified by the relation $p(m) = Gm$. We introduce on M/G a relations topology; that is, $V \subset M/G$ is open if and only if $p^{-1}(V)$ is open on M. In general, M/G can be a rather poor space.

If G acts freely and properly on M, then M/G is a smooth manifold and $p : M \to M/G$ is the principal lamination with Lie group G.

We introduce the following constraints on the principal lamination.

1. p is a smooth full-rank function.

2. $p : M \to M/G$ has a cross-section (i.e. a smooth mapping $\sigma : M/G \to M$ such that $p \cdot \sigma$ a is the identity mapping on M/G if and only if M is equivalent to $M/G \times G$.

3. The topological conditions that guarantee the existence of a section, that is, if M/G or G is a contraction mapping, a cross-section must exist, are specified.

We formulate a theorem, which is necessary for obtaining a global factorization of the DITS state space.

Let $Q_m : G \to G \cdot m$ be specified by $g \to Q(G, m)$. The following result about the global structure of a DITS with symmetries holds.

Theorem 4.9. *We assume that $\hat{F}(M \times U, M, \psi)$ is a controlled DITS with a state space symmetry (C, Q). Then, if Q is free and proper, and $p : M \to M/G$ has a cross-section σ, then $\hat{F}$ is isomorphic to the system*

$$\dot{y} = \Psi(y, u),$$

$$\dot{g} = (T_e L_g)(T_e Q_{\sigma(y)})^{-1} \left[\Psi(\sigma(y), u) - (T_y \sigma)\Psi(y, u) \right], \qquad (4.53)$$

defined on $M/G \times G$.

We formulate an assertion on feedback equivalence of DITSs with symmetries.

Assertion 4.1. *Let the DITS* $F(M \times U, M, \psi)$ *have a symmetry* (G, θ, Q) *such that* Q *is free and proper. Then, there exists a system* F *with symmetry* (G, Q) *to which* F *is feedback-equivalent under the condition that* $p : M \to M/G$ *has a cross-section* σ.

Combining Theorem 4.9 and Assertion 4.1, we obtain the following corollary.

Corollary 4.1. *Let DITS* $\hat{F}(M \times U, M, \psi)$ *have a symmetry* (G, θ, Q), Q *be free and proper, and* $p : M \to M/G$ *have a cross-section. Then, there exists a model of DITS* F *with state space symmetry* (G, Q) *to which* $\hat{F}$ *is feedback-equivalent. Consequently,* F *has a global structure.*

The feasibility of applying the results to the investigation of agents. It is of interest to investigate the decomposability of DITSs composed of neurallike agents that are described by the system of equations

$$\dot{x}(t) = \psi(x(t), u(t)). \qquad (4.54)$$

One can define for (4.54) a decomposed system L as a nontrivial cascade of subsystem L^1 and L^2. If the Lie algebra $\hat{L}(L)$ is the semidirect sum of finite-dimensional subalgebra L^1 and the ideal of L^2, it has a nontrivial cascade decomposition into subsystems L^1 and L^2 such that $\hat{L}(L^1) = L^1$, and $\hat{L}(L^2) = L^2$. Using this fact and Levy's theorem one can demonstrate that if $\hat{L}(L)$ is finite-dimensional, the DITS admits a nontrivial decomposition into a parallel cascade of L^i systems with simple Lie algebras followed by a cascade of one-dimensional systems L^j. As a result, the basic informational transformation is done in subsystems with simple Lie algebras. The state space M of the original system L is adopted here as the state space of these systems. Therefore, despite the fact that the system has been partitioned into simpler parts, the overall dimensionality of these parts is, in general, larger than that of the original system. (One can reduce at the local level this dimensionality by replacing the L^i system by matrix equivalents defined on the exponential functions of the Lie algebras that correspond to them.)

These results can be compared with the conditions for decomposability obtained by analyzing the DITS symmetries described in this section for which the subsystem dimensionality equals that of the original system. No assumptions about the finite dimensionality of the Lie algebra are required here. We consider a class of neural nets described by the linear-analytic equations

$$\dot{x}(t) = f(x) + \sum_{i=1}^{k} u_i g_i(x). \qquad (4.55)$$

One can formulate for it the necessary and sufficient conditions for parallel-cascade decomposability by Lie algebras. In doing so, one can pose the condition that each component of the input action be applied to only one of the subsystems; that is, the decomposition procedure partitions the inputs into disjoint subsets. However, such an approach cannot be applied to the decomposition of DITS with scalar input.

If DITS $\hat{F}(B, M, \psi)$ has an infinitesimal symmetry (G, θ, Q), local commutativity of the diagram means that $\psi * \varepsilon_B = \varepsilon_m$ and $\pi * \varepsilon_B = \varepsilon_n$. Let $\Delta_B = \mathrm{span}\{\varepsilon \| \varepsilon_B \in T_e G\}$ and the same hold for Δ_m. Then, $\psi * \Delta B \subset \Delta_m$ and $\pi * \Delta_B = \Delta$, and Δ_m is a controlled invariant distribution. Models of neural networks, including affine ones, have invariant distributions that induce decompositions of the system into simpler subsystems. However, because the symmetry conditions are constraints, the decompositions are obtained as more detailed and structured.

A class of multiple agents that are described by a commutative diagram is examined in this section. Constraints on systems with symmetry under which one can explicitly expose the hidden structure of the controlled process are formulated. We show that the effect of the DITS on the information-transforming process depends substantially on the type of system symmetry. The informational process is subject here to the action of the cascade group, transformations, or the action of a dynamic-transformation operator with feedback. The obtained results can be expanded to adaptive learning systems by introducing the corresponding optimization models. When doing so, one can expect that DITS of which the quality functional is invariant in symmetry-conserving transformations will be described adequately by a nonlinear system with optimal feedback and will have a differential-geometric structure, which is of interest from the point of view of applications. The results of the investigations presented here can be used in the study of a synergetic model of a neural network on the basis of potential-dependent ion channels in biomembranes.

5. Fiber Bundles, Multiple Agents, and Observability

In the last decade, important work has been done on a differential-geometric approach to nonlinear input state-output systems, which in local coordinates have the form

$$\dot{x} = g(x, u), \quad y = h(x), \tag{4.56}$$

where x is the state of the system, u is the input and y is the output. Most of the attention has been directed to the formulation in this context of fundamental system-theoretic concepts such as controllability, observability, minimality, and realization theory.

In spite of some very natural formulations and elegant results which have been achieved, there are certain disadvantages in the whole approach, from which we summarize the following points.

(a) Normally the equations

$$\dot{x} = g(x, u) \tag{4.57}$$

are interpreted as a family of vector fields on a manifold parameterized by u; that is, for every fixed $\overline{u}$, $g(\cdot, \overline{u})$ is a globally defined vector field. We propose another framework by looking at (4.57) as a coordinatization of the diagram

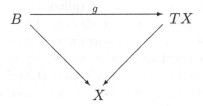

where B is a fiber bundle above the state space manifold X and the fibers of B are the state-dependent input spaces, whereas TX is as usual the tangent bundle of X (the possible velocities at every point of X).

(b) The "usual" definition of *observability* has some drawbacks. In fact, observability is defined as *distinguishability*; that is, for every x_1 and x_2 (elements of X) there exists a certain input function (in principle dependent on x_1 and x_2) such that the output function of the system starting from x_1 under the influence of this input function is different from the output function of the system starting from x_2 under the influence of the same input function. Of course, from a practical point of view this notion of observability is not very useful, and also is not in accord with the usual definition of observability or reconstructibility for general systems.

Hence, despite the work of Susmann (1983) on universal inputs (i.e., input functions) which distinguish between every two states x_1 and x_2, this approach remains unsatisfactory.

(c) In the class of nonlinear systems (4.56), memoryless systems

$$y = h(u) \tag{4.58}$$

are not included. Of course, one could extend the system (4.56) to the form

$$\dot{x} = g(x, u), \quad y = h(x, u), \tag{4.59}$$

but this gives, if one wants to regard observability as distinguishability, the following rather complicated notion of observability. As can be seen, distinguishability of (4.59) with $y \in \mathbb{R}^p$, $u \in \mathbb{R}^m$, and $x \in \mathbb{R}^n$ is equivalent to distinguishability of

$$\dot{x} = g(x, u), \quad \overline{y} = \overline{h}(x), \tag{4.60}$$

where $\overline{h} : \mathbb{R}^n \to (\mathbb{R}^p)^{\mathbb{R}^m}$ is defined by $\overline{h}(x)(u) = h(x, u)$.

Checking the Lie algebra conditions for distinguishability for the system (4.60) is not very easy.

(d) It is often not clear how to distinguish a priori between inputs and outputs. Especially in the case of a nonlinear system, it could be possible that a separation of what we call external variables in input variables and output variables should be interpreted only locally. An example is the (nearly) ideal diode given by the *I-V* characteristic in Figure 4.8. For $I < 0$ it is natural to regard I as the input and V as the output, whereas for $V > 0$ it is natural to see V as the input and I as the output. An input-output description should be given in the scattering variables $(I - V, I + V)$. Moreover, in the case of nonlinear systems it can happen that a global separation of the external variables in inputs and outputs is simply not possible! This results in a definition of a system,

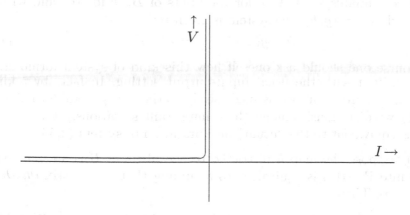

Figure 4.8. Current-voltage characteristic of the ideal diode.

that is a generalization of the usual input-output framework. It appears that various notions such as the definitions of autonomous (i.e., without inputs), memoryless, time-reversible, Hamiltonian, and gradient systems are very natural in this framework.

5.1   Smooth Nonlinear Systems

The smooth (say C^∞) systems can be represented in the commutative diagram

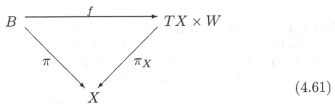

$$(4.61)$$

where (all spaces are smooth manifolds) B is a fiber bundle above X with projection π, TX is the tangent bundle of X, π_x the natural projection of TX on X, and f is a smooth map. W is the space of external variables (think of the inputs and the outputs). X is the state space and the fiber $\pi_{-1}(x)$ in B above $\in X$ represents the space of inputs (to be seen initially as dummy variables), which is state-dependent (think of forces acting at different points of a curved surface).

This definition formalizes the idea that at every point $x \in X$ we have a set of possible velocities (elements of TX) and possible values of the external variables (elements of W), namely the space

$$f(\pi^{-1}(x)) \subset T_x X \times W.$$

We denote the system (4.61) by $\Sigma(X, W, B, f)$. It is easily seen that in local coordinates x for X, v for the fibers of B, w for W, and with f factored in $f = (g, h)$, the system is given by

$$\dot{x} = g(x, v), \quad w = h(x, v). \tag{4.62}$$

Of course one should ask oneself how this kind of system formulation is connected with the usual input-output setting. In fact, by adding more and more assumptions successively to the very general formulation (4.61) we distinguish among three important situations, of which the last is equivalent to the "usual" interpretation of system (4.56).

(i) Suppose the map h restricted to the fibers of B is an immersive map into W (this is equivalent to assuming that the matrix $\partial h / \partial v$ is injective). Then:

Lemma 4.1. *Let h be restricted, $(\overline{x}, \overline{u})$ and $\overline{w}$ be points in B and W, respectively, such that $h(\overline{x}, \overline{v}) = \overline{w}$. Then locally around $(\overline{x}, \overline{v})$ and $\overline{w}$*

there are coordinates (x, v) *for* B *(such that* v *are coordinates for the fibers of), coordinates* (w_1, w_2) *for* W, *and a map* $\overline{h}$ *such that* h *has the form*

$$(x, v) \gg h > (w_1, w_2) = (\overline{h}(x, v), v). \tag{4.63}$$

Proof: The lemma follows from the implicit function theorem.
Hence locally we can interpret a part of the external variables, that is, w_1, as the outputs, and a complementary part, that is, w_2, as the inputs! If we denote w_1 by y and w_2 by u, then system (4.62) has the form (of course only locally)

$$\dot{x} = y(x, u), \quad y = \overline{h}(x, u). \tag{4.64}$$

(ii) Now we not only assume that $\partial h / \partial v$ is injective, which results in a local input-output parameterization (4.64), but we also assume that the output set denoted by Y is globally defined. Moreover, we assume that W is a fiber bundle above Y, which we call $p : W \to Y$, and that h is a bundle morphism (i.e., maps fibers of B into fibers of W). Then:

Lemma 4.2. *Let* $h : B \to W$ *be a bundle morphism, which is a diffeomorphism restricted to the fibers. Let* $\overline{x} \in X$ *and* $\overline{y} \in Y$ *be such that* $h(\pi^{-1}(\overline{x})) = p^{-1}(\overline{y})$. *Take coordinates* x *around* $\overline{x}$ *for* X *and coordinates* y *around* $\overline{y}$ *for* Y. *Let* $(\overline{x}, \overline{v})$ *be a point in the fiber above* $\overline{x}$ *and let* $(\overline{y}, \overline{u})$ *be a point in the fiber above* $\overline{y}$ *such that* $h(\overline{x}, \overline{v}) = (\overline{y}, \overline{u})$. *Then there are local coordinates* v *around* $\overline{v}$ *for the fibers of* B, *coordinates* u *around* $\overline{u}$ *for the fibers of* W *and a map* $\overline{h} : X \to Y$ *such that* h *has the form*

$$(x, v) \gg h > (y, u) = (\overline{h}(x), v). \tag{4.65}$$

Proof: Choose a locally trivializing chart $(0, \phi)$ of W around $\overline{y}$. Then $\phi : p^{-1}(0) \to 0 \times U$, with U the standard fiber of W. Take local coordinates u around $\overline{u} \in U$. Then (y, u) forms a coordinate system for W around $(\overline{y}, \overline{u})$. Because h is a bundle morphism, it has the form

$$(x, \overline{v}) \gg h > (y, u) = (\overline{h}(x), h'(x, \overline{v})).$$

where (x, v) is a coordinate system for B around $(\overline{x}, \overline{v})$. Now adapt this last coordinate system by defining

$$v = (h')^{-1}(x, u) \quad \text{with } x \text{ fixed.}$$

Because h restricted to the fibers is a diffeomorphism, v is well defined and (x, v) forms a coordinate system for B in which h has the form

$$(x, v) \gg h > (y, u) = (\overline{h}(x), u).$$

Hence under the conditions of Lemma 4.2 our system is locally (around $\overline{x} \in X$ and $\overline{y} \in Y$) described by

$$\dot{x} = g(x, u), \quad y = \overline{h}(x). \tag{4.66}$$

This input-output formulation is essentially the same as the one proposed by Brockett (1980) and Takens (1981), who take the input spaces as the fibers of a bundle above a globally defined output space Y. In fact, this situation should be regarded as the normal setting for nonlinear control systems.

(iii) Take the same assumptions as in (ii) and assume moreover that W is a trivial bundle (i.e., $W = Y \times U$), and that B is a trivial bundle (i.e., $B = X \times V$). Because h is a diffeomorphism on the fibers, we can identify U and V. In this case the output set Y and the input set U are globally defined, and the system is described by

$$\dot{x} = g(x, u), \quad y = \overline{h}(x), \tag{4.67}$$

where for each fixed $\overline{u}$, $g(\cdot, \overline{u})$ is a globally defined vector field on X. This is the "usual" interpretation of (4.56).

5.2   Minimality and Observability

Minimality. We want to give a definition of minimality for a general (smooth) nonlinear system

Definition 4.19. *Let $\Sigma(X, W, B, f)$ and $\Sigma'(X', W, B', f')$ be two smooth systems. Then we say $\Sigma' \leq \Sigma$ if there exist surjective submersions $\phi : X \to X', \Phi : B \to B'$ such that the diagram commutes.*

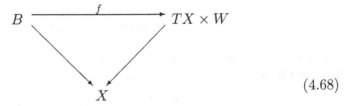

$$\tag{4.68}$$

Σ is called *equivalent to* Σ' (denoted $\Sigma \sim \Sigma'$) if ϕ and Φ are diffeomorphisms.

We call Σ *minimal* if $\Sigma' \leq \Sigma \Rightarrow \Sigma' \sim \Sigma$.

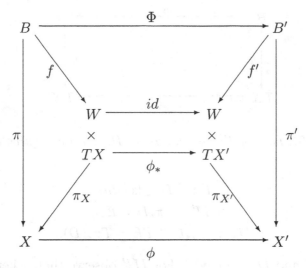

Of course, Definition 4.19 is an elegant but rather abstract definition of minimality. From a differential geometric point of view it is very natural to see what these conditions of commutativity mean locally. In fact, we show in Theorem 4.11 that locally these conditions of commutativity do have a very direct interpretation. But first we have to state some preparatory lemmas and theorems.

Let us look at (4.68). Because Φ is a submersion it induces an involutive distribution D on B given by

$$D := \{Z \in TB | \Phi_* \dot{Z} = 0\}$$

(the foliation generated by D is of the form $\Phi^{-1}(c)$ with c constant). In the same way ϕ induces an involutive distribution E on X. Now the information in diagram (4.68) is contained in three subdiagrams (we assume $f = (g, h)$ and $f' = (g', h')$):

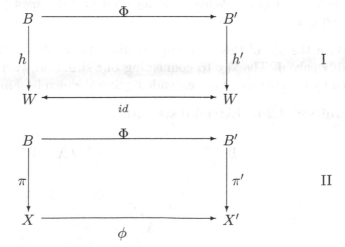

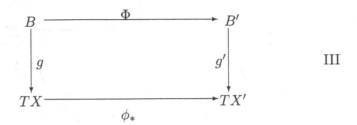

Lemma 4.3. *Locally the diagrams I, II, III are equivalent, respectively, to*

$$I' : \quad D \subset \ker dh,$$
$$II' : \quad \pi_* D = E,$$
$$III' : \quad g_* D \subset TE = T\pi_*(D). \qquad (4.69)$$

Proof: I' and II' are trivial. For III' observe that, when ϕ induces a distribution E on X, then ϕ_* induces the distribution TE on TX.

Now we want to relate conditions I', II', III' with the theory of nonlinear disturbance decoupling. Consider in local coordinates the system

$$\dot{x} = f(x) + \sum_{i=1}^{m} u_i g_i(x) \quad \text{on a manifold } X.$$

We can interpret this as an affine distribution on manifold.

Theorem 4.10. *Let $D \in A(\Delta_0)$. Then the condition*

$$[\Delta, D] \subseteq D + \Delta_0 \qquad (4.70)$$

(we call such a $D \in A(\Delta_0)\Delta(\mathrm{mod}\,\Delta_0)$ invariant) is equivalent to the two conditions: (a) there exists a vector field $F \in \Delta$ such that $[F, D] \subseteq D$; (b) there exist vector fields $B_i \in \Delta_0$ such that the span $\{B_i\} = \Delta_0$ and $[B_i, D] \subset D$.

With the aid of this theorem the disturbance decoupling problem is readily solved. The key to connecting our situation with this theory is given by the concept of the extended system, which is of interest in itself.

Definition 4.20. (Extended system). *Let*

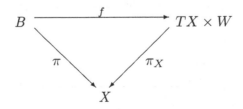

Then we define the extended system of $\Sigma(X, W, B, f)$ *as follows: We
define* Δ_0 *as the vertical tangent space of B; that is,*

$$\Delta_0 := \{Z \in TB | \pi_* Z = 0\}.$$

Note that Δ_0 is *automatically involutive.*

Now take a point $(\overline{x}, \overline{v}) \in B$. Then $g(\overline{x}, \overline{v})$ is an element of $T_{\overline{x}} X$. Now
define

$$\Delta(\overline{x}, \overline{v}) := \{Z \in T_{(\overline{x}, \overline{v})} | \pi_* Z = g(\overline{x}, \overline{v})\}.$$

So $\Delta(\overline{x}, \overline{v})$ consists of the possible lifts of $g(\overline{x}, \overline{v})$ in $(\overline{x}, \overline{v})$. Then it is
easy to see that Δ is an affine distribution on B, and that $\Delta - \Delta = \Delta_0$. We call the affine system (Δ, Δ_0) on B constructed in this way,
together with the output function $h : B \rightarrow W$, the extended system
$\Sigma^e(X, W, B, f)$.

We have the following.

Lemma 4.4. *(a) Let D be an involutive distribution on B such that
$D \cap \Delta_0$ has constant dimension. Then $\pi_* D$ is a well-defined and involutive distribution on X if and only if $D + \Delta_0$ is an involutive distribution.*

*(b) Let D be an involutive distribution on B and let $D \cap \Delta_0$ have
constant dimension. Then the following two conditions are equivalent:
(i) $\pi_* D$ is a well-defined and involutive distribution on X, and $g_* D \subset T\pi_* D$.
(ii) $[\Delta, D] \subset D + \Delta_0$.*

Proof: (a) Let $D + \Delta_0$ be involutive. Because D and Δ_0 are involutive this is equivalent to $[D, \Delta_0] \subset D + \Delta_0$. Applying Theorem 4.10 to
this case gives a basis $\{Z_1, \dots, Z_k\}$ of D such that $[Z_i, \Delta_0] \subseteq \Delta_0$. In co-
ordinates (x, u) for B, the last expression is equivalent to $Z_i(x, u) = (Z_{ix}, Z_{iu}(x, u))$, where Z_{ix} and Z_{iu} are the components of Z_i in the
x- and u-directions, respectively. Hence $\pi_* D = \text{span} \{Z_{1x}, \dots, Z_{kx}\}$ and
is easily seen to be involutive. The converse statement is trivial.

(b) Assume (i); then there exist coordinates (x, u) for B such that
$D = \{\partial/\partial x_1, \dots, \partial/\partial x_x\}$ (the integral manifolds of D are contained in
the sections $u = \text{const}$). Then $g_* D \subset T\pi_* D$ is equivalent to

$$\left(\frac{\partial g}{\partial x_i} \right)_{j^e \text{comp}} = 0$$

with $i = 1, \dots, k$ and $j = k + l, \dots, n$ (n is the dimension of X). From
these expressions $[\Delta, D] \subset D + \Delta_0$ readily follows. The converse statement is based on the same argument.

Now we are prepared to state the main theorem of this section. First
we have to give another definition.

Definition 4.21. (Local minimality). *Let* $\Sigma(X, W, B, f)$ *be a smooth system. Let* $\bar{x} \in X$. *Then* $\Sigma(X, W, B, f)$ *is called locally minimal (around* $\bar{x}$*) if when* D *and* E *are distributions (around* $\bar{x}$*) that satisfy conditions* I', II', III' *of Lemma 4.3, then* D *and* E *must be the zero distributions.*

It is readily seen from Definition 4.19 that minimality of $\Sigma(X, W, B, f)$ locally implies local minimality (locally every involutive distribution can be factored out).

Combining Lemma 4.3, Definition 4.20, and Lemma 4.4 we can state:

Theorem 4.11. $\Sigma(X, W, B, f) = (g, h)$ *is locally minimal if and only if the extended system* $\Sigma^e(X, W, B, f) = (g, h)$ *satisfies the condition that there exists no nonzero involutive distribution* D *on* B *such that*

$$(i) \quad [\Delta, D] \subset D + \Delta_0,$$
$$(ii) \quad D \subset \ker dh. \tag{4.71}$$

Then we can define the sequence $\{D^{mu}\}, \mu = 0, 1, 2, \ldots$ as follows:

$$D^0 = \ker dh,$$
$$D^\mu = D^{\mu-1} \cap \Delta^{-1}(\Delta_0 + D^{\mu-1}), \quad \mu = 1, 2, \ldots.$$

Then $\{D^\mu\}$, $\mu = 0, 1, 2, \ldots$, is a decreasing sequence of involutive distributions, and for some $k \geq \dim(\ker dh) D^k = D^\mu$ for all $\mu \geq k$. Then D^k is the maximal involutive distribution that satisfies

$$(i) \, [\Delta, D^k] \subset D^k + \Delta_0,$$
$$(ii) \, D^k \subset \ker dh.$$

From Theorem 4.11 it follows that $\Sigma(X, W, B, f)$ is locally minimal if and only if $D^k = 0$.

Observability. It is natural to suppose that our definition of minimality has something to do with controllability and observability. However, because the definition of a nonlinear system (4.61) also includes autonomous systems, (i.e., no inputs), minimality cannot be expected to imply, in general, some kind of controllability. In fact an autonomous linear system

$$\dot{x} = Ax, \quad y = Cx$$

is easily seen to be minimal if and only if (A, C) is observable. Moreover, it seems natural to define a notion of *observability* only in the case that the system (4.61) has at least a local input-output representation; i.e., we make the standing assumption that $(\partial h/\partial v)$ is injective (see Lemma 4.1). Therefore, locally we have as our system

$$\dot{x} = g(x, u), \quad y = \overline{h}(x, u) \tag{4.72}$$

for every possible input-output coordinatization (y, u) of W. For such an input-output system local minimality implies the following notion of observability, which we call local distinguishability.

Proposition 4.2. *Choose a local input-output parameterization as in (4.72). Then local minimality implies that the only involutive distribution E on X that satisfies*

(i) $[g(\cdot, u), E] \subset E$ for all u (E is invariant under $g(\cdot, u)$),

(ii) $E \subset \ker d_x h(\cdot, u)$ for all u ($d_x \overline{h}$ means differentiation with respect to x) is the zero distribution.

Proof: Let E be a distribution on X that satisfies (i) and (ii). Then we can lift E in a trivial way to a distribution D on B by requiring that the integral manifolds of D be contained in the sections $u = $ const. Then one can see that D satisfies $[\Delta, D] \subset D + \Delta_0$ and $D \subset \ker dh$. Hence $D = 0$ and $E = 0$.

Corollary 4.2. *Suppose there exists an input-output coordinatization*

$$\dot{x} = g(x, u), \quad y = \overline{h}(x). \tag{4.73}$$

Then minimality implies local weak observability.

Proof: As can be seen from Proposition 4.2, local minimality in this more restricted case implies that the only involutive distribution E on X that satisfies

$$i) \quad [g(\cdot, u), E] \subset E \quad \text{for all } u,$$

$$ii) \quad E \subset \ker d\overline{h}$$

is the zero distribution. It can be seen that the largest distribution that satisfies (i) and (ii) is given by the null space of the codistribution P generated by elements of the form

$$L_{g(\cdot, u^1)} L_{g(\cdot, u^2)} \cdots L_{g(\cdot, u^k)} d\overline{h}, \quad \text{with } u^j \text{ arbitrary.}$$

Because this distribution has to be zero, the codistribution P equals $T_x^* X$, in every $\in X$. This is, apart from singularities (which we don't want to consider), equivalent to local weak observability.

Moreover, let (4.73) be locally weakly observable. Then all feedback transformations $u \mapsto v = \alpha(x, u)$ that leave the form (4.73) invariant (i.e., y is only the function x) are exactly the output feedback transformations $u \mapsto v = \alpha(y, u)$. It can be easily seen in local coordinates that after such output feedback is applied the modified system is still locally weakly observable.

In Proposition 4.2 and its corollary we have shown that local minimality implies a notion of observability, which generalizes the usual notion of local weak observability. Now we define a much stronger notion. Let us denote the (defined only locally) vector field $\dot{x} = y(x, \overline{u})$ for fixed $\overline{u}$ by $g^{\overline{u}}$ and the function $\overline{h}(x, \overline{u})$ by $h^{\overline{u}}$ (with g and $\overline{h}$ as in (4.72)).

Definition 4.22. *Let $\Sigma(X, W, B, f) = (g, h)$ be a smooth nonlinear system. It is called strongly observable if for every possible input-output coordinatization (4.72) the autonomous system*

$$\dot{x} = g^{\overline{u}}(x), \quad y = h^{\overline{u}}(x) \tag{4.74}$$

with $\overline{u}$ constant is locally weakly observable, for all $\overline{u}$.

Proposition 4.3. *Consider the Pfaffian system constructed as follows:*

$$P = dh^{\overline{u}} + L_{g^{\overline{u}}} dh^{\overline{u}} + L_{g^{\overline{u}}}(L_{g^{\overline{u}}} dh^{\overline{u}}) + \cdots + L_{g^{\overline{u}}}^{n-1} dh^{\overline{u}},$$

with n the dimension of X and $L_{g^{\overline{u}}}$ the Lie derivative with respect to $g^{\overline{u}}$. As is well known, the condition that the Pfaffian system P as defined above satisfies the condition $P_x = T_x^* X$ for all $x \in X$ (the so-called observability rank condition) implies that the system

$$\dot{x} = g^{\overline{u}}(x), \quad y = h^{\overline{u}}(x)$$

is locally weakly observable. Hence, when the observability rank condition is satisfied for all u, the system is strongly observable.

We call the Pfaffian system P the observability codistribution.

Controllability. The aim of this section is to define a kind of controllability which is "dual" to the definition of local distinguishability (Proposition 4.2) and which we use in the following section. The notion of controllability we use is the so-called "strong accessibility."

Definition 4.23. *Let $\dot{x} = g(x, u)$ be a nonlinear system in local coordinates. Define $R(T, x_0)$ as the set of points reachable from x_0 in exactly time T; in other words,*

$$R(T, x_0) := \{x_1 \in X \mid \exists \text{ state trajectory } x(t) \text{ generated by } g$$
$$\text{such that } x(0) = x_0 \text{ and } x(T) = x_1\}.$$

We call the system *strongly accessible* if for all $x_0 \in X$, and for all $T > 0$ the set $R(T, x_0)$ has a nonempty interior.

For systems of the form (in local coordinates)

$$\dot{x} = f(x) + \sum_{i=1}^{m} u_i g_i(x) \qquad (4.75)$$

(i.e., affine systems) we can define A as the smallest Lie algebra that contains $\{g_1, \ldots, g_m\}$ and which is invariant under f (i.e., $[f, A] \subset A$). It is known that $A_x = T_x X$ for every $x \in X$ implies that the system (4.75) is strongly accessible. In fact, when the system is analytic, strong accessibility and the rank condition $A_x = T_x X$ for every $x \in X$, are equivalent. We call A the *controllability distribution* and the rank condition the *controllability rank condition*. Now it is clear that for affine systems (4.75) this kind of controllability is an elegant "dual" of local weak observability.

It is well known that the extended system (see Definition 4.20) is an affine system. Hence for this system we can apply the rank condition described above. This makes sense because the strong accessibility of $\Sigma(X, W, B, f)$ is very much related to the strong accessibility of $\Sigma^e(X, W, B, f)$, which can be seen from the following two propositions.

Proposition 4.4. *If $\Sigma^e(X, W, B, f = (g, h))$ is strongly accessible, then $\Sigma(X, W, B, f = (g, h))$ is strongly accessible as well.*

Proof: In local coordinates the dynamics of Σ^e and Σ are given by

$$I \quad \dot{x} = g(x, u) \quad (\Sigma),$$
$$II \quad \dot{x} = g(x, v) \quad (\Sigma^e),$$
$$\dot{v} = u.$$

It is easy to show that if for Σ^e one can steer to a point x_1 then the same is possible for Σ (even with an input that is smoother).

The converse is harder.

Proposition 4.5. *Let $\Sigma(X, W, B, f = (g, h))$ be strongly accessible. In addition if the fibers of B are connected, then $\Sigma^e(X, W, B, f = (g, h))$ is strongly accessible.*

Proof: Consider the same representation of Σ and Σ^e as in the proof of Proposition 4.4. Let $x_0 \in X$ and x_1 be in the (nonempty) interior of $R_\Sigma(x_0, T)$ (the reachable set of system Σ). Then it is possible to reach x_1 from x_0 by an input function $v(t)$ that cannot be generated by the differential equation $\dot{v} = u$. However, we know that the set of the v generated in this way is dense in L^2. (For this we certainly need that the fibers of B are connected.) Because we only have to prove that the interior of a set is nonempty, this makes no difference. Now it is obvious from the equations

$$\dot{x} = g(x, v), \quad \dot{v} = u$$

that if we can reach an open set in the x-part of the (extended) state, then it is surely possible in the (x, v)-state.

In this chapter, the problem of geometrical description of control systems is studied. The connection of the optimal control and Yang–Mills fields has been established. A geometric model of a controlled object as a dynamic information-transforming system is examined. A description of the information-transforming system within the framework of the geometric formalism is also proposed. We suppose that our approach can be applied to multiple agents. After a classification of the fiber bundle types of conflict and conflict-free maneuvers, a weighted energy can be proposed as the cost function to select the optimal one. Various local and global controllability and observability conditions are derived. For the general multiagent case, a convex optimization algorithm is proposed to find the optimal multilegged maneuvers. To completely characterize the optimal conflict-free maneuvers, many issues remain to be addressed. Possible directions of future research include the analysis of the proposed mathematical models in terms of its performance and its robustness with respect to uncertainty of the agents' positions and velocities, and a more realistic study for the agent dynamics.

6.   Notes and Sources

The concept of the unified geometrical control theory is taken from Butkovskiy's work (Butkovskiy and Samoilenko, 1990). The connection of the optimal control problem and Yang–Mills fields in this chapter have been established by Yatsenko (1985). The material contained in Section 4.4 and 4.5 is based on the papers of Van der Shaft (1987). A geometrical model of a neural network is discussed by Yatsenko and Rakitina (1994).

Chapter 5

SUPERCONDUCTING LEVITATION AND BILINEAR SYSTEMS

1. Introduction

The suspension of objects with no visible means of support is a fascinating phenomenon (Moon, 1992; Kozorez, Kolodeev and Kryukov, 1975; Yatsenko,1989; Hull, 2004, Kozorez, et al., 2006). To deprive objects of the effects of gravity is a dream common to generations of thinkers from Benjamin Franklin to Robert Goddard, and even to mystics of the East. This modern fascination with superconducting levitation stems from four singular technical and scientific achievements:(i) the creation of superconducting gravity meters; (ii) the creation of high-speed vehicles to carry people at 500 km/hr; (iii) the creation of a digitally controlled magnetic levitation turbo molecular pump; and (iv) the discovery of new superconducting materials.

The modern development of super high-speed transport systems, known as maglev, started in the late 1960s as a natural consequence of the development of low-temperature superconducting wire, the transistor and chip-based electronic control technology. Maglev provides high-speed running, safety, reliability, low environmental impact, and minimum maintenance. In the 1980s, maglev matured to the point where Japanese and German technologists were ready to market these new high-speed levitated machines.

At the same time, Paul C. W. Chu of the University of Houston and co-workers in 1987 discovered a new, higher-temperature superconductivity (HTS) in the non inter metallic compounds (nyttrium–barium–copper oxide). Those premature promises of superconducting materials have been tempered by the practical difficulties of development. First, bulk YBCO (yttrium–barium–copper oxide) was found to have a low

P.M. Pardalos, V. Yatsenko, *Optimization and Control
of Bilinear Systems*, doi: 10.1007/978-0-387-73669-3,
© Springer Science+Business Media, LLC 2008

current density, and early samples were found to be too brittle to fabricate into useful wire. Scientists are interested in YBCO because when it is cooled below around 90 Kelvin, which can be accomplished with liquid nitrogen, it becomes a superconductor. The two most important properties of YBCO are that it has no electrical resistance and that it expels a magnetic field.

However, from the very beginning, the hallmark of these new superconductors was their ability to levitate small magnets. This property, captured on the covers of both scientific and popular magazines, inspired a group of engineers and applied scientists to envision a new set of levitation applications based on superconducting magnetic bearings.

In the past few years, the original technical obstacles of YBCO have gradually been overcome, and new superconducting materials such as bismuth–strontium–calcium–copper oxide (BSCCO) have been discovered. Higher-current densities for practical applications have been achieved, and longer wire lengths have been produced with good superconducting properties. At this juncture of superconducting technology, we can now envisage, in the coming decade, the levitation of large machine components as well as the enhancement of existing maglev transportation systems with new high-temperature superconducting magnets.

A levitation phenomenon is created by opposing magnetic fluxes. Commonly it refers to levitated high-speed trains equipped with superconducting magnets, proposed by James R. Powell and Gordon T. Danby of Brookhaven National Laboratory in the late 1960s. It has been pursued since 1970 by the Japan Railway Technical Research Institute, which is presently building a second maglev test track 40-km long. In the 1980s demonstration maglevs were built in Germany. We can imagine the relative velocity of 100–200 m/sec between moving bodies with no contact, no wear, no need for fluid or gas intervention, and no need for active controls.

The superconductivity phenomenon was a significant step to improve suspensions. Most, but not all, conductors of electrical current, when cooled sufficiently in the direction of absolute zero, become superconductors. The superconducting state itself is one in which there is zero electrical resistance and perfect diamagnetism. Free suspension of a probe of a superconducting gravimeter is realized by the Braunbeck–Meisner phenomenon. Here we concentrate on a new highly sensitive cryogenic-optical sensor and a method of estimation of the gravitational perturbation acting on the levitated probe.

In this chapter we describe basic properties of a magnetic levitation, theoretical background, and control algorithms of a probe stability.

Bilinear control schemes of the static and dynamic types are proposed for the control of a magnetic levitation system. The proposed controllers guarantee the asymptotic regulation of the system states to their desired values. We also describe a simple superconducting gravity meter, its mathematical model, and design of nonlinear controllers that stabilize it at an equilibrium state. Furthermore, an accurate mathematical model of asymptotically stable estimation of a weak noisy signal using the stochastic measurement model is proposed.

2. Stability and Levitation

Levitation can be achieved using electric or magnetic forces or by using air pressure, although some purists would argue whether flying or hovering is levitation. However, the analogy of magnetic levitation with the suspension of aircraft provides insight into the essential requirements for levitation; that is, lift alone is not levitation. The success of the Wright machine in 1903 was based, in part, on the invention of a mechanism on the wings to achieve stable levitated flight. The same can be said about magnetic bearing design, namely, that an understanding of the nature of mechanical stability is crucial for the creation of a successful levitation device.

Simple notions of stability often use the paradigm of the ball in a potential well or on the top of a potential hill. This idea uses the concept of potential energy, which states that physical systems are stable when they are at their lowest energy level.

The minimum potential energy definition of stability is good to begin with, but is not enough in order to understand magnetic levitation. Not only must one consider the stability of the center of mass of the body, but it is also necessary to achieve the stability of the orientation or an angular position of the body. If the levitated body is deformable, the stability of the deformed shape may also be important.

The second difficulty with the analogy with particles in gravitational potential wells is that we have to define what we mean by the magnetic or electric potential energy (Kozorez and Cheborin, 1977; Bandurin, Zinovyev, and Kozorez, 1979). This is straightforward if the sources of the levitating magnetic or electric forces are fixed. But when magnetization or electric currents are induced due to changes in the position or orientation of our levitated body, then the static concept of stability using potential energy can involve pitfalls that can yield the wrong conclusion regarding the stability of the system.

To be really rigorous in magneto-mechanics, one must discuss stability in the context of dynamics. For example, in some systems one can have static instability but dynamic stability. This is especially true in the case of time-varying electric or magnetic fields as in the case of actively controlled magnetic bearings. However, it is also important when the forces (mechanical or magnetic) depend on generalized velocities.

In general, the use of concepts of dynamic stability in the presence of modeling error due to uncertainties, rooted in modern nonlinear dynamics, must be employed in order to obtain a robust position control of a magnetic levitation system. This theory not only requires the knowledge of how magnetic forces and torques change with position and orientation (i.e., magnetic stiffness), but also the knowledge of how these forces change with both linear and angular velocities.

Earnshaw's Theorem. It is said that a collection of point charges cannot be maintained in an equilibrium configuration solely by the electrostatic interaction of the charges. Early in the nineteenth century (1839) a British minister and natural philosopher, Samuel Earnshaw (1805–1888), examined this question and stated a fundamental proposition known as Earnshaw's theorem. The essence of this theorem is that a group of particles governed by inverse square law forces cannot be in stable equilibrium. The theorem naturally applies to charged particles and magnetic poles and dipoles. A modern statement of this theorem can be found in Jeans (Jeans, 1925; Brandt, 1989, 1990; Braunbeck, 1939: "A charged particle in the field of a fixed set of charges cannot rest in stable equilibrium"). This theorem can be extended to a set of magnets and fixed circuits with constant current sources. To the chagrin of many a would-be inventor, and contrary to the judgment of many patent officers or lawyers, the theorem rules out many clever magnetic levitation schemes. This is especially the case of levitation with a set of permanent magnets as any reader can verify. Equilibrium is possible, but stability is not.

Later on we address the question of how and why one can achieve stable levitation of a superconducting ring using an active feedback. However, here we try to motivate why superconducting systems appear to violate or escape the consequences of Earnshaw's theorem. One of the first to show how diamagnetic or superconducting materials could support stable levitation was Braunbeck (1939).

Earnshaw's theorem is based on the mathematics of inverse square force laws. Particles that experience such forces must obey a partial differential equation known as Laplace's equation. The solutions of this equation do not admit local minima or maxima, but only saddle-type equilibria. However, there are circumstances under which electric and magnetic systems can avoid the consequences of Earnshaw's theorem:

- Time-varying fields (e.g., eddy currents, alternating gradient)

- Active feedback

- Diamagnetic systems

- Ferrofluids

- Superconductors

The theorem is easily proved if the electric and magnetic sources are fixed in space and time, and one seeks to establish the stability of a single free-moving magnet or charged particle. However, in the presence of polarizable, magnetizable, or superconducting materials, the motion of the test body will induce changes in the electric and magnetic sources in the nearby bodies. In general, magnetic flux attractors such as ferromagnetic materials still obey Earnshaw's theorem, whereas for flux repellers such as diamagnetic or Type I superconductors, stability can sometimes be obtained. Superconductors, however, have several modes of stable levitation:

- Type I or Meissner repulsive levitation based on complete flux exclusion

- Type II repulsive levitation based on both partial flux exclusion and flux pinning

- Type III suspension levitation based on flux pinning forces

- Type IV suspension levitation based on magnetic potential well

In the case of Meissner repulsive levitation superconducting currents in the bowl-shaped object move in response to changes in the levitated magnet. The concave shape is required to achieve an energy potential well.

In the case of Type II levitation, both repulsive and suspension (or attractive) stable levitation forces are possible without shaping the superconductor. Magnetic flux exclusion produces equivalent magnetic pressures that result in repulsive levitation whereas flux attraction creates magnetic tensions (similar to ferromagnetic materials) which can support suspension levitation. Flux penetration into superconductors is different from ferromagnetic materials, however.

In Type III superconductors, vortexlike supercurrent structures in the material create paths for the flux lines. When the external sources of these flux lines move, however, these supercurrent vortices resist motion or are pinned in the superconducting material. This so-called *fluxpinning* is believed to be the source of stable levitation in these materials (Brandt, 1989, 1990).

Type IV suspension levitation is described in detail in the next section.

Finally, from a fundamental point of view, it is not completely understood why supercurrent-based magnetic forces can produce stable attractive levitation whereas spin-based magnetic forces in ferromagnetic materials produce unstable attractive or suspension levitation. Given the restricted assumptions upon which Earnshaw's theorem is based, the possibility that some new magnetic material will be discovered, which supports stable levitation, cannot be entirely ruled out.

3. Dynamics of Magnetically Levitated Systems

This section considers mathematical models of a sensor based on the principle of magnetic levitation (Kozorez and Cheborin, 1977; Hull, 2004). The sensor consists of two superconducting current rings and a levitated probe placed between them (see Section 5.1). The stability is provided by a set of superconducting short-circuited loops placed around the floating ring. A novel method using short-circuited superconducting loops as stabilizers has been proposed. We showed that for a given magnetic configuration there exists a minimum current in the levitated ring below which the system is unstable.

The newly developed superconducting gravimeter represents a spring-type device. An analogue of the mechanical spring of our device accomplishes the magnetic returning force acting on a superconducting probe in a nonuniform magnetic field of superconducting rings or a permanent magnet (in another variant). Due to the high stability of superconducting currents of rings a highly stable nondissipative spring is created.

As shown by White (1959), and Abraham and Marsden (1978), a set of variables uniquely defining an energy state can be determined for any electromechanical system possessing a power function and storing energy in the form of magnetic field energy. In this case such variables will be mechanical displacements (mechanical degrees of freedom) q_j, $j = 1, 2, \ldots, l$ (l is the number of degrees of freedom), as well as total magnetic fluxes Ψ_m and currents I_m, $m = 1, 2, \ldots, n$ (n is the number of superconducting rings).

There are inner couplings between magnetic variables

$$\Psi_m = \sum_{i=1}^{n} L_{im} I_i, \quad m = 1, 2, \ldots, n, \tag{5.1}$$

where L_{im} are mutual inductances, and L_{ii} are internal inductances. In the case of superconducting current rings, magnetic-flux linkages Ψ_m retain constant value independently of variations of a ring position. This circumstance allows us to consider the relations (5.1) as a system of n equations for currents I_i, $i = 1, 2, \ldots, n$, where Ψ_m ($m = 1, 2, \ldots, n$) are constants.

We assume that the determinant Δ of (5.1) is not equal to zero. Then it can be solved for currents

$$I_m = \frac{\Delta_m}{\Delta}, \quad m = 1, 2, \ldots, n, \tag{5.2}$$

where Δ_m is the determinant of the current I_m.

If we place the solution of (5.2) into the following formula for the energy of magnetic field of the current loop system

$$W = \frac{1}{2} \sum_{i,m=1}^{n} L_{im} I_i I_m, \tag{5.3}$$

then the energy will be expressed in terms of magnetic-flux linkages Ψ_i, Ψ_m, and the inductance L_{im},

$$W = \frac{1}{2} \Delta^{-2} \sum_{i,m=1}^{n} L_{im} \Delta_i(\Psi_i, \Psi_m, L_{im}) \Delta_m(\Psi_i, \Psi_m, L_{im}), \tag{5.4}$$

$$\Delta = \Delta(\Psi_i, \Psi_m, L_{im}), \quad i, m = 1, 2, \ldots, n. \tag{5.5}$$

It follows from (5.4) that the energy W depends only on the mechanical coordinates q_j, which are incorporated in mutual inductances L_{im} ($i \neq m$). Because of this, in "pure mechanical" terms it is either a power function or potential energy. The formula for magnetic force (White, 1959) prompts precisely which mechanical function will be the energy of the magnetic field. This formula appears as

$$\Delta = \Delta(\Psi_i, \Psi_m, L_{im}), \quad i, m = 1, 2, \ldots, n;$$

that is, the magnetic force is a partial derivative with respect to magnetic energy expressed in terms of magnetic-flux linkages and coordinates, taken with opposite sign.

But this is precisely the definition of force as a function of potential energy of any power field. Therefore, energy of a magnetic field in form (5.4), where $\Psi_i, \Psi_m = \text{const}$, is the potential energy of the magnetic interaction of n ideal currents; that is,

$$W = U_m = U_m(L_{im}(q_j)), \quad i, m = 1, 2, \ldots, n. \tag{5.6}$$

If the system is located in an external power field, for example, in gravitational one, gravitational energy U_G should be added to the magnetic potential energy. Then the total energy of the system is

$$U = U_m + U_G. \tag{5.7}$$

For circuit system's inner linkages between magnetic variables take the form

$$\Psi_1 = LI_1 + L_{12}I_2 + L_{13}I_3 + L_{14}I_4,$$
$$\Psi_2 = L_{12}I_1 + LI_2 + L_{23}I_3 + L_{24}I_4,$$
$$\Psi_3 = L_{13}I_1 + L_{23}I_2 + LI_3 + L_{24}I_4,$$
$$\Psi_4 = L_{14}I_1 + L_{24}I_2 + L_{34}I_3 + LI_4, \tag{5.8}$$

and energy of the magnetic field (5.3) with respect to (5.8) can be written as

$$W = U_m = \frac{1}{2}(LI_1^2 + L_{12}I_1I_2 + L_{13}I_1I_3 + L_{14}I_1I_4 + L_{12}I_1I_2$$
$$+ LI_2^2 + L_{23}I_2I_3 + L_{24}I_2I_4 + L_{13}I_1I_3 + L_{23}I_2I_3L_{14}I_1I_4$$
$$+ LI_3^2 + L_{34}I_3I_4 + L_{24}I_2I_4 + L_{34}I_3I_4 + LI_4^2)$$
$$= \frac{1}{2}(\Psi_1 I_1 + \Psi_2 I_2 + \Psi_2 I_3 + \Psi_1 I_4). \tag{5.9}$$

All the coils of the sensor are modeled by thin short-circuited ring-shaped loops of similar radius, therefore internal inductances of the loops are $L_1 = L_{22} = L_{33} = L_{44} = L$.

By solving the system of equations (5.8), we find expressions for currents and substitute them into (5.9) thus defining dependence of magnetic potential energy on mechanical coordinates:

$$U_m = \Psi_1^2 (2L)^{-1}\{2(1 - y_{14})(1 - y_{23}^2) - (y_{13} - y_{24})^2 - (y_{13} - y_{34})^2$$
$$+ 2y_{23}(y_{12} - y_{24})(y_{13} - y_{34})$$
$$+ 2p[(y_{12}y_{34} - y_{13}y_{24})(y_{12} - y_{13} - y_{24} + y_{34})$$
$$- (1 - y_{14})(1 - y_{23})(y_{12} + y_{13} + y_{24} + y_{34})] + p^2[2(1 - y_{14}^2)(1 - y_{23})$$
$$- (y_{12} - y_{13})^2 - (y_{24} - y_{34})^2 + 2y_{14}(y_{12} - y_{13})(y_{24} - y_{34})]\}$$
$$\times [(1 - y_{14}^2)(1 - y_{23}^2) - y_{12}^2 - y_{13}^2 - y_{24}^2 - y_{34}^2$$
$$+ (y_{12}y_{34} - y_{13}y_{24})^2 + 2y_{14}(y_{12}y_{24} - y_{13}y_{34})$$
$$+ 2y_{23}(y_{12}y_{13} - y_{24}y_{34}) - 2y_{14}y_{23}(y_{12}y_{34} - y_{13}y_{24})]^{-1}$$
$$= \Psi_1^2 (2L)^{-1}(M + Np + Qp^2)D^{-1}, \tag{5.10}$$

where $y_{im} = L_{im}L^{-1}$; $p = \Psi_2\Psi_1^{-1}$, and relative mutual inductances $y_{im} = y_{im}(q_1, \ldots, q_0)$ are functions of coordinates.

In order to define the explicit relation $y_{im}(q)$ we introduce the inertial coordinate system $O\xi\eta\zeta$, whose $O\eta$ axis coincides with the axis of stationary loops 1, 4 of the sensor; i_1, i_2, i_3 are basis vectors of the system $O\xi\eta\zeta$. We place in the center of mass of the sensor the origin of the coordinate system associated with it, with basis vectors i_{11}, i_{12}, i_{13}, and with its O_1, ζ_1 axis coinciding with the axis of the loops 2, 3. We describe the position of the center of mass of the probe by cylindrical coordinates ρ_2, α, ζ and orientation of trihedron $O_1\xi_1\eta_1\zeta_1$ with respect to system $O\xi\eta\zeta$ is described by Euler angles (v is a nutation angle, ψ is a precession angle, and φ is a proper rotation angle).

As is seen from formula (5.10), the potential energy depends on all six mutual inductances y_{im}, but y_{14}, $y_{23} = \text{const}$. Therefore, only four inductances y_{12}, y_{13}, y_{24}, y_{34} are to be determined. All of them are calculated in a similar way.

Let us define the following notation. $\mathbf{R}_i$ ($i = 1, 2$) are radius-vectors of centers of rings' mass in the system $O\xi\eta\zeta$; dl_1, dl_2 are elements of arcs of rings 1, 2; $\mathbf{R}_{12}$ is a radius-vector connecting the center of the ith Ring with the respective element dl_i; $\mathbf{e}$ is a radius-vector of the center of the ring 2 in the system $O_1\xi_1\eta_1\zeta_1$. Then the mutual inductance can be calculated by the Neumann formula

$$y_{12} = \frac{L_{12}}{L} = \frac{1}{20\pi} \oint \oint \frac{dl_1 dl_2}{|\mathbf{R}_{12}|}, \tag{5.11}$$

$$\mathbf{R}_{12} = \mathbf{R}_2 + \mathbf{e} + \mathbf{a}_2 - \mathbf{a}_1 - \mathbf{R}_1, \tag{5.12}$$

where

$$\mathbf{R}_i = \rho_i \cos \alpha i_1 + \rho_i \sin \alpha i_2 + \zeta_i i_3,$$

$$\mathbf{a}_i = a(\cos \lambda_i i_{11} + \sin \lambda_i i_{12}), \quad i = 1, 2,$$

$$\mathbf{e} = e i_{13}, \quad dl_i = a(\sin \lambda_i i_{i1} - \cos \lambda_i i_{i2}) d\lambda_i.$$

Because Ring 1 is fixed and the coordinate system $O\xi\eta\zeta$ is selected such that its axis $O\eta$ coincides with the axis of Ring 1, all coordinates describing the position of Ring 1 have the following values:

$$\rho'\rho_i, \quad \alpha_1 = \frac{\pi}{2}, \quad \zeta_1 = 0, \quad \vartheta_1 = \frac{\pi}{2}, \quad \Psi_1 = 0, \quad \phi_1 = 0. \tag{5.13}$$

Then for fixed Ring 1

$$\mathbf{R}_1 = \rho_1 i_2,$$

$$\mathbf{a}_1 = a(\cos \lambda_1 i_1 + \sin \lambda_1 i_3),$$

$$dl_1 = a(\sin \lambda_1 i_1 - \cos \lambda_1 i_3) d\lambda_1,$$

and for SE:

$$\mathbf{R}_2 = \rho_2 \cos \alpha i_1 + \rho_2 \sin \alpha i_2 + \zeta i_3,$$

$$\mathbf{a}_z = \{[\cos(\lambda_2 + \lambda) \cos \Psi - \sin(\lambda_2 + \phi) \sin \Psi$$
$$- \sin(\lambda_2 + \phi) \sin \Psi \cos \vartheta]i_2 + \sin(\lambda_2 + \phi) \sin \vartheta i_3\},$$

$$\mathbf{e} = e \sin \phi \sin \vartheta i_1 - e \cos \phi \sin \vartheta i_2 + e \cos \vartheta i_3,$$

$$dl_2 = a\{[\sin(\lambda_2 + \phi) \cos \Psi + \cos(\lambda_2 + \phi) \sin \Psi \cos \theta]i_1$$
$$- [\sin(\lambda_2 + \phi) \sin \Psi - \cos(\lambda_2 + \phi) \sin \Psi \cos \vartheta]i_2$$
$$- \cos(\lambda_2 + \phi) \sin \vartheta i_3\}.$$

By performing elementary transformations, we obtain

$$y_{12} = \frac{1}{40\pi} \int_0^{2\pi} d\lambda_1 \int_0^{2\pi} [\sin x_4 \cos \lambda_1 \cos(\lambda_2 + x_6)$$
$$+ \sin \lambda_1 \sin(\lambda_2 + x_6) + \cos x_4 \sin x_5 \sin \lambda_1 \cos(\lambda_2 + x_6)]$$
$$\times \left\{ \frac{1}{2} + e^2 + \rho_1^2 + x_1^2 + x_3^2 + \frac{1}{2}[\cos x_4 \sin x_5 \right.$$
$$- \cos \lambda_1 \sin(\lambda_2 + \varphi_6) - \cos x_5 \cos \lambda_1 \cos(\lambda_2 + x_6)]$$
$$- 2\rho_1 x_1 \sin x_2 + x_1[\cos(x_2 - x_5) \cos(\lambda_2 + x_6)$$
$$+ \sin(x_2 - x_5) \cos x_4 \sin(\lambda_2 + x_6) - \cos x_2 \cos \lambda_1]$$
$$- \rho_1[\sin x_5 \cos(\lambda_2 + x_6) + \cos x_4 \cos x_5 \sin(\lambda_2 + x_6)]$$
$$- e(\sin x_4 \sin x_5 \cos \lambda_1 + \cos x_4 \sin \lambda_1)$$
$$- 2e x_1 \sin(x_2 - x_5) \sin x_4 + 2e\rho_1 \sin x_4 \cos x_5$$
$$\left. - x_3[\sin \lambda_1 - \sin x_4(\lambda_2 + x_6)] + 2e x_3 \cos x_4 \right\}^{-1/2} d\lambda_2,$$

where dimensionless variables are introduced

$$x_1 = \frac{\rho_2}{2a}; \quad x_2 = \alpha; \quad x_3 = \frac{\zeta}{2a}; \quad x_1 = v; \quad x_5 = \psi; \quad x_6 = \varphi. \quad (5.14)$$

Thus, the collection of formulae (5.10) and (5.11) determines dependence of magnetic potential energy on coordinates of SE, and total potential energy

$$U = U_m - mg\rho \qquad (5.15)$$

provided that direction of a gravitational force coincides with the direction of the $O\eta$ axis.

The integrals in the formula (5.11) are not taken in general form. Only in the case where the axes of fixed loops coincide, can the integral relationship be reduced to linear combinations of complete elliptic integrals. In our case, where magnetic forces are large as compared with perturbing ones, the potential energy can be expanded into the following power series

$$U = U_0 + \sum_{j=1}^{6} \left(\frac{PU}{Pq} \right) \bigg|_0 (q_j - q_{j0})$$ (5.16)

$$+ \frac{1}{2} \sum_{j,n=1}^{6} \frac{\partial^2 U}{\partial q_j \partial q_n} \bigg|_0 (q_j - q_{j0})(q_n - q_{n0}),$$

where derivatives are calculated at the point q_{j0}:

$$x_{10} = x_{10}; \quad x_{20} = \frac{\pi}{2}; \quad x_{30} = 0; \quad x_{40} = \frac{\pi}{2}; \quad x_{50} = x_{60} = 0. \quad (5.17)$$

After simple manipulations the final expression for potential energy can be rewritten as

$$U = \sum_{\substack{i,m=1 \\ i \neq m}} \left(\frac{\partial U_m}{\partial y_{im}} \frac{\partial y_{im}}{\partial x_1} \bigg|_0 - mg \right) (x_2 - x_{10})$$

$$+ \frac{1}{2} \sum_{j,n=1}^{5} \sum_{\substack{r,s=1 \\ r \neq s}}^{4} \sum_{\substack{i,m=1 \\ i \neq m}}^{4} \left(\frac{\partial^2 U_m}{\partial y_{im} \partial y_{rs}} \frac{\partial y_{im}}{\partial x_j} \frac{\partial y_{rs}}{\partial x_n} + \frac{\partial U_m}{\partial y_{im}} \frac{\partial^2 y_{im}}{\partial x_j \partial x_n} \right) \bigg|_0$$

$$\times (x_j - x_{j0})(x_n - x_{n0}),$$ (5.18)

where

$$\frac{\partial U_m}{\partial y_{im}} = \frac{\Psi_1}{2L} [(M_{im} + N_{im}p + Q_{im}p^2)D - (M + Np + Qp^2)D_{im}]D^{-2};$$

$$\frac{\partial U_m^2}{\partial y_{im} \partial y_{rs}} = \frac{\Psi_1}{2L} [(M_{im,rs}D^2 - M_{im}DD_{rs} - M_{rs}DD_{im} - MDD_{im,rs}$$

$$+ 2MD_{im}D_{rs}) + p(N_{im}DD_{rs} - N_{rs}DD_{im}ND D_{im,rs} + 2ND_{im}D_{rs})$$

$$+ p^2(Q_{im,rs}D^2 - Q_{im}DD_{rs} - Q_{rs}DD_{im} - QDD_{im,rs} + 2QD_{im}D_{rs})]D^{-3}$$

(expressions for M, N, Q, D are clear from formula (5.10), symbol M_{im} denotes the derivative of M with respect to y_{im});

$$M_{12} = -M_{24} = [y_{12} - y_{24} - y_{23}(y_{13} - y_{34})];$$
$$M_{13} = -M_{34} = [y_{13} - y_{34} - y_{23}(y_{12} - y_{24})];$$
$$N_{12} = 2[y_{34}(y_{12} - y_{13} - y_{24} + y_{34}) + (y_{12}y_{34} - y_{13}y_{24})$$
$$- (1 - y_{14})(1 - y_{23})];$$
$$N_{13} = -2[y_{24}(y_{12} - y_{13} - y_{24} + y_{34}) + (y_{12}y_{34} - y_{13}y_{24})$$
$$+ (1 - y_{14})(1 - y_{23})];$$
$$N_{24} = -2[y_{13}(y_{12} - y_{13} - y_{24} + y_{34}) + (y_{12}y_{34} - y_{13}y_{24})$$
$$+ (1 - y_{14})(1 - y_{23})];$$
$$N_{34} = 2[y_{12}(y_{12} - y_{13} - y_{24} + y_{34}) - (y_{12}y_{34} - y_{13}y_{24})$$
$$+ (1 - y_{14})(1 - y_{23})];$$
$$Q_{12} = -Q_{13} = [y_{12} - y_{13} - y_{14}(y_{24} - y_{34})];$$
$$Q_{24} = -Q_{34} = [y_{24} - y_{34} - y_{14}(y_{12} - y_{13})];$$
$$D_{12} = -2[y_{12} + y_{34}(y_{12}y_{34} - y_{14}y_{24}) - y_{14}y_{24} - y_{13}y_{23} + y_{14}y_{23}y_{34}];$$
$$D_{13} = -2[y_{13} + y_2(y_{12}y_{34} - y_{13}y_{24}) - y_{14}y_{34} - y_{12}y_{23} + y_{14}y_{23}y_{24}];$$
$$D_{24} = -2[y_{24} + y_{13}(y_{12}y_{34} - y_{13}y_{24}) - y_{12}y_{14} - y_{23}y_{34} + y_{13}y_{14}y_{23}];$$
$$D_{34} = -2[y_{34} + y_{12}(y_{12}y_{34} - y_{13}y_{24}) - y_{13}y_{14} - y_{23}y_{24} + y_{12}y_{14}y_{23}];$$
$$M_{12,12} = M_{13,13} = M_{34,34} = Q_{12,12} = Q_{13,13} = Q_{24,24} = Q_{34,34} = -2;$$
$$N_{12,12} = 4y_{34}; \quad N_{13,13} = 4y_{24}; \quad N_{24,24} = 4y_{13}; \quad N_{34,34} = 4y_{12};$$
$$D_{12,12} = -2(1 - y_{34}^2); \quad D_{13,13} = -2(1 - y_{24}^2);$$
$$D_{24,24} = -2(1 - y_{13}^2); \quad D_{34,34} = -2(1 - y_{12}^2);$$
$$M_{12,13} = 2y_{14}; \quad M_{12,24} = 2;$$
$$N_{12,13} = -2(y_{24} + y_{34}); \quad N_{12,24} = -(y_{13} + y_{34});$$
$$Q_{12,13} = 2; \quad Q_{12,24} = 2y_{23};$$
$$D_{12,13} = 2(y_{14} - y_{24}y_{34}); \quad D_{12,24} = 2(y_{23} - y_{13}y_{34});$$
$$M_{12,34} = -2y_{24}; \quad M_{13,24} = -2y_{14};$$
$$N_{12,34} = 2[2(y_{12} + y_{34}) - y_{13} - y_{24}];$$
$$N_{13,24} = 2[2(y_{13} + y_{24}) - y_{12}y_{34}];$$
$$Q_{12,34} = 2y_{23}; \quad Q_{13,24} = -2y_{23};$$
$$D_{12,34} = 2(2y_{12}y_{34} - y_{13}y_{24} - y_{14}y_{23});$$
$$D_{13,24} = 2(2y_{13}y_{24} - y_{12}y_{34} - y_{14}y_{23});$$
$$M_{13,34} = 2; \quad M_{24,34} = 2y_{14};$$

$$N_{13,34} = -2(y_{12} + y_{24}); \quad N_{24,34} = -2(y_{12} + y_{13});$$
$$Q_{13,34} = 2y_{23}; \quad Q_{24,34} = 2;$$
$$D_{13,34} = 2(y_{23} - y_{12}y_{24}); \quad D_{24,34} = 2(y_{14} - y_{12}y_{13});$$
$$y_{14} = y_{23} = \text{const}; \quad y_{im} = \frac{1}{5k}[(1 + k'^2)\mathbf{K}(k) - 2\mathbf{E}(k)];$$
$$\frac{\partial y_{im}}{\partial x_1} = \frac{1}{5b}[2k'^2\mathbf{K}(k) - (2 - k^2)\mathbf{E}(k)];$$
$$\frac{\partial^2 y_{im}}{\partial x_1^2} = -2\frac{\partial^2 y_{im}}{\partial x_3^2} = \frac{k^3}{5k'^2}[1];$$
$$\frac{\partial^2 y_{im}}{\partial x_2^2} = -\frac{1}{10k'^2}(k\rho_1 + kd + b)\{b[2] + k^2(\rho_1 + d)[1]\};$$
$$\frac{\partial^2 y_{im}}{\partial x_4^2} = \frac{\partial^2 y_{im}}{\partial x_5^2} = \frac{k}{40k'k}\{k'^2[(2 - 3k^2 + 2k^4)\mathbf{E}(k)$$
$$- k'^2(2 - k^2)\mathbf{K}(k)] - 4kd(d + b)[1]\};$$
$$\frac{\partial^2 y_{im}}{\partial x_3 \partial x_4} = \frac{1}{20k'^2}\{b[2] + 2k^3d[1]\};$$
$$\frac{\partial^2 y_{im}}{\partial x_2 \partial x_5} = \frac{1}{20k'^2k}(k\rho_1 + kd + b)\{b[2] + 2k^3d[1]\};$$
$$[1] = [k'^2\mathbf{K}(k) - (1 - 2k^2)\mathbf{E}(k)];$$
$$[2] = [k'^2(4 + k^2)\mathbf{K}(k) - (4 - k^2 - 2k^4)\mathbf{E}(k)];$$

$$\text{at } im = 12 \quad b = k', \quad d = e, \quad \rho_1 = 0;$$
$$\text{at } im = 13 \quad b = k', \quad d = -e, \quad \rho_1 = 0;$$
$$\text{at } im = 24 \quad b = -k', \quad d = e, \quad \rho_1 = 2h;$$
$$\text{at } im = 34 \quad b = -k', \quad d = -e, \quad \rho_1 = 2h;$$

$\mathbf{K}(k)$, $\mathbf{E}(k)$ are complete elliptical integrals of the first and second kind of the absolute value of k_{im}, and $k_{im}^2 = [1 + (x_1 - \rho_1 - d)^2]^{-1}$; $k'^2 = 1 - k^2$.

The zero term $U_m(q_0)$ is omitted in decomposition (5.16) because the potential energy is determined accurately to a constant, and equality to zero of the coefficient is the necessary condition of equilibrium of $(x_1 - x_{10})$, the system with a gravitational force. Using the following condition

$$\frac{\partial U_m}{\partial y_{12}}\frac{\partial x_{12}}{\partial x_1}\bigg|_0 + \frac{\partial U_m}{\partial y_{13}}\frac{\partial y_{13}}{\partial y_1}\bigg|_0 + \frac{\partial U_m}{\partial y_{25}}\frac{\partial x_{24}}{\partial y_1}\bigg|_0 + \frac{\partial U_m}{\partial y_{34}}\frac{\partial y_{34}}{\partial x_1}\bigg|_0 = mg \quad (5.19)$$

we can find the value of x_{10} at which gravitational force of SE is balanced by magnetic interaction forces, and which is placed into expression (5.16).

In order to obtain the dynamic equations of the sensor, we use the results discussed above and the formula for kinetic energy of a free body in the form (5.16):

$$T = \frac{1}{2}m(\dot{\rho}^2 + \rho^2\dot{\alpha}^2 + \dot{\zeta}^2) + \frac{1}{2}A(\dot{v}\sin\varphi - \dot{\psi}\sin v\cos\varphi)^2$$
$$+ \frac{1}{2}B(\dot{v}\cos\varphi + \dot{\psi}\sin v\sin\varphi)^2 + \frac{1}{2}C(\dot{\varphi} + \dot{\psi}\cos v)^2. \tag{5.20}$$

Let us suppose that the principal moments of inertia of the sensor with respect to axes rigidly bound to coordinate system A, B, C are equal to $4ma^2$ and let us go over to dimensionless coordinates $x_1, \ldots, x_6$ and dimensionless time $\tau = t\omega$ by introducing characteristic frequency of sensor oscillations ω. Then dimensionless kinetic energy $\tilde{T}$ will be

$$2\tilde{T} = 2T(4ma^2\omega^2)^{-1} = \dot{x}_1^2 + x_1^2\dot{x}_2^2 + \dot{x}_4^2 + \dot{x}_5^2 + \dot{x}_6^2 + 2\dot{x}_5\dot{x}_6\cos x_4. \tag{5.21}$$

Applying Lagrange equations of the first kind

$$\frac{d}{dt}\frac{\partial L}{\partial q_i} = \frac{\partial L}{\partial q_i}, \tag{5.22}$$

where $L = T - U$ is a Lagrange function, we obtain the required dynamic equations of the sensor:

$$\ddot{x}_1 = x_1\dot{x}_2^2 - \gamma\frac{\partial U}{\partial x_1} - x_1\dot{x}_2^2 - \gamma\frac{\partial^2 U}{\partial x_1^2}\bigg|_0 (x_1 - x_{10});$$

$$\ddot{x}_2 = -x_1^{-2}\left[2x_1\dot{x}_1\dot{x}_2\,\gamma\frac{\partial^2 U}{\partial x_2^2}\bigg|_0\left(x_2 - \frac{\pi}{2}\right) + \gamma\frac{\partial^2 U}{\partial x_2\partial x_3}\bigg|_0 x_5\right];$$

$$\dot{x}_3 = -\gamma\frac{\partial^2 U}{\partial x_3^2}\bigg|_0 x_3 - \gamma\frac{\partial^2 U}{\partial x_3\partial x_4}\bigg|_0\left(x_4 - \frac{\pi}{2}\right);$$

$$\ddot{x}_4 = -\dot{x}_5\dot{x}_6\sin x_4 - \gamma\frac{\partial^2 U}{\partial x_4^2}\bigg|_0\left(x_4 - \frac{\pi}{2}\right) - \gamma\frac{\partial^2 U}{\partial x_3\partial x_4}\bigg|_0 x_3;$$

$$\ddot{x}_5 = -\sin^{-2} x_4 \left[\dot{x}_4 \dot{x}_5 \sin x_4 \cos x_4 - \dot{x}_4 \dot{x}_6 \sin x_4 + \gamma \frac{\partial^2 U}{\partial x_5^2}\bigg|_0 x_5 \right.$$

$$\left. + \gamma \frac{\partial^2 U}{\partial x_2 \partial x_5}\bigg|_0 \left(x_2 - \frac{\pi}{2} \right) \right];$$

$$\ddot{x}_6 = -\sin^{-2} x_4 \left[\dot{x}_4 \dot{x}_6 \sin x_4 \cos x_4 - \dot{x}_4 \dot{x}_5 \sin x_4 + \gamma \frac{\partial^2 U}{\partial x_5^2}\bigg|_0 x_5 \cos x_4 \right.$$

$$\left. + \gamma \frac{\partial^2 U}{\partial x_2 \partial x_5}\bigg|_0 \left(x_2 - \frac{\pi}{2} \right) \cos x_4 \right]. \tag{5.23}$$

Here the numeric value $\gamma = \Psi_1^2 (8Lma^2\omega^2)^{-1}$ is determined from condition

$$\gamma \sum_{\substack{i,m=1 \\ i \neq m}} \frac{\partial U_m}{\partial y_{im}} \frac{\partial y_{im}}{\partial x_1}\bigg|_0 = \frac{g}{2a\omega^2}. \tag{5.24}$$

Thus, the expressions for potential and kinetic energies and differential equations of motion of SE in the form convenient for numerical analysis of stability and dynamics of gravity-inertial devices are obtained.

4. Controlled Levitation and Bilinear Dynamics

The estimation of the signal acting on a macroscopic probe was the subject of numerous studies (Menskii, 1983; Vick, 1970; Braginskii, 1970). So far, however, no work has been done on estimation of a limiting weak noisy signal with unknown parameters acting on a probe in a controlled potential well. This section is intended to fill this gap to a certain extent. We propose a more accurate mathematical model of asymptotically stable estimation of a limiting weak noisy signal using the stochastic measurement model first proposed by Yatsenko (1989). Here we describe magnetic levitation of a superconducting probe in a chaotic state. We propose a sensor for optimal estimation of a signal acting on a levitated body.

4.1 Statement of the Problem

Consider the mathematical model of the sensitivity element of a gravity-inertial device

$$\dot{y} = f(y, u) = f_0 + \sum_{i=1}^{6} f_i(y) u_i,$$

$$z_1 = \frac{1}{T} \int_0^T \frac{1}{c} \left(\int_0^t g(y, \tau) d\tau \right) dt, \tag{5.25}$$

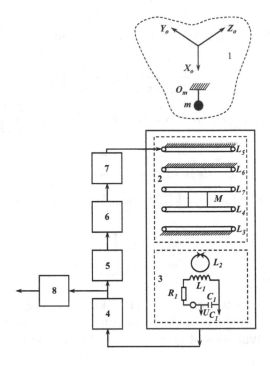

Figure 5.1.

where

$$
f_0 = \begin{bmatrix} y_2 \\ a_11y_1 + a_{12}y_5 + a_{13}y_5^2 + a_{14}y_5y_1 + a_{15}y_5y_1 \\ y_4 \\ a_{21}y_3 + a_{22}y_4 + a_{23}y_5 + a_{24}y_4 + a_{26}\sin(k_1 + k_2y_3 + k_3y_5) \\ y_6 \\ a_{31}y_5 + a_{32}y_6 + a_{33}y_1 + a_{34}y_4 + a_{36}\sin(k_1 + k_2y_3 + k_3y_5) \end{bmatrix} ;
$$

$$
f_0(y) = \begin{bmatrix} 0 \\ a_{17} + a_{18}y_1 \\ 0 \\ 0 \\ 0 \\ 0 \end{bmatrix} ; \quad
f_2(y) = \begin{bmatrix} 0 \\ a_{19} + a_{110}y_1 \\ 0 \\ 0 \\ 0 \\ 0 \end{bmatrix} ; \quad
f_3 = \begin{bmatrix} 0 \\ a_{16} \\ 0 \\ 0 \\ 0 \\ 0 \end{bmatrix} ;
$$

$$
f_4 = \begin{bmatrix} 0 \\ a_{111} \\ 0 \\ 0 \\ 0 \\ 0 \end{bmatrix} ; \quad
f_5 = \begin{bmatrix} 0 \\ 0 \\ 0 \\ a_{25} \\ 0 \\ a_{35} \end{bmatrix} ; \quad
f_6 = \begin{bmatrix} 0 \\ 0 \\ 0 \\ a_{26} \\ 0 \\ a_{37} \end{bmatrix} ;
$$

$y = (y_1, \ldots, y_6)$ is the state vector, $y(0) = 0$; $u_1(t)$ is the scalar control; $u_2(t) = u_1^2(t)$; $u_3(t) = r(t) + s(t)$ is the additive mixture of the signal and gravitational noise acting on the probe; $u_4(t)$ is a perturbation represented by a stationary stochastic process; $u_5(t)$ is δ-correlated noise; $u_6(t) = A \sin \omega t$ is a deterministic function; $(a_{1j}, \; j = 1, \ldots, 6)$, k_1, k_2, k_3, c, T are constant parameters; $(a_{ij}, \; i = 2, 3, \; j = 1, \ldots, 6)$ are parameters continuously dependent on y_1; z_1 is the one-dimensional model output.

Equation (5.25) describes a controlled high-sensitivity measuring instrument utilizing some nonlinear effects, such as the Josephson effect or the magnetic potential well effect (for more details, see Likharev and Ul'rikh, 1978 and Barone and Paterno, 1982). Investigation of the properties of the system (5.25) is therefore relevant both for the construction of efficient estimation algorithms and for the solution of control problems.

It is required to construct a perturbation-invariant mathematical model of asymptotically stable estimation of the signal $r(t)$ given the observations z_1.

The problem is solved in several stages.

1. Synthesis of a control algorithm u_1 that ensures asymptotic stability of the unperturbed motion $y_1 = 0$, $y_2 = 0$

2. Optimization of a magnetic levited system

3. Synthesis of an adaptive filtering algorithm

4. Numerical analysis of a mathematical estimation model

4.2   Optimal Synthesis of Chaotic Dynamics

In the optimization methods described by Yatsenko (2003) sensor dynamics throughout the state space are represented either with a single set of coupled maps

$$y_i(n + 1) - f_i[y(n), \gamma, a], \quad i - 1, \ldots, N, \quad y \subset \mathbb{R}^N \qquad (5.26)$$

or a set of ordinary differential equations

$$\dot{y}_i(t) = f_i[y(t), \gamma, a], \quad i = 1, \ldots, N, \quad y \in \mathbb{R}^N, \qquad (5.27)$$

where $y \in \Gamma \subset \mathbb{R}^n$ is a state vector; $a \in \mathbb{R}^n$ is a parameter vector; γ represents a noise term. If $\gamma(t) = 0$ and $a = \text{const}$ equation (5.27) defines a deterministic dynamical system. The time series of sensor measurement is then a sequence of observations $\{s_n\}$, $M = 1$, where $s_n = h[y(t = n \triangle t)]$, with a measurement function h and a sampling $\triangle t$. The number of observed variables is assumed to be sufficient to embed the dynamics. The functions $\{f_i\}$ may be of any form, but are usually taken to be a series expansion. This method has been successfully tested with Taylor- and Fourier-series expansion. In this manner, the modeling is done by finding the best expansion coefficients to reproduce the experimental data. Often, the form of the functions $\{f_i\}$ is known, but the coefficients are unknown. For example, this situation occurs frequently with rate equations for measurement processes. The added information greatly reduces the number of undetermined parameters, thus making the modeling computationally more efficient.

The modeling procedure begins with the step of choosing some trial coefficients. The error in these parameters can be computed by taking each data point $x(t_n)$ as an initial condition for the model equations. The predicted value $y(t_{n+1})$ can then be calculated for CMs as

$$y_i(n+1) = f_i[x(n), a], \quad i = 1, \ldots, N,$$

or for ODEs as

$$y_i(t_{n+1}) = x_i(t_n) + \int_{t_n}^{t_{n+1}} f_i[y(t'), a]dt, \quad i = 1, \ldots, N \qquad (5.28)$$

and compared to the experimentally determined value. It is well known that more stable models can often be obtained by comparing the prediction and the experimental data several time steps into the future. For the present analysis, we predict the value only to the time of the first unused experimental data point. The error in the model is thus obtained by summing these differences

$$F = \frac{1}{N(M-1) - N_c} \sum_{i=1}^{M} \sum_{j=1}^{N} \frac{1}{\sigma_{ij}^2} [y_j(t_i) - s_j(t_i)]^2, \qquad (5.29)$$

where N_c is the number of free coefficients a_i, M is the number of data points, and σ_{ij} is the error in the jth vector component of the ith calibration measurement. The task of finding the optimal model parameters has now been reduced to a minimization problem. Thus, the best parameters are determined by

$$\min_a F(a, y), \quad \alpha_i^{\min} \le \alpha_i \le \alpha_i^{\max}, \quad i = 1, \ldots, r, \qquad (5.30)$$

where α_i are the system characteristics of the sensor (fractal dimension, pointwise dimension, information dimension, generalized dimension, embedding dimension, Lyapunov dimension, metric entropy, etc.)

A minimal embedding of dimension N is determined by means of Hausdorff dimension d or any other generalized dimension. The essence of the present method is as follows. We consider relatively slow parameter *a*. As a result, we should solve the corresponding constrained optimization problem. The constrained optimization algorithm was implemented as a function in MATLAB 5.3.1 running on a UNIX Computer.

Therefore the ability to determine these coefficients rests upon the strength of the algorithm employed to search through the space of parameters. Because this has been formulated as a standard F_ν^2 identification problem, the normal statistical tests can be applied. Typically, $F_\nu \simeq 1$ implies that the modeling was successful; however, more sophisticated tests can be applied as well, for example, the F test. If the experimental errors σ_{ij} are unavailable, the normalization factor can simply be removed from equation (5.29). This means that the F_ν tests cannot be applied, but the best possible model can still be determined by locating the global minimum of F_ν in the parameter space.

4.3 Chaotic Dynamics of Levitated Probes

The connection of the displacement y_1 of the probe in the magnetic field by means of u_3 and an output signal of the sensor can be described by an equation for state variables $y_3, \ldots, y_6$ and some functional z (a model of the quantum interferometer S). This model admits (Yatsenko, 1989) the following bilinear model (BM),

$$\dot{y} = Ay + (Bu_1 + Cu_1^2)y + Du_1 + Eu_1^2 + Fu_3 + Gu_4, \quad z = Ly, \quad (5.31)$$

where A, B, C, D, E, F, G, L are the matrix; $y \in \widehat{Y} \subset \mathbb{R}^2$; $z \in \mathbb{R}^1$.

Then there exist some possibilities for optimization of information characteristics of the measurement using the parameter matrix a and control $u(\widehat{n})$. On the basis of these characteristics we can provide a matrix and topological behaviors of discrete approximation of the BM $\{T, \widehat{Y}, S, \Psi\}$ using symbolic dynamic methods. Here $\{\widehat{T}^n; \ n \in \mathbb{Z}\}$ is a cascade; $T : \widehat{Y} \to \widehat{Y}$, $\Psi : \widehat{Y} \to \mathcal{L}$ is the map "input-output" of the system S; $\mathcal{L}$ is a finite alphabet. A further optimization of the sensor can be reached near a Smale's horseshoe of additional Lebesgue measure of the dynamical system $\{T, \widehat{Y}, S, \Psi\}$.

The requirement of the equilibrium of the probe is provided by a feedback

$$\widehat{u}_1 + \widehat{\alpha}\widehat{u}_1 = \widehat{\alpha}r(y - u_0)$$

in the simplified model

$$\ddot{y} - \widehat{u}_4 = \widehat{u}_1 + \widehat{u}_3.$$

Here $u_0(t)$ is a fixed relation of the time of a probe; $\widehat{u}_1 = d_2u_1$, $\widehat{u}_3 = f_2u_3$, $\widehat{u}_4(y) = g_2u_4 = \delta y + \widehat{K}(y)y$, δ, r, $\widehat{\alpha}$ are constants, $\widehat{K}(y) = (1/\widehat{B})(y^2 - 1)(y^2 - B)$, $\widehat{B} > 1$. Under the parameter $r = 0$, the feedback realizes the three stable states $y = 0, \pm\sqrt{\widehat{B}}$ and two saddle points $y = \pm 1$. Under some values of the parameters $(\delta, \widehat{\alpha}, r, \widehat{B})$ the probe u_0 will be moved from one point to another.

However, under different values of the parameters we will have a limit cycle and a chaotic mode. If $u_0 = \widehat{u}_3 = 0$ and $r = r_0$ then the origin of the coordinate system will be an unstable saddle point of the spiral type. A numerical model of a measurement has chaotic properties (strange attractor) that can be used for constructing a better sensitivity measurement.

Using the linear model near a stable point of the probe

$$\dot{x} = Ax + Du_1 + Fu_3 + Gu_4, \quad z = Lx, \quad x \in \mathbb{R}^2, \qquad (5.32)$$

a normalized polynomial $\Theta(\lambda) = \alpha_1\lambda^2 + \alpha_2\lambda + \alpha_3$ with a negative real part of roots, and the method of synthesis of a nonperturbed motion under control $y = 0$, $u_1 = 0$, we can find a stabilizing control

$$u_1 = -(\alpha_3 + a_{11})a_{17}^{-1}\alpha^{-1}z. \qquad (5.33)$$

The next section explains these results.

4.4   Asymptotic Stability of Measurements

The system of equations for the state variables $y_3, \ldots, y_6$ and the functional z_1 describe the superconducting interferometer (Likharev, 1978) mapping $y_1 \to z_1$ which admits a linear representation. Introducing the function $z = \alpha y_1 + \beta y_2$, where β is a constant, and setting $a_{1j} = 0$, $j = 2, \ldots, 5$, we obtain the bilinear system

$$\begin{bmatrix} \dot{y}_1 \\ \dot{y}_2 \end{bmatrix} = \begin{bmatrix} 0 & 1 \\ a_1 & 0 \end{bmatrix} \begin{bmatrix} y_1 \\ y_2 \end{bmatrix} + \begin{bmatrix} 0 \\ a_2 + a_3 y_1 \end{bmatrix} u_1 + \begin{bmatrix} 0 \\ a_5 + a_4 y_1 \end{bmatrix} u_1^2$$

$$+ \begin{bmatrix} 0 \\ a_6 \end{bmatrix} u_3 + \begin{bmatrix} 0 \\ a_7 \end{bmatrix} u_4, \quad z = \alpha y_1 + \beta y_2. \qquad (5.34)$$

Here $a_1 = a_{11}$, $a_2 = a_{17}$, $a_3 = a_{18}$, $a_4 = a_{110}$, $a_5 = a_{19}$, $a_6 = a_{16}$, and $a_7 = a_{111}$. Let $a_3(t) = 0$ and $u_4(t) = 0$. We derive the conditions of asymptotic stability of the bilinear system from the linear measurement model for which the existence conditions are given in Yatsenko (1989). Linearizing (5.34) in the neighborhood of the stable equilibrium $y_1 = 0$, $y_2 = 0$ of the probe for $u_1(t) = 0$, we obtain

$$\begin{bmatrix} \dot{x}_1 \\ \dot{x}_2 \end{bmatrix} = \begin{bmatrix} 0 & 1 \\ a_1 & 0 \end{bmatrix} \begin{bmatrix} x_1 \\ x_2 \end{bmatrix} + \begin{bmatrix} 0 \\ a_2 \end{bmatrix} u_1,$$

$$z = [\alpha \ \beta][y_1 \ y_2]^T. \qquad (5.35)$$

The system (5.35) can be rewritten in a more general form as

$$\dot{x} = Ax + Bu_1, \quad z = Cx. \tag{5.36}$$

We need to determine the control

$$u_1 = K(z), \quad K(0) = 0, \tag{5.37}$$

which ensures asymptotic stability of the unperturbed motion $x = 0$.

Controllability of the linearized system

$$\dot{x} = Ax + Bu_1 \tag{5.38}$$

implies stabilizability of the bilinear system (5.34) by the linear control $u_1 = Fx$.

Let $\theta(\lambda) = \alpha_1 \lambda^2 + \alpha_2 \lambda + \alpha_3$ be an arbitrary unnormalized characteristic polynomial whose roots have negative real parts. Find a matrix K such that the roots of the characteristic polynomial of the matrix $A + BKC$ coincide with the roots of the polynomial $\theta(\lambda)$.

By controllability of the system (5.38), there exists a matrix $F = \|f_j\|$ $(j = 1, 2)$ such that $A + BF$ has the specified spectrum. Thus the existence of the sought matrix K is equivalent to solvability of the equation

$$KC = F. \tag{5.39}$$

Let

$$C_1 = \alpha, \quad C_2 = \beta, \quad E_1 = [1\ 0]^T, \quad E_2 = [0,\ 1]. \tag{5.40}$$

For the existence of a matrix K satisfying equation (5.39) it is necessary and sufficient that

$$FQ = 0, \quad Q = E_2 - E_1 C_1^{-1} C_2, \tag{5.41}$$

where, using (5.35),

$$Q = \begin{bmatrix} -\alpha^{-1}\beta \\ 1 \end{bmatrix}. \tag{5.42}$$

If the elements of the matrix (5.42) are treated as the coordinates of a vector q in two-dimensional space, then the condition (5.41) implies that the vector $s = (f_j)$, $j = 1, 2$, should be collinear with the vector $q = (1, \alpha^{-1}\beta)$. Therefore, for a fixed alignment of the sensitivity axis of the quantum interferometer, equation (5.41) is satisfied by a one-parametric family of matrices F of the form

$$F = [f_1 \alpha^{-1}\beta f_1]. \tag{5.43}$$

For the characteristic polynomial of the matrix $A + BF$ to coincide with the given polynomial $\theta(\lambda)$, it is necessary that $f_1 = -(a_3 + a_1)a_2^{-1}$, $\alpha_1 = 1$, and $\alpha_2 = -a_2\alpha^{-1}\beta f_1$. Thus, for $\alpha_1 = 1$, $a_2 \neq 0$, $\alpha \neq 0$, the output feedback matrix $K = FE_1C_1^{-1}$ is given by

$$K = [f_1\alpha^{-1}] \tag{5.44}$$

and the control $u_1 = Kz = f_1\alpha^{-1}z$ ensures asymptotic stability of the equilibrium (5.34) in some neighborhood H of x (Seraji, 1974; Lebedev, 1978).

4.5   Synthesizing the Adaptive Filter

Let the observer input be an additive mixture of gravitational signal and noise,

$$v_c(t) = r(t) + s(t), v_c = v_3. \tag{5.45}$$

Assume that the signal $r(t)$ and the noise $s(t)$ on the output of the measuring instrument have the spectral densities

$$S_R(\omega) = S_r^{in}(\omega)|W(j\omega)|^2, \tag{5.46}$$

$$S_Q(\omega) = S_s^{in}(\omega)|W(j\omega)|^2, \tag{5.47}$$

where

$$W(p) = \frac{a_6\alpha}{p^2 - a_1} = \frac{a}{p^2 - a_1}.$$

Let

$$S_r^{in}(\omega) = \frac{R^2(t)}{g(\omega)}, \quad S_s^{in}(\omega) = \frac{Q^2(t)}{(\omega^2 + p_1^2)(\omega^2 + p_2^2)g(\omega)}. \tag{5.48}$$

Using the controller, we obtain the following expression for the spectral density of the signal and the noise on the gravimeter output,

$$S_R(\omega) = R^2(t), \quad S_S(\omega) = \frac{Q^2(t)}{(\omega^2 + p_1^2)(\omega^2 + p_2^2)}. \tag{5.49}$$

Here

$$g(\omega) = |W_p(j\omega)|^2, \quad W_p(p) = \frac{a}{p^2 + a_2f_1\beta\alpha^{-1}p - (a_1 + a_2f_1)}. \tag{5.50}$$

Assume that the parameters Q and R fall between the bounds

$$Q_{min} \leq Q \leq Q_{max}, \quad R_{min} \leq R \leq R_{max}, \tag{5.51}$$

p_1, p_2 are located in the complex plane so that the dispersion is bounded. Let us synthesize an adaptive filter minimizing the mean square error,

$$I = \min_{\hat{y}(t)} M[r_1(t) - \hat{y}(t)]^2. \tag{5.52}$$

Here $\hat{y}(t)$ is the signal on the output of the adaptive filter connected to the output of the quantum interferometer, $r_1(t)$ is the useful signal on the output of the quantum interferometer. The adaptive filter satisfies the following requirements.

1. The adaptive filter transfer function coincides with the transfer function of the optimal filter for large values of the ratio Q/R. This property holds if for (5.49), given (5.51), we have for any ω

$$|G(j\omega)|^2 = O\left(\frac{Q^2}{R^2} + \omega^4\right), \tag{5.53}$$

where $O(\cdot)$ indicates smallness of a quantity relative to another quantity, and

$$|G(j\omega)|^2 = \frac{|B(j\omega)|^2|G(j\omega)|^2}{|A(j\omega)|^2|D(j\omega)|^2} - \omega^4. \tag{5.54}$$

2. The adaptive filter transfer function is very close to the transfer function of the optimal filter for other values of the ratio Q/R compared to 1.

The filter is synthesized following the general Kolmogorov–Wiener filtering theory (Kalman and Bucy, 1961; Bose and Chen, 1995; Diniz, 1997). When (5.54) holds, we have the following proposition, which is proved in Zagarii and Shubladze (1981).

Proposition 5.1. *For the observable signal (5.45) with spectral densities (5.46)–(5.46), when (5.51)–(5.54) hold, the coefficient of the transfer function* $W_{\text{opt}}(j\omega)$ *ensuring (5.52) is independent of* p_i, $i = 1, 2$, *and is determined by the ratio* Q/R. *Also*

$$W_{\text{opt}}(j\omega) = \frac{Q/R}{(j\omega)^2 + 2\sqrt{\frac{Q}{R}}\cos(\pi/2N)j\omega + Q/R}. \tag{5.55}$$

From (5.55) it follows that for small Q/R property 2 also holds, because as Q/R goes to zero both the optimal and the adaptive filter become open-loop.

4.6   Estimation of Gravitational Signals

Estimation of the useful signal level involves the following. Given the absolute value of the observed signal $v_{1r} = R_1 + s_1$, generate the function $\varphi(t)$,

$$\varphi(t) = \begin{cases} |v_{1r}(t^*) + k(t - t^*), & q_1 > 0, \\ \max |v_{1r}(t)|, & q_1 < 0, \quad q_2^0, \\ |v_{1r}(t^*)| - k_1(t - t'' - \Delta l), & q_2 > 0, \end{cases} \tag{5.56}$$

where t^* is the time moment when

$$v_{1r}(f^*) = \max_{t \in (t^* - \Delta t, t^*)} |v_{1r}(t)|; \tag{5.57}$$

$$k, k_1 \geq \max |r_1(t)|; \tag{5.58}$$

$$\Delta t = \frac{2\pi \int_0^\infty S(Q, \omega) d\omega}{\int_0^\infty \omega S(Q, \omega) d\omega}; \tag{5.59}$$

$$q_1(t) = |v_{1r}(t)| - |v_{1r}(t^*) + k(t - t^*)|, \quad t > t^*; \tag{5.60}$$

$$q_2(t) = |v_{1r}(t - \Delta t)| - |v_{1r}(t) - k_1(t - t^* - \Delta t)|, \quad t > t^* + \Delta t. \tag{5.61}$$

If the function $\varphi(t)$ is known, then the useful signal level is given by $Q = \varphi(t) - \psi$. Here it is defined by

$$k_1 \left[1 - \frac{\Delta}{2\pi} \left(\frac{\lambda_2}{\lambda_0} \right) \exp \left(-\frac{\psi^2}{2\lambda_0} \right) \right] = k \left[1 - \Phi \left(\frac{\psi}{\lambda_0} \right) \right],$$

$$\lambda_{2i} = \int_0^\infty \omega^{2i} S_R(R, \omega) d\omega;$$

$\Phi(\psi/\lambda_0)$ is the standard normal distribution function. The noise level is estimated using a wideband filter. The absolute value of the filtered signal is integrated with a time constant much larger than the time constant of the integrating filter.

4.7   Numerical Analysis of the Estimation Model

Consider the estimation system shown in Figure 5.2. Let $S_1, S_2, \ldots, S_{14}$ be the signal transformation operators constructed using the mathematical models of the subsystems 1–14. Then the mathematical model of the estimation system can be expressed by the collection of the following operators: the operator S_1 measuring the signal $u_3(t)$, the operator S_2 measuring the state vector y, the adaptive estimation operator S_4, the inverse operator $S_5 = (S_1 S_2)^{-1}$ (Figure 5.3), signal readout operators S_7, S_9, parameter identification operator of the bilinear observation model

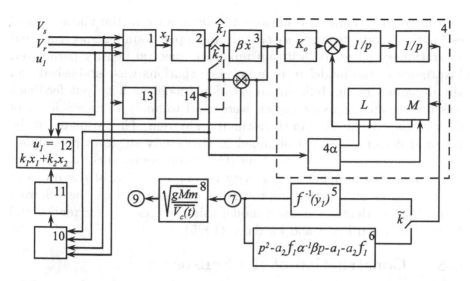

Figure 5.2. The probe in a magnetic potential well; (2) quantum interferometer; (3) differentiator; (4) adaptive filter; (4a) adaptation block; (5, 6) the inverse of models 1, 2; (7) gravimeter readout; (8) determination of the coordinate x_m; (9) displacement sensor readout; (10) bilinear model identifier; (11) synthesis of controller parameters; (12) feedback; (13) compensation of sensor support vibrations; (14) two-channel noise compensation.

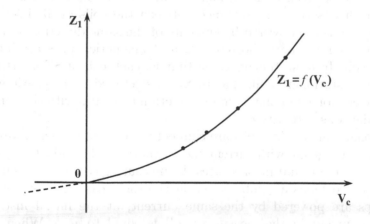

Figure 5.3. The static characteristic of sensor in Figure 5.2.

S_{10}, the operator S_{11} synthesizing the controller parameters, and the operators S_{13}, S_{14} maintaining the signal-to-noise ratio $r_1(t)/s_1(t)$ required for optimal filtering. The operators S_{13}, S_{14} are constructed using the Hamiltonian model of a system of free physical pendulums with a couple of coaxial ideally conducting rings on their end faces and a two-channel noise compensation circuit for $s_1(t)$ (Yatsenko, 2003).

Numerical simulation confirmed the invariance of the mathematical estimation model to noise $u_4(t)$ with an upper bounded spectrum and to stationary noise $s(t)$ with an unknown spectral density parameter. Invariance of the model to impulse perturbations was established and the dynamics of the bilinear observation model with output feedback was investigated. Specific signals were used to analyze the efficiency of digital adaptive filtering in the estimation system. The estimates of the minimal detectable signal obtained in this study suggest applicability of the technique to some fundamental experiments, such as checking the equivalence principle in relativity theory, detection of gravitational waves, and so on. Further improvement of sensitivity can be achieved by optimizing the measurement model using the results of Ermoliev and Wets (1984) and Horst and Pardalos (1995).

4.8   Construction of the Sensor

Mechanically, the sensor represents a free body or sensing element (SE) designed as a rigid pack of two coaxial short-circuited superconducting coils, and suspended in a magnetic field of two stationary superconducting current coils whose axes in nonperturbed state coincide with the axis of SE. The sensing element is positioned between stationary coils, with the distance between each stationary coil and the SE coil nearest to it being much less than the distance between the coils of SE itself. The stationary coils are powered by currents of the same direction and same strength, then they are shorted out, and magnetic fluxes are induced into SE coils from stationary coils; that is, each coil of SE is attracted to the next stationary coil; that is, SE is stretched by magnetic forces. The acceleration component whose direction coincides with the common axis of the coils is registered.

Stationary suspension coils are connected in series in such a way that after their energizing with current they form two independent loops with a common section that incorporates the measuring coil. Currents flowing in the formed loops are subtracted on the the measuring coil. Because the loops are powered by the same current, at the initial instant of time the current in the measuring coil is equal to zero. When SE is displaced along its axis under the effect of acceleration, the current in one stationary coil increases by the value ΔI_2 whereas in the other it decreases by ΔI_2. Current $\Delta I_1 + \Delta I_2$ will flow through the measuring coil in this case.

However, increments of currents of opposite signs arise only in the case of displacement along the axis. But if perturbations arise in the direction perpendicular to the axis or along the angle of inclination of the SE axis, currents in both loops either increase or decrease. Current in the measuring coil will not vary. As this takes place, the greater the symmetry of the magnetic systems, the more invariant will be the circuit with respect to the mentioned perturbations.

5. Nonlinear Dynamics and Chaos

As we discussed in the introduction to this chapter, the term "linear" refers to the dependence of the magnetic forces to the first power of the state variables. In general, however, most magnetic phenomena are nonlinear in position or angular state variables, in velocity or angular velocity variables, or in the magnetic field, or electric circuit variables. In spite of the reality of nonlinear forces, most of the techniques for nonlinear analysis use linear models in order to simplify the mathematics. However, these models fail to capture important physical phenomena. Nonlinear phenomena include the following: amplitude-dependence of natural frequencies; jump and hysteretic behavior in forced vibration problems; limit-cycle periodic motions; subharmonic generation; and the most recently observed phenomena of chaotic dynamics and unpredictable motion. The differences between linear and nonlinear models are summarized in Table 5.1. Introductory books in nonlinear dynamics include Hageronn (1988) and the classic book by Stocker (1950). An introduction to chaotic phenomena in nonlinear dynamics can be found in the book by Moon (1992).

In spite of the obvious nonlinear properties of magnetically levitated systems, very little analytical or experimental work is reported on nonlinear dynamics of levitated systems. One would think that safety considerations would demand a more realistic study of the dynamics of such systems.

A few studies of chaotic dynamics of levitated bodies have been reported, however. These include several papers (Moon, 1988; Yatsenko, 1989), and a few from Japan (Gafka and Tani, 1992; Kuroda, Tanada, and Kikushima, 1992). In this section, we discuss two simple cases:

Table 5.1. Comparison of linear and nonlinear phenomena

Linear Dynamics	Nonlinear Dynamics
Resonance	Subharmonics
Instability	Limit cycle
Periodic motion	Chaos
Robust with respect to initial conditions	Sensitive to initial conditions
Predictable	Unpredictable
Unique solution	Multiple solutions

1. The vertical motion of a levitated coil moving over a conducting guideway

2. The lateral vibration of a magnet over a YBCO superconducting bearing (Stocker, 1950)

When the coil is close to the sheet conductor, the force is inversely proportional to the height of the coil above the conducting sheet z; that is,

$$F = \frac{\mu_0 I^2 \beta}{2\pi z}. \tag{5.62}$$

In this model we have assumed that the force acts principally on the coil elements transverse to the horizontal velocity. We also assume that the guideway has a vertical wavelike deformation pattern of amplitude $A_0 \cos kx$ and wavelength Λ, where $\Lambda \ll z$. If the coil moves with a horizontal velocity (i.e., $x = V_0 t$) and a wave number $k = 2\pi/\Lambda$, then the wavelike track will produce a sinusoidal forcing term on the levitated coil proportional to $\cos \omega t$, where $\omega = 2\pi V_0/\Lambda$.

Vertical Heave Dynamics of a Maglev Vehicle. Consider a coil element of length β carrying constant current I moving over a continuous-sheet guideway. In the high-speed limit, the force on the coil will be given by the field due to an image coil below the guideway of opposite current direction.

Under these very generous assumptions we can derive an equation for the vertical motion (called heave) of the form

$$m\ddot{z} + \delta\dot{z} - \frac{\mu_0 I^2 \beta}{2\pi z} = -mg + m\frac{V_0^2 4\pi^2 A_0}{\Lambda^2} \cos \omega t, \tag{5.63}$$

where $\omega = 2\pi V_0/\Lambda$ and an arbitrary damping term has been added. This system can be written in the form of a third-order autonomous system of first-order differential equations:

$$\dot{z} = v,$$

$$\dot{v} = -cv + \frac{b}{z} - a + f\cos\phi,$$

$$\dot{\phi} = \omega. \tag{5.64}$$

This system of equations can easily be numerically integrated in time using a Runge–Kutta or other suitable algorithm. The trajectory is easily projected onto the phase plane of z versus v. As the amplitude of the guideway waviness is increased, one can see a change in the geometry of the motion from elliptical to a distorted ellipse to chaotic motion. The chaotic motion is better viewed by looking at a stroboscopic view of the dynamics by plotting (z_n, v_n) at discrete values of the phase $\phi = \omega t$ or $t_n = 2\pi n/\omega$. This picture is called a *Poincaré map* (Moon, 1992). In contrast to unordered continuous time the Poincaré map shows a fractal like structure. This type of chaotic motion with fractal structure is called a *strange attractor*. It indicates that the dynamics are very sensitive to the initial conditions.

Chaoticlike dynamics in a levitated model moving over a rotating-wheel guideway have been observed by Moon (1988,1992).

Chaotic Lateral Vibration of a YBCO Magnetic Bearing. In this chapter we saw that the magnetic force between a permanent magnet and a high-temperature superconductor such as YBCO is hysteretic near the critical temperature. Hysteretic forces are both nonlinear and dissipative and can produce complex nonlinear dynamics. A permanent magnet is restrained to move laterally over a YBCO superconductor. As the gap between the magnet and the superconductor is decreased, the dynamics of the magnet becomes increasingly complex in a pattern called *period doubling*. Subharmonic frequencies appear in the spectrum of the form $m\omega/n$, where $n = 2, 4, \ldots, 2k$. This bifurcation behavior is shown in the Poincaré map as a function of the magnet–YBCO gap. At a critical gap, the motion becomes chaotic. Another tool for observing chaotic motions is to plot a *return map* on one of the state variables, say X_{n+1}, versus X_n, where X_n is the displacement of the magnet at discrete times synchronous with the driving amplitude; that is, $t_n = 2\pi n/\omega$. The return map shows a simple parabolic shape. This map is similar to a very famous equation of chaos known as the *logistic map* (Moon, 1992,):

$$X_{n+1} = aX_n(1 - X_n). \tag{5.65}$$

For $a > 3.57$ the dynamics may become chaotic, and this equation generates a probability density function.

This simple experiment again indicates that although magnetically levitated bodies are governed by deterministic forces, the nonlinear nature of the forces can generate complex and sometimes unpredictable dynamics that are sensitive to initial conditions and changes in other system parameters. Thus, care in the design of such systems should include exploration of the possible nonlinear behavior of levitation devices.

6. Notes and Sources

In this chapter, we described a superconducting gravity meter, its mathematical model, and a nonlinear controller that stabilize a probe at the equilibrium state. We have also presented a mathematical model of the superconducting suspension which is based on a magnetic levitation. A nonlinear control algorithm has been implemented for the purpose of maintaining chaotic behavior in the sensor.

The present chapter proposed the mathematical models used to estimate the weak signal based on continuous bilinear observations. The proposed algorithms are more effective as compared with the known ones. Also, the algorithms that are considered here make it possible to find the value of a minimally detectable signal. However, it is necessary to use effective global optimization procedures. There exist more effective global optimization algorithms that could be used to solve this problem (Pardalos et al., 2001). The obtained estimates are directly applied to the estimation of the value of a gravity signal present in controlled cryogenic sensors.

The material contained in Chapter 5 is based on early papers of Yatsenko (1989), Samoilenko and Yatsenko (1991b), Kozorez and Cheborin (1977), and Moon (1992). The bilinear model of controlled magnetically levitated systems has been proposed by Yatsenko (1989). The mathematical model of a gravity-inertial sensor is due to Kozorez and Cheborin (1977). A chaotic phenomenon in nonlinear dynamics can be found in the papers of Moon (1988, 1992).

Chapter 6

OPTIMIZATION AND CONTROL OF QUANTUM-MECHANICAL PROCESSES

In the last century, society was transformed by the conscious application of modern technologies to engines of the industrial revolution. It is easy to predict that in the twenty-first century it will be quantum and biomolecular technologies that will influence all our lives. There is a sense in which quantum technology already has a profound effect (Lloyd, 1993). A large part of the national product of the industrialized countries stems from quantum mechanics. We are referring to transistors, the 'fundamental particle' of modern electronics. It is now that this information is found to be profoundly different, and in some cases, much more powerful than that based on classical mechanics.

In quantum computing, information is manipulated not discretely, in the classical way, as a series of zeroes and ones (bits), but as a continuous superposition (qubits) where the number of possibilities is vastly greater. In effect, many computations are performed simultaneously, and calculations that would be intractable classically become feasible quantally. In this way, computational theory is becoming a branch of physics rather than mathematics.

Useful optimization and control of quantum physical processes has been the subject of many investigations (Butkovskiy and Samoilenko, 1990; Samoilenko and Yatsenko, 1991a; Butkovskiy, 1991; D'Alessandro, 2000). Problems of controlling microprocesses and quantum ensembles were posed and solved for plasma and laser devices, particle accelerators, nuclear power plants, and units of automation and computer technology. Although quantum processors have not yet been developed, there is no doubt that their production is near at hand. Some quantum effects have already been applied in optoelectronic computer devices, but their integration into diverse computer systems still belongs to the future (Ozava, 1984).

P.M. Pardalos, V. Yatsenko, *Optimization and Control of Bilinear Systems*, doi: 10.1007/978-0-387-73669-3,
© Springer Science+Business Media, LLC 2008

Recently there has been an intense interplay among methods of quantum physics, optimization, and the theory of control. The problem of controlling quantum states was in fact brought forth due to the rise of quantum mechanics. For instance, many experimental facts of quantum mechanics were established with the use of macroscopic fields acting on quantum ensembles, which from the modern viewpoint can be considered as the control. As the technology of experimentation evolved, new problems in controlling quantum systems arose (Butkovskiy and Samoilenko, 1990; Yatsenko, 1993), and their solution required special methods (Brockett, 1973; Yatsenko, 1984; Agrachev and Sachkov, 2004).

Optimization problems in quantum and biomolecular computing are also very important. Some exciting results and developments (experimental as well as theoretical) have emerged (Hendricson, 1995; Horst amd Pardalos, 1995; Xue, 1996; Pardalos, Liu, and Xue, 1997; Pardalos et al., 1995; Pardalos, Tseveendorj, and Enkhbat, 2003). Interesting global optimization problems have been formulated and developed for processes of nonlinear transformations of information in quantum systems (Butkovskiy and Samoilenko, 1990; Petrov et al., 1982; Yatsenko, Titarenko, and Kolesnik, 1994; Yatsenko, 1995; Grover, 1996) and in biomedicine (Haken, 1996; Yatsenko, 1996). An important question regarding the optimization of those systems is how to construct efficient controlled mathematical models.

However, many applied optimization problems have not been considered yet. It is necessary to use optimization methods of quantum and biomolecular systems because of the practical importance of the implementation of physical processes satisfying the required quality criteria (Pardalos, 1988; Pardalos and Li 1990; Pardalos, Gu, and Du, 1994; Pardalos, Liu, and Xue, 1997).

Most of the attention is focused on the following problems.

1. Mathematical modeling of controlled quantum and biomolecular systems

2. Mathematical modeling of controlled physical and chemical processes in the brain; consider the brain as a quantum macroscopic object (Jibu and Yassue, 1995; Gomatam, 1999)

3. Optimal construction of a set of states accessible from a given initial state

4. Optimization of the set of controls steering the system from a given initial state to a desired accessible state with the greatest or specified probability

5. Identification of a control that is optimal with respect to a given criterion, such as the response time or the minimum number of switches (in bang-bang control)

6. Identification of the measuring operator and the method of its implementation by means of programmed control providing the most reliable measurement at a given instant T

7. Construction of a system of microscopic feedback providing the possibility of control with data accumulation

This chapter deals with the progress made in the optimal bilinear system control of quantum systems. It concentrates on applying the geometric technique in order to investigate a finite control problem of a two-level quantum system, resonance control of a three-level system, simulation of bilinear quantum control systems, and optimal control using the Bellman principle. We show that a quantum object described by a Schrödinger equation can be controlled in an optimal way by electromagnetic modes. We also demonstrate an application of these techniques and an algebra-geometric approach to the study of dynamic processes in nonlinear systems.

The analytical method for solving problems of finite control of quantum systems by modulated electromagnetic fields is considered in Section 1. This method is based on dynamic symmetry properties and makes it possible to calculate in an explicit form the evolution matrix, which is a representation of the correspondent dynamic group of systems. The formal solutions of problems of finite control of occupancies of two-level and three-level atoms, which are controlled by lasers with arbitrary admissible laws of modulation for the emission parameters, are considered as examples.

Sections 2 and 3 concentrate on a simulation of quantum control systems. An algebraic approach to the study of bilinear quantum systems is proposed. Mathematical models of controlled quantum objects are considered.

Section 4 considers the Bellman principle for quantum systems. We show an analogy between the optimal principle for controlled systems and quantum mechanics. For nonlinear stochastic systems an association with the Shrödinger equation is established.

In Section 5, we present new mathematical models of classical (CL) and quantum-mechanical lattices (QML). System-theoretical results on the observability, controllability, and minimal realizability theorems are formulated for CL. The cellular dynamaton (CD) based on quantum oscillators is presented. We investigate the conditions when stochastic resonance can occur through the interaction of dynamical neurons with the intrinsic deterministic noise and external periodic control. We find a chaotic motion in the phase-space surrounding the dynamaton separatrix. The suppression of chaos around the hyperbolic saddle arises only for a critical external control field strength and phase.

1.   Control of Quantum Systems
1.1   Evolution of Quantum Systems

Contemporary engineering makes manipulation and monitoring of separate atoms possible (Dirac, 1958; Allen and Eberly, 1975; Carrol and Hioe, 1988; Prants, 1990; Sakurai, 1994). Fundamental questions of quantum theory such as the superposition and indeterminancy principles, reduction of wave function, and quantum skips are studied now not only within theoretical study, but also within experimental analysis. Both statements and interpretation of experiments demand an analytical description on the theoretical level for the evolution of a nonstationary quantum dynamical system (QDS) without resort to different perturbation methods. It is known that the evolution matrix, which is used for determination of the control vector-function for realization of the given controllability of the system, gives the complete picture of QDS. Intensities, frequencies, and phases of external electromagnetic fields usually perform the role of controls. The goal of this experiment is using the controlled transition of QDS from the given initial state to the desirable terminal state in a definite time. Generally speaking, controls that achieve their objective in minimum time are desired to minimize dissipative effects associated with residual couplings to the system environment. From a mathematical perspective, many of these problems reduce to time-optimal control of bilinear systems evolving on finite- or infinite-dimensional Lie groups. Although bilinear control problems have previously been studied in great detail, rich new mathematical structures can be found in quantum problems. The added structure enables complete characterization of time-optimal trajectories and reachable sets for some of these systems. Another class of quantum optimal control problem is steering in the presence of relaxation. Recent work of this type has shown, for example, that significant improvements can be made in the sensitivity of multidimensional nuclear magnetic resonance (NMR) experiments.

It is a typical problem of finite control under the condition that the controlled value satisfies the quantum motion equations. To generalize the analysis it is expedient to consider the control processes for QDS on the Lie groups for all unitary transformations of domains of admissible states into themselves (Butkovskiy and Samoilenko, 1990; Samoilenko and Yatsenko, 1991a; D'Alessandro, 2000). Mathematical foundations of the theory of control for quantum-mechanical processes were created by Butkovskiy and Samoilenko (1990). Models of the harmonic oscillator and the spin of an electron are usually considered as controlled quantum systems.

Timely evolution of QDS is described by the continuous one-parameter group G of unitary operators S, which are acting as both transformations of the dynamic variables in the Heisenberg representation

$$A(t) = S^+(t)S(t),$$ (6.1)

and transformations of a pure

$$|\Psi(t)\rangle = S(t)|\Psi(0)\rangle$$ (6.2)

or of mixed state of QDS

$$\rho(t) = S(t)\rho(0)S^+(t)$$ (6.3)

in the Shrödinger representation. The goal of the control is to get the given values of certain characteristics of QDS α_u for a definite time instant $t = T$. Then the problem of finite control is stated as follows. It is required to find on the interval $0 \le t \le T$ certain control functions $u(t) = \{u_j(t)\}$, $j = 1, 2, \ldots, n$ from the set of the admissible controls U, which make it possible to synthesize the evolution operator $S_u(T, 0)$ realizing the mapping

$$\alpha_u(0) \xrightarrow{\;S_u(T,0)\;} \alpha_u(T)$$ (6.4)

under the constraint

$$i\frac{d}{dt}S = H(u, t)S, \quad S(0) = I, \quad h = 1.$$ (6.5)

Arbitrary values (6.1)–(6.3) as well as their mean values, moments of a distribution, and so on can serve as α_u depending on the problem statement. The evolution generator $H(t)$ is identified in quantum mechanics with the Hamiltonian, the group G is called the dynamic group, and the corresponding Lie algebra to it is called the L-dynamic algebra of QDS with corresponding dynamic symmetry (Jurdjevic, 1997). Let us assume that H generates a finite-dimensional dynamic Lie algebra; that is,

$$H(t) = \sum_{i=1}^{n} u_i(t)L_i.$$ (6.6)

The totality of operators $\{L_i\}$ generates a basis of an n-dimensional Lie algebra, and the $u_i(t)$ are scalar functions of time with complex values. It follows from the classical Frobenius theorem (Arnold, 1983; Boothby, 1975; Jurdjevic, 1997) that the solution (6.5) is at least local and can be represented as

$$S = \prod_{i=1}^{n} \exp(g_i(u, t) L_i). \tag{6.7}$$

By substituting (6.7) in (6.5) we obtain a system of nonlinear first-order differential equations for parameters of a dynamic group g_i:

$$u_k(t) = N_{k\ell}(g)(t), \quad k, \ell = 1, \ldots, n, \tag{6.8}$$

where $N_{k\ell}(g)$ is an $(n \times n)$ matrix, the elements of which are analytical functions of g_i. The multiplicative variant of parameterization of G (6.7) essentially simplifies the calculation of probability amplitudes, which are observed in quantum theory. For every abstract dynamic algebra L the system (6.8) is invariant with respect to the set of representations and realizations of this algebra and its form is determined by the structure of L. It is possible to draw the following conclusions on the basis of structure attributes only. If L belongs to the class of solvable algebras, then it is evident that its basis can be organized to possess the triangular form for the matrix $N_{k\ell}$ and, therefore, to reduce the solution of (6.8) to n successive integrations. The solution in the form (6.8) is valid for all t; that is, global (Brockett, 1979) for solvable algebras. According to the Levy–Maltsev theorem an arbitrary Lie algebra assumes the expansion

$$L = \mathfrak{R} \oplus R, \tag{6.9}$$

where $\mathfrak{R}$ is a semisimple Lie algebra, and R is a radical. The following decomposition of the solution of equation (6.5) is possible due to (6.9),

$$S = S_{\mathfrak{R}} S_R, \tag{6.10}$$

where

$$i \frac{d}{dt} S_{\mathfrak{R}} = H_{\mathfrak{R}}(t) S_{\mathfrak{R}}, \quad S_{\mathfrak{R}}(0) = I, \tag{6.11}$$

$$i \frac{d}{dt} S_R = S_{\mathfrak{R}}^{+} H_R(t) S_{\mathfrak{R}} S_R, \tag{6.12}$$

$$H(t) = H_{\mathfrak{R}}(t) \oplus H_R(t), \tag{6.13}$$

and the operators $H_{\mathfrak{R}}$ and H_R generate the Lie algebras $\mathfrak{R}$ and R, correspondingly. Because every semisimple algebra $\mathfrak{R}$ can be decomposed uniquely into the direct sum of simple algebras

$$\mathfrak{R} = \mathfrak{R}_1 \oplus \cdots \oplus \mathfrak{R}_k, \tag{6.14}$$

then further decomposition is possible; that is,

$$S_{\mathfrak{R}} = \prod_{i=1}^{n} S_i, \tag{6.15}$$

where each co-factor satisfies the equation of type (6.11) with the Hamiltonian H_i that generates the corresponding ideal $\mathfrak{R}_i$. Moreover,

$$H_{\mathfrak{R}}(t) = H_1(t) + \cdots + H_k(t). \tag{6.16}$$

We reduce our study in this book to the finite control problem of QDS occupancy with a finite number of nonequidistant levels of energy. Such systems are controlled on the whole state space (Hunt, 1972; Isidory, 1995) because their groups are compact. The group $SU(2)$ is the fundamental group in the theory of interaction of radiation with a substance, QDS with $2j + 1$ energy levels, which has the group $SU(2)$. DS is described by the Hamiltonian

$$H(t) = u_0(t)L_0 + u^*(t)L_- + u(t)L_+, \tag{6.17}$$

where u and u_0 are differentiable complex-valued functions of time and the generators $SU(2)$ are written in the spherical basis

$$[L_+, L_-] = 2L_0, \quad [L_0, L_\pm] = \pm L_\pm. \tag{6.18}$$

If we look for the solution of the evolution equation (6.5) with the Hamiltonian (6.17) as

$$S = \exp\left(g_0 - i \int_0^t u_0(\tau)d\tau\right) L_0 \exp g_- L_- \exp g_+ L_+, \tag{6.19}$$

then the system (6.8), which consists of three first-order differential equations, can be reduced to a certain second-order equation for a new variable

$$g \equiv \exp(g_0/2), \tag{6.20}$$

namely,

$$\ddot{g} - \left(\frac{\dot{u}}{u} + iu_0\right)\dot{u} + |u|^2 g = 0, \quad g(0) = 1, \quad \dot{g}(0) = 0. \tag{6.21}$$

The initial parameters are expressed in terms of the parameter g and the controls as follows,

$$g_0 = 2\ln g, \quad g_- = \frac{ig\dot{g}}{u}\exp\left[-i\int_0^t u_0 d\tau\right],$$

$$g_+ = -\frac{iu}{g^2}\exp\left[i\int_0^t u_0 d\tau\right]. \tag{6.22}$$

It is necessary to construct a $(2j+1)$-dimensional irreducible unitary representation of the group for selected noncanonical parameterization (6.19). We calculate the matrix elements of the operator (6.19) in the standard basis

$$|j, m\rangle, \quad m = -j, -j+1, \ldots, j \qquad (6.23)$$

and find

$$S_{m'm}^{(j)} = \exp\left[-ij \int_0^t u_0 d\tau\right] \sum_{n=-j}^{j} \left[\frac{(j-m')!\,(j-m)!}{(j+m')!\,(j+m)!}\right]^{1/2}$$

$$\times \frac{(j+n)!}{(j-n')!\,(n-m)!\,(n-m')!} g^{m+m'} \tilde{g}^{n-m'} (-\tilde{g}^*)^{n-m}. \qquad (6.24)$$

For convenience, we introduced the following parameter into (6.24)

$$\tilde{g} \equiv g - g^{-1}. \qquad (6.25)$$

The following conservation law is valid for an arbitrary representation of $SU(2)$,

$$|g|^2 + |\tilde{g}|^2 = 1. \qquad (6.26)$$

The expressions (6.17)–(6.26) are valid for an arbitrary QDS with $SU(2)$. However, it is necessary to note that the standard basis (6.23) is applied in the physics of magnetic resonance whereas another numeration of states is used in laser physics, namely,

$$|N, n\rangle, \quad n = 1, \ldots, N. \qquad (6.27)$$

The connection between the two bases is given by the relations

$$N = 2j + 1, \quad m = n - j - 1, \qquad (6.28)$$

where N is the total number of QDS levels, and n is the number of the levels.

1.2   Finite Control of Quantum Systems

In this section we consider a nonrelaxing N-level atom with nonsingular nonequidistant spectra as the QDS,

$$H_0|N, n\rangle = \omega_n|N, n\rangle, \quad n = 1, \ldots, N. \qquad (6.29)$$

This atom interacts coherently with the modulated polychromatic laser field

$$E(t) = \sum_{k=1}^{N-1} \sum_{\ell=2}^{N} \operatorname{Re} E_{k\ell}(t) \exp j(\omega_{k\ell} t + \varphi_{k\ell}), \quad k < \ell, \qquad (6.30)$$

which consists of M $(= 1, 2, \ldots, N(N-1)/2)$ components. We assume that the component (k, ℓ) with amplitude $E_{k\ell}$, central frequency $\omega_{k,\ell}$, and phase $\varphi_{k,\ell}$ excites only one transition between the stationary states $|N, k\rangle$ and $|N, \ell\rangle$. The Hamiltonian of the interaction of an atom with the electric field in the dipole approximation has the form

$$H_{\text{int}}(t) = -E(t)d, \tag{6.31}$$

where d is the operator of the electric dipole moment of an atom. The operator H_{int} depends explicitly on time and control and it is finite on the time interval $[0, T]$. Generally speaking, the electric dipole transitions are possible only between states with different parity, and, therefore, some of the matrix elements of the Hamiltonian H_{int} vanish. However, for generality, we suppose that all transitions in the atoms $N(N-1)/2$ are solvable. Therefore, all results further obtained are valid for all transitions of an arbitrary physical nature, but the electrical dipole ones. It is necessary to find such a control, which transits the N-level atom from the given initial stationary state $|\Psi(0)\rangle$ to the desirable terminal stationary state $|\Psi(T)\rangle$; that is, to find the explicit form of the evolution matrix, which realizes the required transformation:

$$|\Psi(T)\rangle = S(u, T)\,|\Psi(0)\rangle. \tag{6.32}$$

The solution of this problem is reduced to the representation of the interaction

$$|\Psi(t)\rangle = \sum_{n=1}^{N} c_n(t)|N, n\rangle \exp(-i\omega_n t) \tag{6.33}$$

and to the application of an approximation of rotating waves. Then, the nonstationary Shrödinger equation

$$i\frac{d}{dt}|\Psi(t)\rangle = \big(H_0 + H_{\text{int}}(t)\big)|\Psi(t)\rangle \tag{6.34}$$

generates the following system of N equations for the probability amplitudes of $c_n(t)$,

$$i\dot{c} = \sum_{\ell \neq k=1}^{N} u_{k\ell}(t) \exp[i\Delta_{k\ell}(t) + i\varphi_{k\ell}]c_\ell \tag{6.35}$$

with controllable real parameters $u_{k\ell}$ and $\Delta_{k\ell}$, which belong to the class U of the piecewise-continuous bounded functions of time. Here the value

$$\Delta_{k\ell}(t) = \omega_k - \omega_\ell + t^{-1}\text{sgn}\,(\ell - k) \int_0^t \omega_{k\ell}(\tau)d\tau \tag{6.36}$$

makes sense for resonance detuning for the $(k\ell)$th transition of an atom, which depends on time for the case of frequency modulation of a laser field (6.30). It is assumed that the phase modulation is absent (i.e., $\varphi_{k\ell} = $ const $\forall\ k,\ell$), and the phase control consists in selecting $\varphi_{k\ell}$ values in order to satisfy the conditions of the DS problem. The values make sense for variable Rabby frequencies and they are equal to

$$u_{k\ell}(t) = -\frac{1}{2}E_{k\ell}(t)d_{k\ell}, \tag{6.37}$$

where $d_{k\ell}$ is the corresponding matrix element of the dipole operator d in the basis (6.29). Let us note it is not assumed to sum by repeating indexes in the expression (6.37). The vector-function of occupancy is the controllable value

$$P = \{P_1(t),\ldots,P_N(t)\}, \quad P_i(t) = |C_i(t)|^2, \quad i = 1,\ldots,N, \tag{6.38}$$

which is transformed according to the evolution matrix

$$P(T) \equiv |S(u,T)|^2 P(0). \tag{6.39}$$

In the general case, when all components of the polychromatic field (6.30) are modulated by both amplitude and frequency the vector-function of control

$$u = \{E_{k\ell}(t), \omega_{k\ell}(t),\ k = 1,\ldots,N-1,\ l = 2,\ldots,N,\ k \neq \ell\} \tag{6.40}$$

has the dimension $N(N-1)$. In the absence of relaxation processes the complete occupancy of an atom is conserved:

$$\sum_{i=1}^{N} P_i(t) = 1. \tag{6.41}$$

As examples of the DS method of application in the control theory of quantum mechanical processes, the problems of finite control for two- and three-level atoms of occupancy are solved in subsequent sections for the practically significant case of excitation by modulated laser fields.

1.3   Amplitude-Frequency Control

The bimodal model of an atom permits the formal solution of the problem of finite control for arbitrary admissible amplitude-frequency control

$$u = \{E(t), \omega(t)\} \in U. \tag{6.42}$$

We rewrite the Shrödinger equation (6.35) in a matrix form

$$i\frac{d}{dt}\begin{pmatrix} c_1 \\ c_2 \end{pmatrix} = \begin{pmatrix} 0 & -\frac{1}{2}E\exp i\Delta t \\ -\frac{1}{2}E\exp(-i\Delta t) & 0 \end{pmatrix}\begin{pmatrix} c_1 \\ c_2 \end{pmatrix}, \tag{6.43}$$

where $E(t)$ is the variable amplitude field intensity, $\Delta(t) = \omega(t) - \omega_0$ is a variable of detuning resonance, and ω_0 is the frequency of transition between stationary states of an atom. Let us decompose the matrix (6.43) in the basis of spinor representation SU(2):

$$L_0 = \frac{1}{2}\begin{vmatrix} 1 & 0 \\ 0 & -1 \end{vmatrix}, \quad L_+ = \begin{vmatrix} 0 & 1 \\ 0 & 0 \end{vmatrix}, \quad L_- = \begin{vmatrix} 0 & 0 \\ 1 & 0 \end{vmatrix} \tag{6.44}$$

and parameterize this group according to (6.19). Then, the solution of (6.43) can be represented as

$$\begin{pmatrix} c_1(t) \\ c_2(t) \end{pmatrix} = \begin{bmatrix} g^* & \tilde{g} \\ \tilde{g}^* & g \end{bmatrix} \begin{pmatrix} c_1(0) \\ c_2(0) \end{pmatrix}, \tag{6.45}$$

where

$$\tilde{g} = -\frac{2ig\dot{g}}{Ed}e^{i\Delta t}. \tag{6.46}$$

In this case the control equation (6.21) has the following form,

$$\ddot{g} - \left(\frac{\dot{E}}{E} - i\Delta\right)\dot{g} + \frac{1}{4}E^2d^2g = 0. \tag{6.47}$$

The formal solution of the problem of the occupancy finite control for a bimodal QDS can be found by an external field with the arbitrary amplitude-frequency modulation (6.42) from (6.45) and (6.39):

$$|g(T)|^2 = \frac{P_1(0) + P_i(T) - 1}{2P_1(0) - 1}. \tag{6.48}$$

Thus, by specifying the input $P_1(0)$ and output $P_1(T)$ values of occupancy of the lower level by means of (6.47) we find such an admissible control (6.42), which provides the required value of the parameter g for the time instant $t = T$. From the expression (6.26) follows that the values g and $\tilde{g}$ are accurate within arbitrary phase multipliers and make sense for probability amplitudes.

In the physics of magnetic and optical resonance the objective of control is often in the attainment of the occupancy inversion, that is, the bimodal system transition into the state $P_2(T) = 1$ by the controls $E(t)$ and/or $\Delta(t)$. Such well-known and practically realizable ways of creating the inversion as a π-impulse excitation (amplitude modulation) and an adiabatic passing through the resonance (frequency modulation) are described by the above-cited formulas (6.47) and (6.48) as particular cases.

1.4   Resonance Control of a Three-Level System

Let us consider a three-level atom with space-tapered spectra, which is controlled by the amplitude by means of a trichromatic laser field with the following vector-function of control,

$$u = \left\{ u_i = -\frac{1}{2} E_i d_i, \ i = 1, 2, 3 \right\} \in U, \qquad (6.49)$$

under the conditions stated in Section 6.1 and the additional resonant condition

$$\omega_{12} = \omega_2 - \omega_1, \quad \omega_{13} = \omega_3 - \omega_1, \quad \omega_{23} = \omega_3 - \omega_2. \qquad (6.50)$$

For the corresponding selection of phases of the components of the laser field the Hamiltonian of the excited atom in the basis (6.29) has the following form,

$$H(t) = \begin{bmatrix} 0 & u_1 & -iu_3 \\ u_1 & 0 & u_2 \\ iu_3 & u_2 & 0 \end{bmatrix}. \qquad (6.51)$$

It is possible to confine to single indexes for numeration of the amplitudes E_i and the matrix elements of the torque dipole d_i for the three-level case. By means of the unitary matrix

$$D = \begin{bmatrix} 1/\sqrt{2} & 0 & 1/\sqrt{2} \\ 0 & 1 & 0 \\ i1/\sqrt{2} & 0 & -i1/\sqrt{2} \end{bmatrix}$$

we transform the Hamiltonian (6.51):

$$H'(t) = D^+ H(t) D. \qquad (6.52)$$

The transformed Hamiltonian $H'(t)$ can be decomposed using the basis of standard three-level irreducible unitary representation of $SU(2)$ with the following nonzero matrix elements of the generators,

$$(L_0)_{11} = -(L_0)_{33} = 1, \quad (L_+)_{12} = (L_+)_{23} = (L_-)_{21} = (L_-)_{32} = \sqrt{2}.$$

We obtain

$$H'(t) = u_3(t)L_0 + \frac{1}{2}\big(u_1(t) + iu_2(t)\big)L_- + \frac{1}{2}\big(u_1(t) - iu_2(t)\big)L_+. \ (6.53)$$

The formula (6.53) for the matrix elements of the $N(= 2j + 1)$-dimensional unitary irreducible representation of $SU(2)$ gives us at once the solution of the nonstationary Shrödinger equation for the three-level QDS ($j = 1$),

$$\begin{bmatrix} c_1'(t) \\ c_2'(t) \\ c_3'(t) \end{bmatrix} = \begin{pmatrix} (g^*)^2 & \sqrt{2}g^*\tilde{g} & \tilde{g}^2 \\ -\sqrt{2}g^*\tilde{g}^* & |g|^2 - |\tilde{g}|^2 & \sqrt{2}g\tilde{g} \\ (\tilde{g}^*)^2 & -\sqrt{2}g\tilde{g}^* & g^2 \end{pmatrix} \begin{bmatrix} c_1'(0) \\ c_2'(0) \\ c_3'(0) \end{bmatrix}, \tag{6.54}$$

where

$$|c_1'(t), c_2'(t), c_3'(t)\rangle^T = D^+|c_1'(t), c_2'(t), c_3'(t)\rangle^T. \tag{6.55}$$

The formal solution of the problem of finite control of occupancies of these systems by means of a trichromatic amplitude modulated laser field has the following form,

$$2|g(T)|^2 - 1 = \frac{P_3'(T) - P_1'(T)}{P_3'(0) - P_1'(0)}. \tag{6.56}$$

The parameter g satisfies the following control differential equation

$$\tilde{g} - \left(\frac{\dot{u}_1 - i\dot{u}_2}{u_1 - iu_2} + iu_3\right)\dot{g} + \frac{1}{4}\left(u_1^1 + u_2^2\right)g = 0. \tag{6.57}$$

2. Simulation of Quantum Control Systems

In this section we develop an algebraic approach to the study of quantum control systems that is based on bilinear dynamical models defined on the orbits of the adjoint representation of a compact Lie group. A mathematically correct model is constructed for the physical statement of the problem; the limits of its physical correctness are then found.

Algebraic and geometric methods developed in modern mathematics are of great interest in finding the solution of the optimal control problems of physical processes in the space of quantum mechanical states. In this case, the theory of right-invariant bilinear control systems on unitary Lie groups (Hirshorn, 1977; Judjevic, 1997) turns out to be a natural tool for the solution. The algebraic-geometrical properties of the quantum bilinear models allow us to reformulate an optimal control problem in terms of the Lie group theory, and hence, in some cases, to obtain explicit solutions for systems that admit a finite-dimensional description. The crucial role here is played by the possibility of an exact embedding of the phase space of the system into a certain Euclidean space; this allows us to reduce the initial problem to the usual and well-studied statement. The verification of the finite-dimensional model of a quantum system and the extraction of an effective part of the infinite-dimensional configuration space are based on the averaging method, which is used in the passage to describe the control process in "real time".

In this section we justify the approach to the study of quantum control systems that is based on models of nonlinear dynamics defined on orbits of the adjoint representation in Lie algebras. On the basis of systems analysis of the physical statement of the problem, we construct the above-mentioned mathematical models and find the limits of their correctness. After that, using these models, we study the main characteristics of quantum control objects.

2.1   Mathematical Models of Quantum Objects

The considered physical system is described by an algebra of observables $\mathcal{L}$ over the state space. Its current state at each instant of time is determined by a unit vector in a complex separable Hilbert space $\mathcal{H}$. A complete as possible description of the quantum system is attained in this way. A state that admits such a description is called pure. It should be noted here that two vectors of a unit length, which differ from each other by the phase factor, describe only one and the same physical state (this is a specific "guade symmetry"). Thus, the space of (pure) states of the quantum mechanical system is isomorphic to the complex projective space over $\mathcal{H}$.

To study the system it is necessary to assign an algebra of corresponding physical values and to describe the assumed values. In the general quantum mechanical case, this algebra is characterized by a certain maximal set of independent observables $\{\widehat{A}_i\}$ that are represented by linear self-adjoint operators acting on $\mathcal{H}$. Because of the self-adjoint property, the domain of each such operator is a dense set of analytical vectors in $\mathcal{H}$ that form a linear subspace $D(\widehat{A}_i) \subset \mathcal{H}$. In the case where an operator $\widehat{A}_i$ is bounded we have $D(\widehat{A}_i) = \mathcal{H}$. Let $D(\mathcal{L}) \subset \mathcal{H}$ be the common invariant domain of the algebra $\mathcal{L}$ of observables. The invariance means that the range $R(\mathcal{H})$ lies within $D(\mathcal{H})$. The set of operators that are observable in the domain $D(L)$ forms a Lie algebra (that is denoted by the same symbol L) using the Lie bracket which has the form $\{\widehat{A}, \widehat{B}\} = -i[\widehat{A}, \widehat{B}] = i(\widehat{B}\widehat{A} - \widehat{A}\widehat{B})$.

Let us consider the space $\mathcal{L}^*$, the dual space of the Lie algebra $\mathcal{L}$, which consists of real linear functions defined on $\mathcal{L}$. The space $\mathcal{L}$ can be considered as a space of operators $\widehat{R}$ defined on $D(\mathcal{L})$. Then the above linear functions can be represented in the form

$$\langle \widehat{A} \rangle = \langle \widehat{A}, \widehat{R} \rangle = \mathrm{Tr}\,(\widehat{A}, \widehat{R}), \quad \widehat{A} \in \mathcal{L}, \quad \widehat{R} \in \mathcal{L}^*. \tag{6.58}$$

The value $\langle \widehat{A} \rangle$ means the expectation of the observables $\widehat{A}$ at a certain quantum mechanical state that is characterized by the operator $\widehat{R}$ when one measures the physical value corresponding to this observable. We note that the pairing of $\mathcal{L}$ and $\mathcal{L}^*$, given by the trace functional $\mathrm{Tr}\,(\cdot)$, can be used for the identification of $\mathcal{L}$ and $\mathcal{L}^*$; furthermore this pairing obeys the property of an inner product; this fact is very important for further considerations. By what has been said above on the interpretation of the operator $\widehat{R}$, it is clear that it should satisfy additional probability constraints. The basis of the space $D(\mathcal{L})$ is defined as an element of the convex set $\mathcal{P}$ of the positive self-adjoint density operators

$$\hat{\rho} = \sum_i \nu_i \, |i\rangle \, \langle i|, \quad \nu_i \geq 0, \quad \sum_i \nu_i = 1. \tag{6.59}$$

The operator $\hat{\rho}$ describes the case where one can assert that the system is a mixture of the states $|i\rangle$ with probabilities ν_i; in contrast to the vector $|i\rangle$, this mixture corresponds uniquely to the physical state of the system. Pure states correspond to extreme points of the convex set of density operators; they also correspond to the case $\nu_i = 1$ for a certain i, and satisfy the condition $\rho^2 = \rho$. Mixed states correspond to all other points.

The convex set $\mathcal{P} \in \mathcal{L}^*$ is no longer a linear space, because the eigenvalues of the operator $\hat{\rho}$ are nonnegative: $\nu_i \geq 0$. However, as is everywhere in $\mathcal{L}^*$, for any $\hat{\rho} \in \mathcal{P}$ and $\widehat{A} \in \mathcal{L}$, the real bilinear form

$$\langle \widehat{A}_\rho \rangle = \langle \widehat{A}, \hat{\rho} \rangle = \mathrm{Tr}\,(\widehat{A}, \hat{\rho}) \tag{6.60}$$

is well defined; this form has the sense of the quantum mechanical mean value of the observable $\widehat{A} \in \mathcal{L}$, at the state $\hat{\rho} \in \mathcal{P}$. Along with the expectation $\langle \rho A \rangle$, the value

$$\widehat{D}_\rho(\widehat{A}) = \langle (\widehat{A} - \langle \widehat{A}, \hat{\rho} \rangle \widehat{I})^2 \rangle = \mathrm{Tr}\,(\widehat{A}^2 \hat{\rho}) - [\mathrm{Tr}\,(\widehat{A}\hat{\rho})]^2 \tag{6.61}$$

is widely used in physics; this value characterizes the variance of the results of the measurement result of the observable $\widehat{A}$ at the state $\hat{\rho}$. Here $\widehat{I}$ is the identity operator on $D(\mathcal{L})$. It can be proved that each moment functional of an arbitrary order, which characterizes the statistics of the observations, can be computed correctly through the trace functional.

2.2   Dynamics of Quantum Systems and Control

We form the law of motion of the system in the representation of states through the density operator and assign it as a linear mapping $L : \mathcal{P} \times T \to \mathcal{P}$. In the system evolution, when there is no interaction with macroscopic measuring devices, this mapping is realized by unitary transformations and represents the Hamiltonian dynamics. Thus, there exists a unitary operator $U(t, t_0)$ such that

$$\hat{\rho}(t_0) \to \hat{\rho} = \widehat{U}(t, t_0) \hat{\rho}(t_0) \widehat{U}^+(t, t_0). \tag{6.62}$$

So, the evolution of the quantum system in time is given by the initial state $\hat{\rho}(t_0) = \hat{\rho}_0$ at the instant of time $t = t_0$ and by a one-parametric subgroup $g(t) = \{\widehat{U}(t)\}_G$ of the group of all unitary operators on $\mathcal{H}$.

In accordance with the Stone theorem, which plays the fundamental role for the whole mathematical apparatus of quantum mechanics, for a similar one-parametric subgroup of unitary transformations, there exists an operator $\widehat{H} \in \mathcal{L}$ (Hamiltonian of the system) and the infinitesimal generator of the shift operator of the system such that

$$\widehat{U}(t, t_0) = \exp[-i(t - t_0)\widehat{H}]. \tag{6.63}$$

In differential form, the differential equation has the form of a Schrödinger operator for the evolution operator on the unitary group

$$i\frac{d}{dt}\widehat{U}(t, t_0) = \widehat{H}\widehat{U}(t, t_0), \quad \widehat{U}(t, t_0) = \widehat{I}. \tag{6.64}$$

The above concepts admit a useful algebraic treatment. The equation (6.64) defines a right-invariant dynamical system on the Lie group $G = U$ of the unitary transformations of the wave function Hilbert space $\mathcal{H}$. The right invariance means that the following assertion holds. If $\widehat{U}(t, t_0)$ is a solution to equation (6.64), then for any unitary operator $\widehat{V} \in U$, $\widehat{U}(t, t_0)\widehat{V}$ is also a solution. The action of the adjoint representation of this group is defined on its Lie algebra; this action coincides with the law of evolution of the density operator of the quantum system:

$$\hat{\rho}(t_0) \rightarrow \hat{\rho}(t) = U(t, t_0)\hat{\rho}(t_0)\widehat{U}^+(t, t_0) = \mathrm{Ad}_{U_t}\hat{\rho}(t_0). \tag{6.65}$$

Thus, the evolution of the state is realized on the orbit of this representation in the Lie algebra. The action of the adjoint representation of the Lie algebra itself generates the following equation in the Liouville form,

$$\frac{d}{dt}\hat{\rho}(t) = \{\widehat{H}, \hat{\rho}\} = \mathrm{ad}_{\widehat{H}}\hat{\rho}, \tag{6.66}$$

which uses the intrinsic operation of taking the Lie bracket.

We use here the following observation. Because the study of the dynamical system (6.64) on the Lie group turns out to be difficult in the general case (the nonlinearity of the group as a differentiable manifold, the complexity of embedding computing, etc.), it can be more convenient to study the corresponding dynamical systems that are generated by its action in a space of one or another representation of the group (in particular, in the space of the adjoint representation). Also, we note that in our case of the algebra of the observables, which is defined as the algebra of bounded self-adjoint operators, $\hat{\rho}$ is an element of this algebra itself. The possibility to reduce an equation on a group to an equation on its Lie algebra is an essential simplification and is realized due to the property of the one-side invariance of the initial system.

The use of the group properties of unitary operators allows the use of the analytical apparatus of the theory of the Lie groups and algebras, to give precise mathematical statements and obtain solutions on a number of basic problems of control of quantum mechanical systems. In order to ensure the physical possibility of control, the Hamiltonian $\widehat{H}$ of the system should be a function of the control parameters $u_\mu(t)$. As control actions, it is natural to consider the electromagnetic field of a high degree of coherence. Lasers, masers, generators of SHF range, and the like can serve as sources of such fields. The Hamiltonian of a quantum system in an exterior field can be represented in the form

$$\widehat{H} = \widehat{H}_0 + \widehat{H}^1(t, u), \qquad (6.67)$$

where $\widehat{H}_0$ is the energy of the quantum system and $\widehat{H}^1(t, u(t))$ is the energy of the interaction of the system with exterior control fields. Thus, the Hamiltonian of the system is directly varied in time in accordance with the required control law; this ensures the physical possibility of control of quantum states and processes.

2.3   Physical Constraints

Further development is based on the refinement of the form of dependencies of the interaction Hamiltonian $H^1(t, u)$ on the control action. Among such types of dependencies, which have a physical sense and significance, the linear dependence is the most interesting. In this case the interaction Hamiltonian is represented in the form of a linear combination of perturbation operators $\widehat{H}_\mu(t)$, where $\mu = 1, \ldots, m$.

We note that the choice of the dipole approximation is sufficiently adequate for the description of the interaction of a system of atoms or molecules with electromagnetic fields in the optical range of waves to which the laser radiation belongs. We may assume the field to be homogeneous with a sufficient accuracy of about 10^{-8} cm. This allows us to neglect multifield effects of interaction. The magnetic dipole moment (of a nonrelativistic system) is less in order than the electric one.

The coherence could also be realized. The interaction of SHF radiation with the dipole magnetic moments (spins) of the atoms of the ferromagnetic crystals is also linear. In the first case, the electric field is the material support for the control action; in the second case, this role is played by the magnetic field; however, both variants are described by one and the same mathematical model and are simultaneously studied in the sequel.

We pass now to the formulation of constraints of a physical-technical character that have appeared when constructing correct mathematical models of quantum control processes. For definiteness, we consider the quantum-optic case. Then, our goal consists of constructing a mathematical model that allows us to study the control resonance laser action on a substance, which leads to a considerable change of the populations of the energetic levels of the substance and to the subsequent change of the properties or structure of the substance when irradiated. An important role is played by the realization of a noninvertible strong change in the substance when irradiated; to this end, the action should be sufficiently intensive (up to tens kV to a square centimeter for irradiation by pulses of a second duration).

The spectrum of possible applications of intensive electromagnetic actions is sufficiently wide. The laser irradiation can change the speed and coordinates of the substance being irradiated, their distribution in coordinates and velocities, and the structure of the particle being irradiated, thus leading to the ionization, dissociation, chemical surgery, isomerization, or polymerization, and can carry phase transitions such as evoparation, condensation, fusion, crystallization, amorphisation, hardening and so on. When such changes are realized by a monochromatic radiation and the result of the action is spectral-dependent, then one speaks about an intensive resonance interaction with the substance.

Due to linearity of perturbation, we can represent the interaction Hamiltonian of resonance of the electromagnetic fields with a substance in the following form,

$$\widehat{H}^1(t, u_\mu) = (D, E) = -\widehat{D} \sum_{\mu=1}^{m} u_\mu(t) \sin(\omega_\mu + f_\mu). \qquad (6.68)$$

Here $\widehat{D}$ is the vector operator of the dipole force moment, $u_\mu(t)$ is a slowly varied vector amplitude of the resonance component of the exterior field (frequency ω_μ), and f_μ is the corresponding vector phase. The vector character of the parameters allows us to take the polarization into account. The control parameters $u_\mu(t)$ belong to the prescribed class of admissible controls

$$U(\Omega, t) = \{u_\mu(t) : T \to \Omega \subset \mathbb{R}_{3n}^+\}, \qquad (6.69)$$

where Ω is the set of the admissible nonnegative values of the controls, which can be specified if necessary.

2.4   Hierarchy of Time Scales

It is very important for quantum systems to select an optimal time interval T for the control function. This requirement is connected with excluding the relaxation processes from consideration. To this end, the control time T should be considerably less than the lifetime of the level being simulated. This lifetime is determined by the time of relaxation of the energy, the so-called longitudinal relaxation time T_j. In many cases, it considerably exceeded by several orders the relaxation time T_2. The longitudinal relaxation time assumes values starting from 10^{-12} sec for radiation-free transitions in the condensed substance up to several hours and even days in the experiments of nuclear magnetic resonance.

For $T \ll T_1$, the radiation itself has a pulse character, and the control action is represented by the envelope of the corresponding wave process. By the very sense of the concept of the envelope, it is given by non-negative functions of time.

The cross-relaxation time arises in extension of the spectral lines. For rarefied gas, the main factors that determine this phenomenon are the interactions with the electromagnetic vacuum, which lead us to the so-called natural extension. The characteristic value of the corresponding cross relaxation time is $T_2 \approx 10^{-6}$ sec. For $T > T_2$, the phase relations are not preserved, and the interaction is of a noncoherent character. The equations of such processes contain relaxation terms and define non-unitary dynamics.

The slowness, which is not of less importance for us, physically means the smallness of change of the amplitude during the period of oscillation of the supporting wave; mathematically, it can be formalized as a requirement that the resonance frequency ω_μ of supporting wave should be considerably greater than the upper bound of the domain, where the spectrum $u_\mu(t)$ is located; that is,

$$\omega_\mu \gg \sup \{\mathrm{Supp}\,(\tilde{u}_\mu(\omega))\}. \tag{6.70}$$

Here *Supp* is the support of a function, that is, the closure of the set on which this function differs from zero; $\tilde{u}_\mu(\omega)$ is the Fourier image of the function $u_\mu(\omega)$. In other words, the resonance exterior field has a narrowband character, and moreover, the width of the band is considerably less than the value taken at its middle value (supporting frequency). The time $T_3 = 1/\omega_\mu$ characterizes the phase coherent property of the transition; the requirement of slowness is expressed by the inequality $T \gg T_3$. This constraint can also be considered as a condition of physical reliability, because the rapid modulation is not possible at present and is intrinsically contradictory in the framework of our consideration. The characteristic values of T_3 in the range of the visual light are of the order of 10^{-15} sec.

Thus, the temporal interval of the admissible control is given by the double inequality

$$T_1, \ T_2 \gg T \gg T_3. \tag{6.71}$$

The presented values of the characteristic times allow us to describe the limits of our consideration. The region covered by them is sufficiently large; this ensures a "space of maneuver" for the elaborator and the correctness of results for the asymptotic study, which is realized in the next section. A more precise choice of T is determined by the technical characteristics of the laser in use.

The necessary relation between the extreme terms of inequality (6.71) has a simple physical sense: the width of energetic levels is considerably smaller than the distance between them. Such a situation is fairly usual for quantum-optical systems in practice.

Thus, we have succeeded in the characterization of the control problems of quantum mechanical systems as control problems in the class of bilinear right-invariant systems on orbits of the Lie algebras of their groups of motion; this stresses the importance of the application of Lie group methods in order to attain the goal of our investigations. We have performed a subject "localization" of mathematical models and indicated an interval test for its adequacy to the physical situation in hand.

3.   Representation of the Interaction

After the analysis of the physical constraints of the problem, which has been based on the available hierarchy of times, we can pass to the simplification of the complete Hamiltonian

$$\widehat{H} = \widehat{H}_0 - \widehat{D} \sum_{\mu=1}^{m} u_\mu(t) \sin(\omega_\mu t + f_\mu) \tag{6.72}$$

of the considered system.

Because $\widehat{H}_0$ does not depend on time, it is convenient to pass to the representation of the interaction. In this passage, the density operator of the system and the Hamiltonian (6.72) are transformed in the following way,

$$\hat{\rho}_{\text{int}}(t) = \widehat{V}_t \hat{\rho}(t) \widehat{V}_t^+,$$
$$\widehat{H}_{\text{int}}(t) = \widehat{V}_t \widehat{H} \widehat{V}_t^{-1},$$
$$\widehat{V}_t = \widehat{V}^t(t) = \exp\{i(t - t_0)\widehat{H}_0\}.$$

Let us consider the set of eigenvectors of the unperturbed Hamiltonian $\widehat{H}_0$ as the basis in the Hilbert space $\mathcal{H}$. The unperturbed dynamics assign the motion defined by the unitary operator $\widehat{V}(t)$. The representation of the interaction means, in fact, the passage to the corresponding "rotational coordinate system." Having the mentioned passage, we obtain the following equation,

$$\frac{d}{dt}\hat{\rho}_{\text{int}}(t) = \{\widehat{H}_{\text{int}}^1, \hat{\rho}_{\text{int}}\} = \text{ad}_{H_{\text{int}}^1}\hat{\rho}_{\text{int}}. \tag{6.73}$$

Thus, the new system turns out to be homogeneous in control actions. Further transformations of the system (6.73) are based on the following relation of scales in the hierarchy of times,

$$T_1, T_2 \gg T \gg T_3. \tag{6.74}$$

They allow us to pass to a simplified description by using the averaging with respect to a physically infinitely small interval Δ that is chosen in such a way that

$$T \gg \Delta \gg T_3. \tag{6.75}$$

The possibility of such a choice is ensured by the values of the characteristic times. The right-hand side of the inequality (6.75) guarantees slow controls preservation and constancy of the slow controls when passing to the slow time; by the averaging theorem, the left-hand side allows us to replace the temporal means by spatial ones. We use this observation for the matrix elements of the perturbation operator in the representation of the interaction. Consider the following matrix element of the interaction Hamiltonian,

$$D_{kl} \exp\{i(\omega_k - \omega_l)(t - t_0)\}u_\mu(t)\sin(\omega_\mu t + f_\mu). \tag{6.76}$$

In the case of resonance of frequencies of the external actions and of a certain passage for a pair of levels $\omega_k - \omega_l = \omega_\mu$, we have

$$D_{kl}\Delta^{-1}\int_{t_0-\Delta}^{t_0} \exp\{i(\omega_k - \omega_l)\tau\}u_\mu(\tau)\sin(\omega_\mu\tau + f_\mu)d\tau = D_{kl}u_\mu(t)\Delta^{-1}$$

$$\times \int_0^{2\pi} (\cos\phi + i\sin\phi)(\sin\phi\cos f_\mu + \sin f_\mu \cos\phi)d\phi + O\left(\frac{1}{\omega_\mu\Delta}\right)$$

$$= \frac{i}{2}\exp(-if_\mu)D_{kl}u_\mu(t) + O\left(\frac{1}{\omega_\mu\Delta}\right) \quad \text{for} \quad \omega_\mu\Delta \to +\infty. \tag{6.77}$$

In the case of absence of resonance of frequencies, that is, when $\omega_k - \omega_l \neq \omega_\mu$, the integrand is a rapidly oscillating function with frequency $\omega_k - \omega_l \pm \omega_\mu$ and zero mean value; as a result, the nonresonant matrix element vanishes.

A small term that will be omitted can be estimated exactly; it is equal to the value of the integral over the part of the period, which remains after extraction of the maximal integral number of periods $T_2 = 2\pi\omega_\mu^{-1}$ from Δ. However, the smallness of the parameter $1/\omega_\mu\Delta$, which is ensured by the corresponding choice of Δ, allows us to take the asymptotic estimation as a satisfactory one.

3.1   Approximation of the Model

It is also useful to note that even when there is no polarization, the quadrature components of the external action (equation (6.77) with $f_\mu = 0$ for the sinusoidal component and with $f_\mu = \pi/2$ for the cosinusoidal one) generate two independent control channels. As we show below, this fact is a necessary condition for ensuring the controllability of the two-level quantum system.

The phase values, which are equal to $-\pi/2$ and π, also yield two new control channels that differ from the quadrature channel by a sign. It is natural not to consider them separately, assuming merely that admissible controls in "quadrature channels" can obtain negative values.

The operation of rapidly omitting the oscillating matrix elements is called by physicists the resonance approximation. The procedure of passing to the "real" time scale presented here is in essence a justification of this approximation in our case. The formal representation of the averaging procedure of the function $f(t)$ on the physically small interval of time Δ is given by the relation

$$\overline{f(t)} = \int_{-\Delta}^{0} f(t+\tau)g(\tau)d\tau,$$

where $g(t)$ is a certain averaging "density" with properties $g(\tau) \geq 0$ and $\int_{-\Delta}^{0} g(\tau)d\tau = 1$.

The simplest case where

$$g(\tau) = \begin{cases} 1/\Delta & \text{for } -\Delta < \tau < 0, \\ 0 & \text{otherwise} \end{cases} \tag{6.78}$$

was used above. The other variants $g(\tau)$ play the role of an "apparatus function of the device" as a formalization of the concept of finite resolution ability. This specific form is essential for a number of mathematical modeling problems. In control theory, their analogues arise as impulse transition functions of various processes.

The above-presented result is based on ideas that are related to the famous functional conjecture of N. N. Bogolyubov, which forms the basis of kinetic theory. The idea of using a smoothing operation on physically infinitely small time scales is contained in Butkovskiy and Samoilenko

(1990), where it is applied for the justification of the deduction of kinetic equations as models of an abbreviated description of the evolution of a statistical ensemble.

So, we have obtained the description of the quantum controlled dynamics in the form

$$
\frac{d}{dt}\hat{\rho}_{\text{int}}(t) = \left\{ \sum_{\mu=1}^{m} \widehat{H}_\mu \hat{u}_\mu(t), \hat{\rho}_{\text{int}} \right\}, \tag{6.79}
$$

where $\widehat{H}_\mu$ denotes a component of the operator of the dipole moment in the representation of the interaction and in the resonance approximation. In the Hilbert space of wave functions, the controlled evolution is described by the equation

$$
\frac{d}{dt}|\psi_{\text{int}}\rangle = \left(\sum_{\mu=1}^{m} \widehat{H}_\mu \hat{u}_\mu(t) \right) |\psi_{\text{int}}\rangle. \tag{6.80}
$$

Thus, as far as control problems of quantum mechanical systems are concerned, we have succeeded in the construction of mathematical models in the class of bilinear right-invariant systems on orbits in finite-dimensional Lie algebras and of groups of motions; this demonstrates the necessity of the use of Lie group theory methods (Brockett, 1972) for their study.

3.2   Quantum Bilinear Dynamics

Let us consider equation (6.80) in more detail. In the representation of the interaction $|\psi_{\text{int}}\rangle$ is a column vector composed of slowly varied components. Only components standing by their vectors that are related to the nonzero matrix elements of the operator of the dipole moment are evolved. We do not consider quantum systems whose spectrum contains infinitely many equidistant levels. An harmonic oscillator belongs to the set of such systems; they are of a considerable theoretical and practical interest, and very effective methods are developed for the study of the corresponding control problems. For a finite number of such levels (including degenerated ones), and also for a finite number of control actions, many components of the wave function are only finitely varied.

Thus, the Hilbert space $\mathcal{H}$ is decomposed into a direct sum of infinite-dimensional subspace that is invariant with respect to the control actions considered, and of an effective finite-dimensional subspace in which the evolution of the system is properly realized. Further consideration is carried out in this effective unitary subspace. We note that this space is formed as a direct sum of linear subspaces each of which corresponds to a certain energy level that is an eigenvalue of the unperturbed Hamiltonian $\widehat{H}_0$; therefore it is not true in general that this space does not contain a part that is control-invariant. For what follows, it is important to preserve this invariant component. Nevertheless, for the case of a non-degenerate spectrum all these stipulations are extra.

Let n be the total dimension of the direct sum introduced above. The wave function is a column vector of dimension n. The density operators and the operators of the physical values are represented by Hermitian matrices of size $n \times n$, and the evolution operator is represented by a unitary matrix of the same size. Unitary operators form the so-called unitary Lie group $U(n)$, and Hermitian matrices form the corresponding Lie algebra $u(n)$ with the Lie bracket

$$\{A, B\} = -i[A, B] = i(BA - AB). \tag{6.81}$$

We obtain a more exact representation if we take into account the following property: the multiplication of the wave function of the system by a phase multiplier of the form $\exp(i\Theta)$ does not change the state described by it. Pure quantum states are uniquely represented by the elements of the projective space CP^n; for the evolution operators, it is appropriate to introduce an equivalence relation between these elements and their equivalent operators which differ from each other by only a phase multiplier. Thus, we can assume that the dynamics of the system is described by $SU(n)$, the special unitary group generated by unitary operators with unit determinant.

The algebra of physical values, the quantum observables of the system, is also reduced if we observe that the trace $\mathrm{Tr}\,(A)$ is a unitary invariant for every observable A; that is, this operator does not participate in the evolution of the system and can be eliminated. Let us demonstrate this by examining the density operator. The unit trace is one of its characteristic properties, which allows us to assign a matrix x to a mixed state of the system so that $\rho = x + 1/n$, where I is the identity matrix of size $n \times n$. A similar operation can be applied to any observable; without loss of generality, we can assume that for each physical value, there is a Hermitian matrix of size $n \times n$ with zero trace. The Lie bracket (6.81) defines the structure of the special unitary algebra $SU(n)$. We obtain the coincidence with the classical definition of $SU(n)$ if each matrix is to be multiplied by the imaginary unit; this matrix becomes skew-Hermitian, and the ordinary matrix commutator can be taken as the Lie bracket.

Returning to the equation of motion for the density matrix (6.79), we note that now (after the above reductions) the evolution of the system can be expressed by the equation

$$\frac{d}{dt}x = \left\{\sum_{\mu=1}^{m}\widehat{H}_\mu\hat{u}_\mu(t), x\right\} = \mathrm{ad}_H(x); \tag{6.82}$$

moreover, x and the components of the vector-operator H_μ are matrices that are elements of the Lie algebra $SU(n)$. Equation (6.82) defines a bilinear system that is homogeneous in controls and is defined on an orbit of the adjoint representation of the group $SU(n)$ in its Lie algebra.

Each element of the orbit of the adjoint representation of $SU(n)$ has the form

$$x(t) = U(t, t_0)x(t_0)U^+(t, t_0), \quad U(t, t_0) \in SU(n). \tag{6.83}$$

Differentiating (6.83) in time, we obtain the following bilinear system on the special unitary group $SU(n)$,

$$i\frac{d}{dt}U(t, t_0) = \left(\sum_{\mu=1}^{m}\widehat{H}_\mu\hat{u}_\mu(t)\right)U(t, t_0), \quad U(t, t_0) = I. \tag{6.84}$$

It is easy to see that this system is right-invariant; that is, if $U(t, t_0)$ is a solution to (6.84), then $U(t, t_0)V$ is also a solution for any $V \in SU(n)$.

It is essential for what follows to mention the properties of reality, compactness, and semisimplicity, which are inherited by the group $SU(n)$. Due to these properties, we can introduce the structure of Euclidean space on the Lie algebra $su(n)$; the inner product in this space is given by the so-called Killing form

$$\langle A, B\rangle - \mathrm{Tr}\,(\mathrm{ad}_A\mathrm{ad}_B), \quad A, B \in su(n), \tag{6.85}$$

which is a positive-definite nondegenerated bilinear form. The invariance is the most important property of this form; this means that it is preserved under the action of the operators of the adjoint representation, and this allows us to make cyclic permutations in the expressions of the form $\langle a, \{b, c\}\rangle = \langle b, \{c, a\}\rangle = \langle c, \{a, b\}\rangle$. It is known that for many Lie algebras (including $su(n)$), the Killing form differs from the trace functional by only a constant factor; that is, we cannot make them differ.

Thus we have embedded the quantum control system into the Lie algebra $SU(n)$; that is, we have revealed implicit group symmetries that are inherited by the initial system. We have carried out the algebraization procedure, which was initiated and developed for mechanical systems without controls.

The algebraization of a dynamical system allows the use of the well-developed tools of the Lie algebras and group theory for the study of this system. These methods for studying a number of problems of mechanics and control, which are developed on this basis, turn out to be very effective.

3.3   Hamiltonian Dynamics

As the first result of the proposed approach, we prove that any dynamical control system of the form (6.62) is Hamiltonian when it is considered on an orbit of the adjoint representation. To do this, we need certain concepts of the symplectic geometry (Dubrovin, Novikov, and Fomenko, 1984); here we restrict ourselves to the most important of them.

A smooth even-dimensional manifold is called symplectic if a closed nondegenerate 2-form ω^2 (symplectic structure) is defined on it. In a natural way, this form assigns a correspondence between the vector fields and the 1-forms on this manifold. Thus, each smooth function on this manifold defines the 1-form (its exterior differential); in turn, this 1-form defines the vector field, which is called Hamiltonian, whose Hamiltonian is the initial function. The phase flow generated by a Hamiltonian vector field preserves the symplectic structure.

A generic orbit of the adjoint representation in a semisimple compact Lie algebra is a symplectic manifold whose symplectic structure is given by the so-called Kirillov form. Each vector ξ that is tangent to an orbit at the point x can be represented in the "bracket form" $\xi = \{a, x\}$. The expression $\omega^2(\{a, x\}, \{b, x\}) = \langle x, \{a, b\}\rangle$, where the Killing form is used as inner product, correctly defines the nondegenerated skew-symmetric closed bilinear form at each point x of the orbit; this form is the Kirillov form.

Consider the dynamical system

$$\frac{d}{dt}x = \{H(t, u), x\} = \mathrm{ad}_H(x) \tag{6.86}$$

The fact that this system preserves the Kirillov form can be proved by a direct computation. Let us show that the function $F(x) = \langle H(t, u), x\rangle$, which is linear on the orbit, is its Hamiltonian. Indeed, its exterior differential $dF(\xi) = \langle H, \xi\rangle$ defines the same 1-form as the Kirillov form when one substitutes the vector field corresponding to (6.86) into it; that is,

$$\omega^1(\xi) = \omega^2(\{a, x\}, \{H, x\}) = \langle H, \{a, x\}\rangle = \langle H, \xi\rangle.$$

The last equation is based on the Killing invariance form. Thus, we have proved the following assertion: each dynamical control system of the form (6.86) is Hamiltonian on an orbit of the adjoint representation of a compact semisimple Lie group.

Note that the Hamiltonian property holds for any choice of admissible controls, and in this case, of course, there is no autonomy property. The assertion is true for any (even non-Markov) character of dependence of the Hamiltonian $H(t, u)$ on control actions.

The invariance of the symplectic form implies the invariance of its exterior powers, and thus, proves the existence of an invariant measure for control systems of the form (6.86) on an orbit of the adjoint representation; this measure is the Liouville form $\wedge_{t=1}^{k} \omega^2 = \omega^2 \wedge \omega^2 \wedge \cdots \wedge \omega^2$; here $2k$ is the dimension of the orbit.

Thus, we have succeeded in characterizing the control problems of quantum mechanic systems as control problems in the class of Hamiltonian systems on orbits in Lie algebras of their groups of motions. This fact stresses the importance of applying Lie group methods, Hamiltonian mechanics, and symplectic geometry for the study of control problems.

It is easy to see that the quantum nature of the equations is not essential for the proof of the Hamiltonian property presented above; this implies that each right-invariant control system on a compact Lie group produces a Hamiltonian system by the action of the adjoint representation in its Lie algebra.

The existence of an invariant measure allows us to hope that it is possible to use ergodic theory for the study of such systems. Also, it is useful to mention the finiteness of the measure constructed, which, in our case, follows from the compactness of an orbit of the adjoint representation. The finiteness implies the possibility of norming; these facts allow us to interpret the measure constructed as a probability one.

4. The Bellman Principle and Quantum Systems

Optimal control theory usually considers the objects, which are described by classical physical laws. In this case, control methods themselves are mainly of a mathematical character and have no certain physical sense. At the same time, it is possible to solve some linear control problems by theoretical physics methods. The present section shows an analogy between the Bellman optimal principle (Agrachev and Sachkov, 2004) for controlled systems and quantum mechanics. For nonlinear stochastic systems, an association is established with a Schrödinger-type equation on the dynamic programming method basis. The importance of establishing quantum-mechanical analogies in order to develop new physical information processing principles was observed in Butkovskiy and Samoilenko (1990), and Gough, Belavkin, and Smolyanov (2005). In addition, quantum features of an atom-molecular structure may be naturally used to solve complicated nonlinear control problems.

As a rule, the present-day problems of optimal control of nonlinear systems are characterized by a considerable complication in their solution. To overcome this difficulty, it is necessary to apply different complicated methods, aimed at decomposing or simplifying the system models. Contrary to such control problem solution methods, an initial problem may be reduced to another problem, solved in the simplest way by contemporary mathematical methods. Also, it is possible to convert the optimal conditions to the Schrödinger-type equation, which can be solved by quantum mechanical methods.

The present section establishes connections between quantum dynamics (Schrödinger equation) and the probabilistic variational principle for dynamical systems (DS). In addition, it is shown that an optimal control problem for quantum systems is a natural generalization of classical dynamical motion laws. For this purpose, the section formulates an optimal equation for the classical dynamical system, and then this equation is used to derive the Hamilton–Jacobi equation integrated, in its turn, by exterior forms. The same control problem is also formulated for a stochastic system as well. If the stochastic control theory is applied for this stochastic system, the Hamilton–Jacobi equation is derived, which is equivalent to some extent to the Schrödinger equation. To obtain the Schrödinger equation, the condition of positiveness for a diffusion constant, which must be physically realizable, is taken into account.

4.1   Deterministic Optimal Control

Let us consider the dynamical system

$$\dot{x} = f(t, x, u), \quad x(t_0) = x_0, \tag{6.87}$$

where $t \in [t_0, t_1] \subset \mathbb{R}$ is time; $x(t) \in \mathbb{R}^n$ is a state vector of a system; $u(t) \in U \subset \mathbb{R}^m$ is a control; f is a sufficiently smooth function $(f : \mathbb{R}^1 \times \mathbb{R}^n \times \mathbb{R}^m \to \mathbb{R}^n)$. Let $\widehat{U}$ be a set of continuous functions $u : [t_0, t_1] \to U$, such that for a fixed function $u(t)$ and some x_0 the equation (6.87) becomes a well-defined dynamical system and its solution $x(t)$ is called a trajectory of the system in relation to the control function $u(t)$. In general, any trajectory $x(t)$ should also fulfill some initial and final conditions.

Let

$$F(t_0, x_0, u) = G(t_0, x_0, t_1, x_u(t_1))$$

$$= \int_{t_0}^{t_1} g_0[x(t), u(t), f] dt + G_0[x(t)], \tag{6.88}$$

be some functional depending on the control $u(t)$ and the initial condition (t_0, x_0). The function $u(t)$ is called a deterministic optimal control if the relevant trajectory $x_u(t)$ given by (6.87) fulfills the boundary conditions and gives a minimum value of $F(t_0, x_0, u)$. Using the Bellman principle of dynamic programming we want to solve the optimal control problem (Kalman, Falb, and Arbib, 1969). This method is based on the function $S(t, x)$, defined as

$$S(t, x) = \inf_{u \in F_{t,x} \subset \tilde{U}} G(t, x, t_1, x_u(t_1))$$

$$= \inf \left\{ \int_{t_0}^{t_1} g_0[x(t), u(t), f]dt + G_0[x(t)] \right\}, \quad t_0 \le t \le t_1, \quad (6.89)$$

where $F_{t,x} \subset \tilde{U}$ denotes a set of all possible $u(t)$ functions for which trajectories $x(t)$ equal x for the initial moment t. The Bellman–Hamilton–Jacobi equation may be proved for this function:

$$\frac{\partial S(t, x)}{\partial t} + \frac{\partial S(t, x)}{\partial x} f(t, x, \tilde{u}(t)) = 0, \quad (6.90)$$

where $\tilde{u}(t)$ is an optimal control function.

We assume that:

(i) No restrictions exist for the values of the control function $u(t)$; that is, $U = \mathbb{R}^n$,

(ii) Equation (6.87) is of the simple form

$$\dot{x} = u, \quad (6.91)$$

(iii) The functional that constitutes the criterion is given as

$$F(t, x, u) = \int_{t_0}^{t_1} L(\tau, x(\tau), u(\tau))d\tau, \quad (6.92)$$

where $L(\tau, x, u)$ is a Lagrange function,

(iv) t_1 and $x(t_1) = x_1$ are fixed values.

Then we have the following equation for the optimal control problem

$$\frac{\partial S(t, x)}{\partial t} + \frac{\partial S(t, x)}{\partial x} \dot{x}(t) + L(t, x, \dot{x}(t)) = 0. \quad (6.93)$$

Taking into account the standard form of the Lagrange function

$$L(t, x, u) = (m/2)u^2 - V(x), \quad (6.94)$$

where $V(x)$ is a potential, we may obtain the equality

$$\frac{\partial S(t, x)}{\partial x} = -m\tilde{u}^T(t). \quad (6.95)$$

Now, applying (6.94) and (6.95), we have, from (6.93),

$$\frac{\partial S(t,x)}{\partial t} - \frac{1}{2m}\left[\frac{\partial S(t,x)}{\partial x}\right]^2 - V(x) = 0. \tag{6.96}$$

It is not difficult to notice that any dynamical system of classical Lagrangian mechanics is of the type considered above. Moreover, we realize that the action of classical mechanics,

$$S_{KM}(t,x) = \int_t^{t_1} L(\tau, \tilde{x}(\tau), \dot{\tilde{x}}(\tau))d\tau, \tag{6.97}$$

treated as a function of the trajectories, initial conditions t, x, and the $S(t,x)$ function, defined for the system considered above, are identical. This helps us to see that $S_{KM}(t,x)$ fulfills equation (6.96). Equation (6.96) for $S_{KM}(t,x)$ is well known in classical mechanics and is called the Bellman–Hamilton–Jacobi equation. Thus, we may really regard classical Lagrangian mechanics as a section of deterministic optimal control.

4.2   The Bellman-Hamilton–Jacobi Theory and Differential Forms

If it is possible to solve optimization problems based on the Bellman–Hamilton–Jacobi equation and the calculus of variations using advanced quantum computers, then it becomes essentially easier to examine non-holonomic problems, especially when they are represented geometrically and invariantly. When using the Hamilton–Jacobi equation and geodesic fields, it is stated that any Euler–Lagrange solution for a positively defined strongly nondegenerate variational problem without pairs of conjugate points delivers a local minimum to the functional (Griffiths, 1983)

$$F : \Theta((G,\omega); [A, B]) \to \mathbb{R}, \tag{6.98}$$

defined by the equation

$$F(N) = \int_N \varphi. \tag{6.99}$$

We introduce the following concepts.

Definition 6.1. *A differential ideal is a graduated ideal $G = \oplus_{q\geq 0}G^q$ in an exterior algebra $A^*(M) = \oplus_{q\geq 0} A^q(M)$ of the form of a class G^∞ on a manifold M, which possesses $dG \subset G$.*

Definition 6.2. *An exterior differential system (G, ω) on a manifold M is specified by a differential ideal $G \subset A^*(M)$ and n-form ω. Let n be the number of independent variables (G, ω).*

Let (G, ω) be the Pfaffian differential system on M, for which the 1-form is its independence condition. Specify the local system (G, ω) by the Pfaffian equations

$$\theta^1 = \cdots = \theta^S = 0, \quad \omega \neq 0, \tag{6.100}$$

specify the two-stage filtration under a fiber bundle

$$W^* \subset L^* \subset T^*(M), \quad \operatorname{rank} L^*/w^* = 1 \tag{6.101}$$

in a cotangent fiber bundle above M, and let $\theta^1 = \cdots = \theta^S = 0$, ω be such locally specified 1-forms, that

$$W^* = \operatorname{span}\{\theta^1 = \cdots = \theta^S\},$$
$$L^* = \operatorname{span}\{\theta^1 = \cdots = \theta^S, \omega\}. \tag{6.102}$$

Let the integral manifold of the system (G, ω) be specified by the mapping

$$f : N \to M, \tag{6.103}$$

where T is a connected 1D manifold, and f is a smooth mapping. Then, the formula (6.103) can be represented as

$$f^*\theta^1 = \cdots = f^*\theta^S = 0, \quad f^*\omega \neq 0. \tag{6.104}$$

Let $N \subset M$ be one, that is $\theta_N^1 = \cdots = \theta_N^S = 0$, $\omega_N \neq 0$. Assume that φ_N is a restriction of φ on N and that $\Theta(G, \omega)$ is a set of integral manifolds of the system (G, ω). Assume also that φ is some 1-form on M, and let

$$F(N, f) = \int_N f * \varphi \tag{6.105}$$

be true for the integral manifold of the system (G, ω).

Definition 6.3. A functional

$$F : \Theta(G, \omega) \to \mathbb{R}$$

is a mapping

$$(N, f) \to \int_N f * \varphi. \tag{6.106}$$

Consider the problem of determining a variation equation for a functional F (i.e., for some manifold Y), determining a Pfaffian system (G, ω) on the integral manifold, which is a one-to-one correspondence with the integral manifold of (G, ω), which meets the Euler–Lagrange functional (6.106).

A variation problem, associated with the functional (6.105), is denoted (G, ω, φ).

Definition 6.4. *A set* (G, ω, φ) *is called here a classical variation problem.*

Pass to the notions of the Hamilton–Jacobi equation and of a geodesic field. They belong to the most important notions of the classical calculus of variations. Consider each of these notions. The following theorem about sufficient conditions for a local minimum is true.

Theorem 6.1. *Let* (G, ω, φ) *be a strongly nondegenerate variational problem on a manifold* M, *and let* $N \subset M$ *be an integral manifold of a system* (G, ω), *and this problem meets the following conditions.*

(i) N *is a solution of the Euler–Lagrange equation,*

(ii) *There exists a quadratic form* $\|A_{\mu\nu}\|$, *positively defined along* N,

(iii) *No two points of the curve* N *are ever conjugated.*

Then N *provides a local minimum of the differential*

$$\delta F(N)L : T_N(\Theta_N(G, \omega; [a, B])) = \mathbb{R} \tag{6.107}$$

for the functional

$$F : \Theta(G, \omega; [a, B]) \to \mathbb{R} \tag{6.108}$$

and this functional is defined by the equation

$$F(N) = \int_N \varphi. \tag{6.109}$$

Definition 6.5. *A function* g, *specified on an open subset* $R \subset Q$ *of a reduced impulse space, is a solution of the Hamilton–Jacobi equation, associated with* (G, ω, φ), *if there exists such a section*

$$s : R \to Y, \quad \omega \circ s = id, \tag{6.110}$$

that

$$dg = s * (\psi_y). \tag{6.111}$$

Definition 6.6. *An action function* $A(q)$ $(q \in U)$ *is defined by the equality*

$$A(q) = \int_{q_0}^{q} s_N * (\varphi_s),$$

and a Hamiltonian of this function is taken along a single curve γ_y, which connects a point q_0 with a point q.

Supposition 6.1. *An action function is a solution to the Hamilton–Jacobi equation.*

In accordance with the Hamilton–Jacobi theorem (Griffiths, 1983), when there exists a general solution of the Hamilton–Jacobi equation, it is possible to integrate the Hamilton equations in quadratures. If there is a strongly nondegenerate variation problem, a solution of the Hamilton–Jacobi equation allows us to integrate the Euler–Lagrange system in quadratures.

4.3 Stochastic Optimal Control and Schrödinger Equations

Consider the case when a dynamic system undergoes both determined and stochastic disturbances. Let stochastic disturbances be represented by a white noise. Then, the natural generalization of the dynamic equation (6.87) of motion is

$$\dot{x} = f(t, x, u) + \sigma(t, x, u)\gamma, \tag{6.112}$$

where f, t, $x(t)$, $u(t)$ are variables identical as in (6.112), γ is the n-dimensional white noise; that is, the generalized derivative of the Wiener process (Gardiner, 1985; Sragovich, 2006), and $\sigma(t, x, u)$ is a function interpreted as a diffusion coefficient.

We restrict ourselves to a stochastic system of the type (6.112) which is the simplest probabilistic generalization of the classical mechanics one. Because the dynamical equation (6.91) in optimal control formulation of classical mechanics is

$$\dot{x} = u, \tag{6.113}$$

its simplest random equivalent, in accordance with (6.112), should have the form

$$\dot{x} = u + a v, \tag{6.114}$$

where a is some constant matrix. For a purpose that is clear later we take $a = (-i/m)^{1/2}I$, where I is the unity matrix. The natural generalization of the criterion (6.92) becomes

$$F(t, x, u) = E_{tx}\left\{ \int_t^{t_1} L(\tau, x(\tau), u(\tau))d\tau \right\}, \tag{6.115}$$

where $E_{tx}\{\ \}$ denotes a kind of mean value, and $u(t)$ is the feedback control function $u(t) = u(t, x(t))$. The mean value operation E_{tx} is uniquely determined by a stochastic process $x(t)$, fulfilling (6.114) and the initial condition $x(t_0) = x_0$ (x is considered here as a common, nonrandom variable). We may define a nonrandom action function $\widetilde{S}(t, x)$ for any stochastic system in the same way as in (6.89). We have

$$\widetilde{S}(t, x) = \inf_u F(t, x, u). \tag{6.116}$$

It may be proved that the function $\widetilde{S}(t, x)$ fulfills, for a general system (6.112) and a criterion (6.115), the equation

$$\frac{\partial \widetilde{S}(t, x)}{\partial t} + B^{\tilde{u}}(t)\widetilde{S}(t, x) + L(t, x, \tilde{u}(t)) = 0, \tag{6.117}$$

where

$$B^{\tilde{u}}(t) = \frac{1}{2} \sum_{i,j=1}^n b_{ij}(t, x, \tilde{u}(t))\frac{\partial^2}{\partial x_i \partial x_j} + \sum_{i=1}^n f_i(t, x, \tilde{u}(t))\frac{\partial}{\partial x_i}; \tag{6.118}$$

$a_{ij}(t, x, u)$ is the i, j element of the $\sigma(t, x^T, u)\sigma^T(t, x, u)$ matrix; $f_i(t, x, u)$ is the ith element of a vector function $f(t, x, u)$: $i, j = 1, 2, \ldots, n$; $\tilde{u}(t) = \tilde{u}(t, \tilde{x}(t)) = \tilde{u}(t, x)$ is a feedback optimal control function. If we consider the simple stochastic generalization (6.114) of classical mechanics we get the following simplified version of equation (6.114),

$$\frac{\partial \widetilde{S}(t, x)}{\partial t} - \frac{i\hbar}{2m} \sum_{i=1}^n \frac{\partial^2 \widetilde{S}(t, x)}{\partial x_i \partial x_i}$$
$$+ \sum_{i=1}^n \tilde{u}(t)\frac{\partial \widetilde{S}(t, x)}{\partial x_i} + L(t, x, \tilde{u}(t)) = 0. \tag{6.119}$$

Let us now consider the following equation, which is equivalent to (6.119),

$$\frac{\partial \widetilde{S}(t, x)}{\partial t} + \min_{u \in \mathbb{R}^n} \left\{ \frac{-i\hbar}{2m} \sum_{i=1}^n \frac{\partial^2 \widetilde{S}(t, x)}{\partial x_i \partial x_i} \right.$$
$$\left. + u\frac{\partial \widetilde{S}(t, x)}{\partial x} + \frac{m}{2}u^2 - V(x) \right\} = 0. \tag{6.120}$$

We easily get that the minimum value of the expression in the bracket above is obtained for:

$$u^T = -\frac{1}{m}\frac{\partial \widetilde{S}(t, x)}{\partial x}, \tag{6.121}$$

where u^T is a transposition of a column vector u. Thus, it is clear from (6.119), (6.120), and (6.121), that

$$\tilde{u}^T(t) = \tilde{u}^T(t, x) = -\frac{1}{m}\frac{\partial \tilde{S}(t, x)}{\partial x}. \tag{6.122}$$

Applying (6.122) to (6.119) we finally get

$$\frac{\partial \tilde{S}(t, x)}{\partial t} - \frac{i\hbar}{2m}\sum_{i=1}^{n}\frac{\partial^2 \tilde{S}(t, x)}{\partial x_i \partial x_i} - \frac{1}{2m}\sum_{i=1}^{n}\left[\frac{\partial \tilde{S}(t, x)}{\partial x_i}\right]^2 - V(x) = 0. \tag{6.123}$$

The equation (6.123) is a kind of Schrödinger equation. Namely, if in a common form for the Schrödinger equation

$$i\hbar\frac{\partial \Psi(t, x)}{\partial t} = -\frac{\hbar^2}{2m}\sum_{i=1}^{n}\frac{\partial^2 \Psi(t, x)}{\partial x_i \partial x_i} + V(x)\Psi(t, x), \tag{6.124}$$

we apply a substitution

$$\Psi(t, x) = \exp\{-(i/\hbar)S_{qm}(t, x)\}, \tag{6.125}$$

and we obtain

$$\frac{\partial S_{qm}(t, x)}{\partial t} + \frac{i\hbar}{2m}\sum_{i=1}^{n}\frac{\partial^2 S_{qm}(t, x)}{\partial x_i \partial x_i}$$

$$-\frac{1}{2m}\sum_{i=1}^{n}\left[\frac{\partial S_{qm}(t, x)}{\partial x_i}\right]^2 - V(x) = 0. \tag{6.126}$$

Thus we see that the last expression is identical to (6.123).

It is worth noticing that for $\hbar \to 0$, the equations (6.123) and (6.126) tend to a classical limit as given in (6.94).

5. Classical and Quantum Controlled Lattices: Self-Organization, Optimization and Biomedical Applications

In this section we discuss developments and optimization of classical and quantum-mechanical cellular dynamatons. Cellular dynamata (Figure 6.1) are complex dynamical systems characterized by two special features: the nodes (component schemes) are all identical copies of a scheme, and they are arranged in a regular spatial lattice (Tsu, 2005; Mahler and Weberruss, 1995).

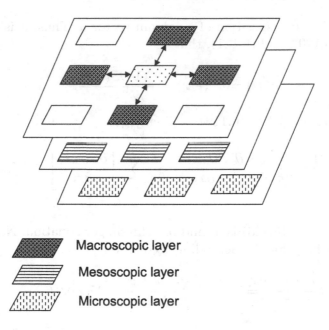

Macroscopic layer

Mesoscopic layer

Microscopic layer

Figure 6.1. Multilevel neural dynamaton.

Definition 6.7. *By a neural dynamical system, (neural dynamaton, or ND), we mean a complex dynamical system in which the nodes are all identical copies of a single controlled Hamiltonian dynamical scheme, the standard cell (Hiebeler and Tater, 1997).*

We consider the cellular dynamaton (CD) (Loscutov and Mikhailov, 1990; Hiebeler and Tater, 1997) in which every cell is considered as the controllable finite-dimensional Hamiltonian system having the form

$$\partial_t x^k = \mathcal{F}_H^k(x^k, a^k, u^k), \quad x^k \in M^k, x^k(0) = x_0^k,$$
$$\partial_t y_j^k = -H_j^k(x^k, a^k), \quad j = 1, \ldots, m, \quad k = 1, \ldots, N,$$
$$u^k = (u_1^k, \ldots, u_m^k) \in \Omega \subset \mathbb{R}^m, \tag{6.127}$$

where $\mathcal{F}_H^k$ is an integral curve of H^k; $H^k(x^k, a^k, u^k)$ is an arbitrary analytical function of states x^k, of cell interaction parameters a^k, and of reconfigurating influences u^k; k is a cell number; $a^k = F^k(x^{k-1}, x^{k+1}, x^k)$. If there exists the Hamiltonian

$$H^k(x^k, a^k, u^k) = H_0^k(x^k, a^k) - \sum_{j=1}^m u_j^k H_j^k(x^k, a^k),$$

then we have the system of equations

$$\dot{x}^k = g_{H_0}^k(x^k, a^k) + \sum_{j=1}^{m} u_j^k g_{H_j}^k(x^k, a^k), \ x^k(0) = x_0^k, \quad x^k \in (M^{2n}, \omega),$$

$$y_j^k = -\frac{\partial H^k}{\partial u_j^k}(x^k, u^k), \quad j = 1, \ldots, m, \quad k = 1, \ldots, N,$$

$$u^k = (u_1^k, \ldots, u_m^k) \in \Omega \subset \mathbb{R}^n. \tag{6.128}$$

Here M is a symplectic manifold with a sympletic form ω and Hamiltonian vector fields $g_{H_j}^k$, $i = 1, \ldots, m$; $\Omega \subset \mathbb{R}^m$ is a control value domain that contains a point 0; $u^k \in \mathcal{U}^k$ are the control vector components that belong to specified classes $\mathcal{U}^k$ of admissible functions. Consider the control u^k as a generalized external macroscopic influence.

5.1   Hamiltonian Models of Cellular Dynamatons

Assume that an elementary cell contained in some CD is described by the Euler–Lagrange equation (Landau and Lifshitz, 1976; Crouch, 1985) with external forces; a cell is a nondissipative system with n degrees of freedom $q_1^k, \ldots, q_m^k$ and a Lagrangian $L^k(q_1^k, \ldots, q_m^k, \dot{q}_1^k, \ldots, \dot{q}_m^k)$. Then, the CD dynamics are defined by the equations

$$\frac{d}{dt}\left(\frac{\partial L^k}{\partial \dot{q}^k}\right) - \frac{\partial L^k}{\partial q^k} = F_i^k, \quad i = 1, \ldots, n, \tag{6.129}$$

where $F^k = (F_1^k, \ldots, F_n^k)$ is a vector of generalized external forces. Let some of the vector components be zero and suppose that the rest of them provide the reconfiguration properties possessed by the CD. Here the Lagrangian cell depends explicitly on the vector parameter of the current value that is determined by the neighboring cell state. Then the representation

$$\dot{q}_k^{l_0} = \frac{\partial H_0^k}{\partial p_i^k}, \quad (\iota - 1, \ldots, n),$$

$$\dot{p}_i^k = -\frac{\partial H_0^k}{\partial q_i^k} + u_i^k, \quad (i = 1, \ldots, m),$$

$$\dot{p}_i^k = \frac{\partial H_0^k}{\partial q_i^k}, \quad (i = 1, \ldots, n), \tag{6.130}$$

where $p_i^k = \partial L^k / \partial \dot{q}^k$ and H_0^k is an internal Hamiltonian for the kth cell, is true.

We have the system

$$\dot{x}^k = g^k_{H_0}(x^k) + \sum_{j=1}^{m} u^k_j g^k_{H_j}(x^k), \quad x^k(0) = x^k_0, \quad x^k \in M^k, \quad (6.131)$$

at the arbitrary coordinates x^k for a smooth manifold M^k ; M^k is a symplectic manifold with a symplectic form ω^k; $g^k_H(x^k)$ and $(j = 0, 1, \ldots, m)$ are Hamiltonian vector fields satisfying the relation $\omega^k(\mathcal{F}_{H_j}) = -dH^K_j$. According to Darboux's theorem (Arnold, 1973; Abraham and Marsden, 1978), there exist canonical coordinates (p^k, q^k) such that

$$\omega^k = \sum_{i=1}^{m} dp^k_i \wedge dq^k_i. \quad (6.132)$$

from the point of view of locality.

If it is assumed that an interaction Hamiltonian H^k_j is equal to q_j $(j = 1, \ldots, m)$, then the system of equations (6.130) is derived. A CD cell "state-output" map may be represented by the expression

$$y^k_j = H^k_j(x^k) \quad (j = 1, \ldots, m); \quad (6.133)$$

that is, an output signal is specified by a disturbance $u^k_1 F^k_H, \ldots, u^k_m F^k_H$ for a natural input, and this circumstance, in its turn, is the reason why the energy $H^k_1, \ldots, H^k_m$ values are changed.

Introduce the following definitions.

Definition 6.8. *A controlled CD model defined by*

$$\partial_t x^k = G^k(x^k, u^k), \quad x^k(0) = x^k_0, \quad x \in M^k,$$

$$y^k_j = h^k_j(x^k, u^k), \quad j = 1, \ldots, r, \quad u^k = (u^k_1, \ldots, u^k_m) \in \Omega \subset \mathbb{R}^n \quad (6.134)$$

is called the Hamiltonian model if it admits the representation in the form of the system

$$\partial_t x^k = \mathcal{F}^k_{H^k}(x^k, u^k), \quad x^k \in M^k, \quad x^k(0) = x^k_0, \quad (6.135)$$

where $H^k(x^k, u^k)$ is the arbitrary analytical function of the kth cell state x^k and of the control parameters u^k; $\mathcal{F}^k_{H^k}$ is integral curve of H^k. If $H^k(x^k, u^k)$ is of the form

$$H^k_0(x^k) - \sum_{j=1}^{m} u^k_j H^k_j(x^k), \quad (6.136)$$

then system (6.128) is obtained. CD cell outputs are specified here by the
expression

$$y_j^k = -\frac{\partial H^k}{\partial u_j^k}(x^k, u^k) \quad (j = 1, \ldots, m). \tag{6.137}$$

Because the state space M^k for system (6.131) is symplectic, it is possible
to yield the Poisson bracket

$$\{F^k, G^k\} = \sum_{j=1}^{n} \left(\frac{\partial F^k}{\partial p_i^k} \frac{\partial G^k}{\partial q_i^k} - \frac{\partial F^k}{\partial q_i^k} \frac{\partial G^k}{\partial q_i^k} \right),$$

$$F^k, G^k \colon M^k \to \mathbb{R}. \tag{6.138}$$

Definition 6.9. *Let L^k be the Lie algebra for the Hamiltonian vector
fields of some kth cell. The linear span for the functions $f^r(H_j^k)$, where
$f^k \in L^k$, is termed the k-cell observation space $\mathcal{H}^k$.*

Because $X_F^k(G^k) = \{F^k, G^k\}$ and $[g_F^k, g_G^k] = g_{\{F,G\}}^k$, $\mathcal{H}_j^k$ is defined by
the functions

$$\{F_1^k, \{F_2^k, \{\ldots \{F_k^k, H_k^k\} \ldots \}\}\}, \tag{6.139}$$

for CD (6.131), (6.133) and here F_r^k, $r = 1, \ldots, k$ is equal to H_i^k, $i = 0, 1, \ldots, m$.

If the theorems considered for the case with nonlinear system controllability and observability (Brockett, 1979) are applied for a CD, then
the following results are obtained.

Proposition 6.1. $\mathcal{H}^k$ *is an ideal formed by $H_1^k, \ldots, H_m^k$ in the Lie algebra (under Poisson bracket) generated, in its turn, by $H_0^k, \ldots, H_m^k$.*

Proposition 6.2. *Let (6.131), (6.133) be a Hamiltonian CD model.
Then:*

(a) *The CD states are strongly accessible and the CD is weakly observable if $\dim d\mathcal{H}^k(x^k) = \dim M^k$, $\forall x^k \in M^k$; otherwise the CD is
quasiminimal.*

(b) *The CD is strongly accessible and observable if $\dim d\mathcal{H}^k(x^k) = \dim M^k$, $\forall x^k \in M^k$, and $\mathcal{H}^k$ allows distinguishing M^k points; otherwise the CD is minimal.*

A nonminimal CD can be reduced to a quasiminimal type under the
same input-output map. If a CD is Hamiltonian, then the Hamiltonian
system is yielded again as the result of the transformation procedure.

It was mentioned earlier that the quasiminimal CDs with one and the same "input-output" map are locally diffeomorphic. In the Hamiltonian case, map equivalence means smooth simplectomorphism; that is, there are local canonical transformation and energy equivalence within the constants (for both CDs). Therefore, the input-output map determines not only the CD state, but also the canonical CD state structure and the internal energy. This approach can be extended to general-type CDs (6.128) and here the observation space $\mathcal{H}$ is used; that is, $\mathcal{H}$ is the linear space that exists for the functions (x, u) and includes $\partial H_1/\partial u, \ldots, \partial H_m/\partial u$ and invariants. A Lie algebra forms by calculations of Poisson brackets of Hamiltonians, Hamiltonian derivatives, and so on.

When the well-known accessibility and observability theorems mentioned in Brockett (1979) and Jurdjevic (1997) are considered for the CD cases, they are formulated as follows.

Theorem 6.2. *Let (6.131), (6.133) be strongly accessible and observable by a CD. Then a CD is minimal if and only if the vector field Lie algebra is self-adjoint.*

Proof: The proof of the theorem follows from the condition of minimality of Hamilton systems (Van Der Shaft, 1982), and the necessary and sufficient conditions of controllability for Lie-determined systems (Brockett, 1979; Jurdjevic, 1997).

Theorem 6.3. *A strongly accessible and observable CD is self-adjoint (and, therefore, Hamiltonian), if the cell output variations $\delta_1 y^k$, $\delta_2 y^k$ are determined on a compact $[0, T]$ and the equality*

$$\int_0^\infty [\delta_1^T u^k(t)\delta_2 y^k(t) - \delta_2^T u^k(t)\delta_1 y^k(t)] = 0.$$

is true for an arbitrary piecewise constant u_j^k and for any two cell input variations $\delta_1 u^k$, $\delta_2 u^k$.

Proof: This result is a synthesis of results from the theory of controlled Hamilton systems put into focus through the maximum principle (Jurdjevic, 1997; Van Der Shaft, 1982).

Besides controllability and observability duality, the Hamiltonian CAs possess a number of other useful features. In particular, the Hamiltonian CD theory is sometimes essential for the quantum models of adaptive computational medium (ACM). The global optimization algorithms (Horst and Pardalos, 1995) can be applied to solve the problem of optimal nonlinear information transforming in ACM.

5.2 Self-Organization of Neural Networks

The phenomenon of self-organization (Loskutov, 1990) that arises due to the interplay of noise and an external control in a bistable elementary cell (EC) has attracted considerable attention in recent years. In a bistable cell characterized by a double-well potential, noise may induce transitions between the two stable states of the system, which behaves stochastically and consequently has a continuous power spectrum. The application of an external time-periodic control has been shown to produce a sharp enhancement of the signal power spectrum about the forcing frequency. This effect has been observed experimentally in simple bistable electronic and optical CD and in more complex systems. Based on this idea, new techniques for extracting periodic signals from the noise background have been used (Klimontovich, 1995). The noise commonly encountered in this context is of external origin in the sense that it appears as a separate term in the appropriate Langeven equations. However, noise or stochasticity also arises in the deterministic dynamical evolution of a bistable cell when coupled to another autonomous or nonautonomous one-degree-of-freedom system.

It is interesting to know the movement property of these systems around the separatrix (Loskutov and Mikhailov, 1990). A characteristic motion of these systems, important in this context, is the motion around the separatrix. It is well known that a generic Hamiltonian perturbation always yields chaotic motion in a phase-space layer surrounding the separatrix. This chaos appearing in the vicinity of the hyperbolic saddle of the bistable potential, although remaining confined within Kolmogorov–Arnold–Moser (KAM) barriers (Arnold, 1973) (which separate the various resonance zones) and being local in character, serves as a precursor to chaotic dynamics. The object of the present section is to examine whether something like stochastic resonance can occur through the interaction of a dynamical bistable EC with intrinsic deterministic noise of this kind and an external periodic forcing. We show that deterministic stochasticity or chaos can be inhibited by critically adjusting the phase and amplitude of the applied resonant driving field.

Our analysis is based on a simple coupled oscillator model describing a nonlinear oscillator with a symmetric double-well potential quadratically coupled to a harmonic oscillator. This dynamical system admits of chaotic behavior and by virtue of having a homoclinic orbit in addition to periodic orbits is amenable to theoretical analysis using Melnikov's technique.

In Melnikov's method one is concerned with the perturbation of the homoclinic manifold in a Hamiltonian system which consists of an integrable part and a small perturbation. It is well known that if Melnikov's function, which measures the leading nontrivial distance between the stable and the unstable manifolds, allows simple zero, then the stable and the unstable manifolds, which for an unperturbed system of oscillators coincide as a smooth homoclinic manifold, intersect transversely for small perturbation generating scattered homoclinic points. This asserts the existence of a Smale's horseshoe (Marcus, 1973) on a Poincaré map and qualitatively explains the onset of deterministic stochasticity around the separatrix. We show that, if an additional external resonant periodic control is brought into play, then, depending on its amplitude and phase, the Melnikov function can be prevented from admitting simple zeroes, which implies that resonance restores the regularity in the dynamics. In other words, deterministic stochasticity is inhibited.

To start with, we consider the following theorem.

Theorem 6.4. *Assume the Hamiltonian dynamaton (HD) is given by*

$$H^k(q^k, p^k, x^k, v^k) = G^k(x^k, v^k)$$
$$+ F^k(q^k, p^k) + \epsilon H^k_{(1)}(q^k, p^k, x^k, v^k), \qquad (6.140)$$

where

$$G^k(x^k, v^k) = \frac{1}{2}\{(v^k)^2 + \omega^2(x^k)^2\}, \qquad (6.141)$$

$$F^k(q^k, p^k) = \frac{1}{2}(p^k)^2 - \frac{1}{2}(q^k)^2 + \frac{1}{4}(q^k)^4, \qquad (6.142)$$

and

$$H^k_{(1)}(q^k, p^k, x^k, v^k) = \frac{\varrho}{2}(x^k - q^k)^2 \qquad (6.143)$$

denote the harmonic oscillator, the nonlinear oscillator with a bistable potential, and the coupling perturbation, respectively, ϱ and ϵ are the coupling and smallness parameters; $I^k = \{1, \ldots, N\}$; N is the number of nodes, equally spaced in the physical space; $\epsilon = \zeta(q^{k+1}, p^{k+1}, x^{k+1}, v^{k+1}, q^{k-1}, p^{k-1}, x^{k-1}, v^{k-1})$ is local bias.

1. *If HD consists of F^k and G^k subsystems, then HD simulates homoclinic and periodic orbits,*

2. *If HD consists of F^k, G^k, and $H^k_{(1)}$ subsystems, and sufficiently small $\epsilon > 0$, then HD simulates the dynamical chaos around the hyperbolic saddle of the bistable potential of each cell,*

3. *If HD consists of F^k, G^k, $H^k_{(1)}$,*

$$H^k_{(2)} = Aq^k \cos(\Omega t + \phi),$$

$\epsilon > 0$ *is sufficiently small,* $A = \varrho(2\bar{h})^{1/2}/\Omega$, $\Omega = \omega$, *and* $\phi = -\pi/2$, *then HD produces a stochastic resonance.*

Proof: The canonically conjugate pairs of coordinates and momenta for G^k and F^k systems are (x^k, v^k) and (q^k, p^k), respectively, and ω is the angular frequency of the harmonic oscillator. The uncoupled system consisting of F^k and G^k systems is integrable. The Hamiltonian perturbation $H^k_{(1)}$ breaks the integrability by introducing horseshoes into the dynamics and thereby making the system chaotic.

Making use of a canonical change of coordinates to action-angle (I^k, θ^k) variables, where θ^k is 2π periodic and $I^k \geq 0$, one obtains G^k as a function of I^k alone as follows,

$$G^k = \omega I^k. \tag{6.144}$$

The action and angle variables for G^k are expressed through the relations

$$x^k = (2I^k/\omega)^{1/2} \sin \theta^k,$$
$$v^k = \omega(2I^k/\omega)^{1/2} \cos \theta^k. \tag{6.145}$$

The integrable equations of motion are ($\epsilon = 0$):

$$\dot{q}^k = \frac{\partial F^k}{\partial p^k}, \quad \dot{p}^k = -\frac{\partial F^k}{\partial q^k},$$
$$\dot{\theta}^k = \omega, \quad \dot{I}^k = 0. \tag{6.146}$$

The Hamiltonian system associated with the F^k system possesses the homoclinic orbit

$$q^k(t) = (2)^{1/2}\mathrm{sech}\,(t - t_0),$$
$$p^k(t) = -(2)^{1/2}\mathrm{sech}\,(t - t_0)\tanh(t - t_0), \tag{6.147}$$

joining the hyperbolic saddle ($q^k = 0, p^k = 0$) to itself. The G^k system (6.144) contains 2π periodic orbits

$$\theta^k = \theta_0 + \omega t, \quad I^k = I_0^k, \tag{6.148}$$

where θ_0^k and I_0^k are determined by the initial conditions. Thus for the uncoupled system $F^k \times G^k$ we have the products of homoclinic and periodic orbits.

Let us now turn toward the perturbed Hamiltonian H^k. The perturbation $H^k_{(1)}$ is smooth. Also, the total Hamiltonian H^k is an integral of motion. The equations of motion are

$$\dot{q}^k = \frac{\partial F^k}{\partial p^k} + \epsilon \frac{\partial H_{(1)}^k}{\partial p^k}, \quad \dot{p}^k = -\frac{\partial F^k}{\partial q^k} - \epsilon \frac{\partial H_{(1)}^k}{\partial q^k},$$

$$\dot{\theta}^k = \omega + \epsilon \frac{\partial H_{(1)}^k}{\partial I^k}, \quad \dot{I}^k = -\epsilon \frac{\partial H_{(1)}^k}{\partial \theta^k}. \tag{6.149}$$

For $\epsilon > 0$, but small, one can show that transverse intersection occurs.

Following Loskutov and Mikhailov (1990) the two-degree-of-freedom autonomous system (6.149) can be reduced to a one-degree-of-freedom nonautonomous system using the classical reduction method. In the process one eliminates the action I^k from equation (6.149) using the integral of motion H^k (6.140). One then further eliminates the time variable t, which is conjugate to H^k, and the resulting equations of motion are written by expressing the coordinate and momentum as functions of the angle variable θ^k.

One need not follow here this procedure explicitly, but can directly use the theorem to calculate Melnikov's function (Melnikov, 1963), which measures the leading nontrivial distance between the stable and unstable manifolds in a direction transverse to dynamic variable θ^k. In practice, the calculation involves the integration of Poisson bracket $\{F^k, H_{(1)}^k\}$ around the homoclinic orbit as follows,

$$\Upsilon(t_0) = \int_{-\infty}^{+\infty} \{F^k, H_{(1)}^k\} \, dt, \quad k = 1, \ldots, N. \tag{6.150}$$

Explicit calculation of the Poisson bracket using equations (6.142), (6.143), and (6.147) yields

$$\{F^k, H_{(1)}^k\} = \frac{\partial F^k}{\partial q^k} \frac{\partial H_{(1)}^k}{\partial p^k} - \frac{\partial F^k}{\partial p^k} \frac{\partial H_{(1)}^k}{\partial q^k} = 2\varrho \operatorname{sech}^2(t - t_0) \tanh(t - t_0)$$

$$+ 2\varrho (I^k/\omega)^{1/2} \sin \omega t \operatorname{sech}(t - t_0) \tanh(t - t_0). \tag{6.151}$$

The Melnikov function is then given by

$$\Upsilon(t_0) = \frac{2\varrho(h)^{1/2}}{\omega} \cos \omega t_0 \left| \pi \omega \operatorname{sech} \frac{\pi \omega}{2} \right|. \tag{6.152}$$

In calculating the relation (6.152) one must take into account that the energy of the homoclinic orbit is zero and $I^k = (h - 0)/\omega$, $h > 0$, where

$$H^k(q^k, p^k, x^k, v^k) = h.$$

Because $\Upsilon(t_0)$ has simple zeroes and is independent of ϵ, we conclude that for $\epsilon > 0$, but sufficiently small, one can have a transverse intersection (and horseshoes on the Poincaré map) on the energy surface

$h > 0$. What follows immediately is that we have simulated the dynamical chaos around the hyperbolic saddle of the bistable potential through the transverse intersection, which is probed by Melnikov's function.

Let us now see the effect of an external time-dependent periodic driving force on this chaos. For this we introduce the perturbation $H_{(2)}^k$, which is of the same order $[O(\epsilon)]$ as $H_{(1)}^k$,

$$H_{(2)}^k = Aq^k \cos(\Omega t + \phi), \tag{6.153}$$

where A, Ω, and ϕ denote the amplitude, the frequency, and the phase of the external field, correspondingly.

It is immediately apparent that the energy function

$$H^k = F^k(q^k, p^k) + G^k(I^k) + \epsilon H_{(1)}^k + \epsilon H_{(2)}^k \tag{6.154}$$

is no longer conserved and one has to consider an equation (Melnikov, 1963) for the time development of H^k in addition to Hamilton's equations of motion

$$\dot{q}^k = \frac{\partial F^k}{\partial p^k} + \epsilon \frac{\partial H_{(1)}^k}{\partial p^k}, \quad \dot{p}^k = -\frac{\partial F^k}{\partial q^k} - \epsilon \frac{\partial H_{(1)}^k}{\partial q^k},$$

$$\dot{\theta} = \omega + \epsilon \frac{\partial H_{(1)}^k}{\partial I^k}, \quad \dot{I}^k = -\epsilon \frac{\partial H_{(1)}^k}{\partial \theta}, \tag{6.155}$$

where

$$f^k = A \cos(\Omega t + \phi). \tag{6.156}$$

One can then use again the classical reduction scheme (Melnikov, 1963) along with an average $\bar{h}$ instead of h (although there are several averaging procedures (Melnikov, 1963) we do not need to have an explicit expression for $\bar{h}$ for our purpose). The relevant Melnikov's function for the problem is as follows,

$$\Upsilon_f^k(t_0) = (1/\omega^2) \left[\Upsilon(t_0) + \int_{-\infty}^{+\infty} \left(-\frac{\partial F^k}{\partial p^k} f^k \right)_{t-t_0} dt \right], \tag{6.157}$$

where $\Upsilon(t_0)$ is given by (6.152) with the replacement of h by some average $\bar{h}$ appropriate for the time-dependent H^k (6.154). The integrand in the expression is a function of the time interval $t - t_0$, where t_0 refers to the time of intersection. On explicit calculation of the integral in (6.157), making use of $\partial F^k / \partial p^k = p^k$ and equation (6.147) we obtain

$$\int_{-\infty}^{+\infty} (\cdots) \, dt = (2)^{1/2} A \sin(\Omega t_0 + \phi) \left[\pi \Omega \operatorname{sech} \frac{\pi \Omega}{2} \right]. \tag{6.158}$$

Also, we have

$$\Upsilon(t_0) = \frac{2\varrho(\bar{h})^{1/2}}{\omega} \cos\left[\pi\Omega \operatorname{sech} \frac{\pi\omega}{2}\right]. \tag{6.159}$$

Let us consider an interesting situation. For $\phi = -\pi/2$ and $\Omega = \omega$, a resonance condition, if one chooses the amplitude of the driving field as

$$A = \varrho(2\bar{h})^{1/2}/\Omega, \tag{6.160}$$

then Melnikov's function $\Upsilon_f^k(t_0)$ vanishes; that is, ceases to have simple zeroes. It implies that the resonance inhibits transverse intersections of the stable and unstable manifolds. As a result, the regularity is restored in the EC and we have a typical situation similar to what is called stochastic resonance. For $A \neq \varrho(2\bar{h})^{1/2}$ Melnikov's function, however, has simple zeroes and the dynamics is chaotic. It has to be noted further that in the absence of any of the perturbation terms the transverse intersections occur and the system becomes chaotic. It is the crucial interplay of both these terms expressed through the condition (6.160) that leads to the inhibition of chaos.

It is pertinent to note the important distinction between stochastic resonance and the suppression of chaos as considered in the present chapter. The stochastic resonance mechanism is related to the oscillating behavior of the signal autocorrelation function for times larger than the relevant decay time (reciprocal of the Kramers rate) in the unperturbed bistable system, and such an effect is apparent even when the perturbation is weak enough not to appreciably affect the rate of the noise-induced switch process (Klimontovich, 1990, 1995, 1999). The suppression of chaos around the hyperbolic saddle, on the other hand, arises only for a critical external field strength and phase. The latter has apparently no role in stochastic resonance. Last we mention that, although we have considered the motion around the separatrix corresponding to a bistable potential, we hope that such an inhibition effect can be seen for other types of potentials where the separatrices do exist.

5.3 Bilinear Lattices and Epileptic Seizures

The dynamics of synapses and global optimization. A simple version of the dynamics of neural synapses can be written (Kandel, 1991) in the form

$$\dot{S}_{kj} = \alpha i_k f_j(t) - H, \tag{6.161}$$

where H defines decay; i is the electric current; $S_{kj} = V_{\text{ion}} s_{kj}$; V_{ion} is the chemical potential; α is the learning rate; s is the action potential; k, j

are cell numbers; $f_j(t)$ is the instantaneous firing rate of neuron j. Decay terms, involving i_k and $f_j(t)$, are essential to forget old information. The learning rate α might also be varied by neuromodulator molecules that control the overall learning process. In the case $S_{kj} = S_{jk}$, there is a Lypunov or "energy" function for equation (6.161),

$$E = -\frac{1}{2}\sum S_{ij}V_iV_j - \sum I_iV_i + \frac{1}{\tau}\sum \int V^{-1}(f')df', \qquad (6.162)$$

and the quantity f_i always decreases in time (Hopfield, 1994). The dynamics then are described by a flow to an attractor where the motion ceases. The existence of this energy function provides a computational tool (Hopfield and Tank, 1985; Hopfield, 1994). In the high-gain limit the system has a direct relationship to physics. It can be started most simply when the asymptotic values are scaled to ±1. The stable points of the dynamic system then each have $V_i = \pm 1$, and the stable states of the dynamic system are the stable points of an Ising magnet with exchange parameters $J_{ij} = S_{ij}$. Many difficult computational problems can be posed as global optimization problems (Pardalos, Floudas, and Klepeis, 1999). If the quantity to be optimized can be mapped onto the form equation (6.161), it defines the connections and the "program" to solve the global optimization problem.

Bilinear lattice model. The bilinear lattice model based on a two-variable reduction of the Hodgkin–Hukley model (Hodgkin and Hukley, 1952) was initially proposed by Morris and Lecar (1981) as a model for barnacle muscle fiber, but is of general utility in modeling the pyramidal cells in network of the CA3 region of the hippocampus (Figure 6.2).

The system of equations for our proposed network model is

$$\dot{x}_1^i = a_1 + b_{11}x_1^i + c_{12}x_2^iu_1^i + u_2^i + b_{13}x_3^i, \qquad (6.163)$$

$$\dot{x}_2^i = a_2 + b_{22}x_2^i, \qquad (6.164)$$

$$\dot{x}_3^i = b_{31}x_1^i + d_3u_2^i, \qquad (6.165)$$

$$y^i(t) = Ex^i(t). \qquad (6.166)$$

This system can by represented by a pair of equations of the form

$$\dot{x}^i(t) = A + \left(B + u_1^i(t)C\right)x^i(t) + Du_2^i, \qquad (6.167)$$

$$y^i(t) = Ex^i(t), \qquad (6.168)$$

with the A, B, C, D, E matrices and the u_j is a scalar function of time and x_j^i. Here

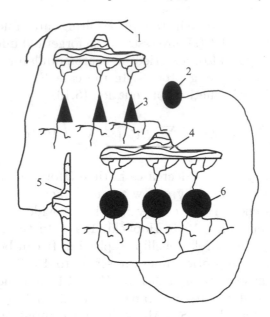

Figure 6.2. A network showing interconnections between an excitatory pathway, a population of pyramidal cells, and a population of inhibitory interneurons (1 is excitatory pathway, 2 is population of excitatory synapses, 3 is pyramidal neurons, 4, 5 are populations of inhibitory synapses, and 6 depicts inhibitory neurons).

$$
A = \begin{bmatrix} a_1 \\ a_2 \\ 0 \end{bmatrix}, \ B = \begin{bmatrix} b_{11} & 0 & b_{13} \\ 0 & b_{22} & 0 \\ b_{31} & 0 & 0 \end{bmatrix}, \ C = \begin{bmatrix} 0 & c_{12} & 0 \\ 0 & 0 & 0 \\ 0 & 0 & 0 \end{bmatrix}, \ D = \begin{bmatrix} 1 \\ 0 \\ d_3 \end{bmatrix},
$$

$a_1 = g_{c_a} m_\infty + g_L V^L$; $a_2 = (\phi \omega_\infty)/\tau_\omega$; $b_{11} = -g_{c_a} m_\infty - g_L$; $b_{13} = -\alpha_{inh}$; $b_{22} = -\phi/\tau_\omega$; $b_{31} = \alpha_{exc}$; $c_{12} = -g_k$; $d_3 = bc$; $m_\infty = f_1(x_1^i, y_1)$; $w_\infty = f_2(x_1^i, y_3, y_4)$; $\alpha_{exc} = f_3(x_1^i, y_5, y_6)$; $\alpha_{inh} = f_4(x_1^i, y_6, y_7)$; $\tau_\omega = f_5(x_1^i, y_3, y_4)$; $u_1^k = (x_1^k - V_k^K)$; $u_2^k = I^k$; x_1^k and x_3^k are the membrane potentials of the pyramidal and inhibitory cells, respectively; x_2^k is the relaxation factor which is essentially the fraction of open potassium channels in the population pyramidal cells; all three variables apply to node k in the lattice. The parameters g_{c_a}, g_k, and g_L are the total conductances for the populations of Ca, K, and leakage channels, respectively. V_k^K is the Nerst potential for potassium in the node. The parameter V_L is a leak potential, τ_ω is a voltage-dependent time constant for W_i, I^k is the applied current, and ϕ and b are temperature scaling factors. The parameter c differentially modifies the current input to the inhibitory interneuron. The parameter V_k^K, the equilibrium potential of potassium for node "k", is taken to be a function of the average extracellular potassium of the six nearest neighbors "k". The value of V_k^K is changed after each interval of simulation time using the following equation,

$$V_k^K = \left(\sum_{j=1}^{6} \frac{V_j^0}{6} \right) - \frac{1}{2},$$

where

$$V_j^0 = \frac{1}{T} \sum_{t_{\text{int}}/6}^{t_{\text{int}}} x_j^k(t)dt;$$

t_{int} determines how long the node's dynamics will remain autonomous without communication with other nearest neighbor nodes, and $T = t_{\text{int}} - t_{\text{int}}/6$ is the interval of time in computing the averages.

This model has a rich variety of relevant dynamical behaviors. The bilinear lattice model can reproduce single-action potentials as well as sustained limit-cycle oscillations for different values of the parameters. Simulations with the proposed bilinear model have shown simple limit cycles as well as periodic state and aperiodic behavior. The periodic state corresponds to a phase-locked mixed-mode state on a torus attractor. As the parameter c is controlled we see a large variety of mixed-mode states interspersed with regions of apparent chaotic behaviour (Larner, Speelman, and Worth, 1997). The dynamics are less complex for smaller values of c; that is, when the current to the inhibitory cells is relatively low compared to the current input to the pyramidal cells. A low degree of inhibition results in a system that is more likely to be periodic, suggesting the type of spatiotemporal coherence that could exist with a seizure. When inhibition is completely absent, the dynamics of a system goes to a fixed point corresponding to a state of total depolarization of the network. An intersection of the transition regions between mixed-mode states undergoes a periodic-doubling sequence as the underlying torus attractor breaks up into a fractal object. When we control the parameters b and y_6 similar behavior is seen in a series of system investigations.

Chaotic lattice model. Bilinear point mappings are self-maintained and a comparatively new part of the theory of dynamic systems, where objects with continuous and discrete time are studied. The following mapping is the particular case of this class (Loskutov and Mikhailov, 1990),

$$x_{n+1} = \sum_{i=1}^{m} a_i u_i(n)(1 - x_n), \tag{6.169}$$

where a_i is a constant parameter; u_i is a scalar control, which can be chosen in a feedback form.

For $a_\infty < a \leq 4$, where $a_\infty = 3.5696\ldots$ the equation has cycles with arbitrary periods including aperiodic trajectories. For $a = 4$ the system dynamics has the property of ergodicity and mixing with exponential divergence of close trajectories.

For (6.169) mappings and

$$y_{n+1} = bu(n)(1 - x_n) \tag{6.170}$$

it is possible to construct a formal mathematical model of the neuron

$$\begin{pmatrix} x_i(n+1) \\ y_i(n+1) \end{pmatrix} = \frac{1}{1 + 2D_i(n)} \begin{pmatrix} 1 + D_i(n) & D_i(n) \\ D_i(n) & 1 + D_i(n) \end{pmatrix} \begin{pmatrix} f[x_i(n)] \\ g[y_i(n)] \end{pmatrix}. \tag{6.171}$$

Here D is the interrelation coefficient between systems (6.169) and (6.170), $f(x) = ax(1-x)$, $g(y) = by(1-y)$, and i is an order number of the neuron.

For $a = b$ and for large D (the synchronism regime) we get the solution represented in Figure 6.3.

For $a \approx b$ ("almost synchronism") the possibility of realization of the regime, which is characteristic for neuron dynamics under the conditions

$$\Delta = |x_n - y_n|, \tag{6.172}$$

$$r(n) = \begin{cases} 1, & \text{for } \Delta \leq \varepsilon, \\ 0, & \text{for } \Delta > \varepsilon, \end{cases} \tag{6.173}$$

where ε is a small parameter.

Figure 6.4 shows the dependence of Δ and r with respect to n ($D = 10^4$, $a = 3.9001$, $b = 3.9$).

The first 300 points were omitted in order to obtain a stable process. For constructing 30 points a steady-state process was used.

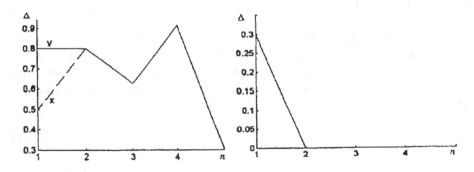

Figure 6.3. Synchronism regime.

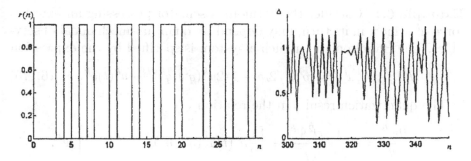

Figure 6.4. "Almost synchronism" regime.

5.4   Quantum Model of Neural Networks

We proceed from the conventional H-Hamiltonian quantization scheme according to which generalized coordinates $q_1, \ldots, q_n$ and momenta $p_1, \ldots, p_n$ are quantized first of all and then H is expressed with respect to these quantum values. Pursuant to the concepts stated in Dirac (1958), when a function set on M is quantized, such quantization means that the Hermitian operator f that acts on a complex Hilbert space is brought into correspondence with each function contained in this set. In addition to this, such operators must satisfy some commutative relations which, in their turn, correspond to Poisson brackets of the classical functions. Assume $M = \mathbb{R}^{2n}$, and let $q_1, \ldots, q_n$ and $p_1, \ldots, p_n$ be the variables that must be quantized. Then the following commutation relations are true,

$$[\hat{q}_i, \hat{q}_j] = 0, \quad [\hat{p}_i, \hat{p}_j] = 0, \quad [\hat{q}_i, \hat{p}_j] = i\hbar\delta_{ij}\hat{I}. \tag{6.174}$$

Let us reduce the value $\hbar$ to 1 in order to simplify our further discussion. Because a skew-Hermitian operator $-i\hat{f}$ corresponds to every Hermitian operator $\hat{f}$, then relations (6.174) mean that the map $\hat{f} \rightarrow -i\hat{f}$ is the morphism of Lie algebra (which contains the Poisson bracket and which is built up on the basis of the classical functions $(q_1, \ldots, q_n, p_1, \ldots, p_n)$ into the skew-Hermitian operator Lie subalgebra. Here the Poisson bracket [f,g] is mapped into the bracket $-i[\hat{f}, \hat{g}] = i[-i\hat{f}, -i\hat{g}]$.

The traditional scheme according to which the values $q_1, \ldots, q_n$ and $p_1, \ldots, p_n$ are quantized and which satisfies expressions (6.174) means that the Hilbert space $L^2(\mathbb{R}^n, C)$ is introduced, that the operator $\hat{q}_j$ is assigned to a coordinate q_j (multiplication by q_j), and that the operator $\hat{p}_j = i\partial/\partial q_j$ is assigned to p. The same scheme may also be applied for some simple Hamiltonian CAs.

Example 6.1. Consider the harmonic oscillator possessing an external force u^k which, in its turn, may depend on quantum-mechanical observables. The Hamiltonian for such a system is specified by the expression

$$H^k(q^k, p^k, u^k) = (p^k)^2/2m + 1/2a^k(q^\ell)(q^k)^2 - u^k(t)q^k. \qquad (6.175)$$

The quantization results in the relation

$$i\hbar \frac{\partial \Psi^k}{\partial t} = -\frac{1}{2m} \frac{\partial^2 \Psi^k}{\partial (q^k)^2} + \left[\frac{1}{2} v^k(t)a^k(q^\ell)(q^k)^2 - u^{l_k}(t)q^k \right] \Psi^k,$$

$$\Psi^k \in L^2(\mathbb{R}, C), \qquad (6.176)$$

which describes, in particular, the dynamics of a particle contained in a single well (bipotential) and this particle is affected, in its turn, by a homogeneous classical external field, and here a field value and a field direction are the arbitrary time functions. Generally, when the Hamiltonians $H_0, \ldots, H_m$ are transformed to the Hermitian operators present in some complex Hilbert space $\mathcal{H}$, the quantized system then looks like

$$i\frac{\partial \Psi^k}{\partial t} = \left(\hat{H}_0^k - \sum_{j=1}^m u_j^k \hat{H}^k \right) \Psi^k, \quad \Psi^k \in \mathcal{H} \qquad (6.177)$$

(Schrödinger representation). Here $\hat{H}^0$ is the Hamiltonian for an isolated quantum system. The second term contained in this expression represents the interaction with the external sources through the system of the interaction Hamiltonians $\hat{H}_1^k, \ldots, \hat{H}_m^k$.

Note that if there is a complex Hilbert space $\mathcal{H}$, then the imaginary part of the inner Hermitian product $\langle \rangle$ defines a simplectic form on $\mathcal{H}$. With respect to this simplectic form, the skew-adjoint operators $-i\hat{H}k_j$ are the linear Hamiltonian vector fields on $\mathcal{H}$. Such fields are the expected observables H_j^k; that is, they correspond to the quadratic Hamiltonians

$$\langle H_j^k \Psi^k, \Psi^k \rangle = \langle \Psi^k \mid H_j^k \mid \Psi^k \rangle, \quad j = 0, \ldots, m. \qquad (6.178)$$

That is why expression (6.176) provides the Hamiltonian system

$$\frac{\partial \Psi^k}{\partial t} = -i\hat{H}_0^k \Psi^k + \sum_{j=1}^m u_j(-i\hat{H}_j^k)\Psi^k, \qquad (6.179)$$

$$y_j^k = \langle \Psi^k \mid H_j^k \mid \Psi^k \rangle, \quad j = 1, \ldots, m \qquad (6.180)$$

with the macroscopic controls $u_j^k, \ldots, u_m^k$ and outputs $y_1^l, \ldots, y_m^k$. The outputs on the infinite-dimensional state space $\mathcal{H}$ are equal to the expected observable H_j^k (Dirac, 1958). Note that the essence of the fact that y_j^k is equal to the expected values of $\hat{H}_j^k$ does not arise from the measurement problem. In addition, although the $\hat{H}_j^k$ measurement process introduces disturbances into a system, these disturbances propagate along the channels that correspond, in their turn, to the inputs $\hat{u}_j^k$. Because H^k is the simplectic space, the Poisson bracket for two observable values, namely, $\hat{H}_1^k$ and $\hat{H}_2^k$, is calculated in the following way

$$\langle \hat{\Psi}^k \mid \hat{H}_1^k \mid \Psi^k \rangle, \quad \langle \Psi^k \mid \hat{H}_2^k \mid \Psi^k \rangle = -\langle \Psi^k \mid [\hat{H}_1^k, \hat{H}_2^k] \mid \Psi^k \rangle. \quad (6.181)$$

The sign "−"arises if the commutators existing for $-i\hat{H}_1^k$ and $-i\hat{H}_2^k$ are taken into account. Therefore, the observation space for a quantum-mechanical CD is specified by the ideal generated by the expected values of $\hat{H}_1^k, \ldots, \hat{H}_m^k$ within the Lie algebra generated by the expected values of $\hat{H}_0^k, \ldots, \hat{H}_m^k$ under the Poisson bracket. Let us compare this observation space with the observation space existing for system (6.128). If the quadratic-linear Hamiltonian is present, then according to the Erenfest theorem, these two observation spaces are equal (in the sense that they are the averaged values which satisfy the classical equations). Therefore, the transputer quantum-mechanical CD can never be minimal. Finally, note that CD quantization may be started not with the variables $q_1^k, \ldots, q_n^k$ and $p_1^k, \ldots, p_n^k$, but with the observation space existing for system (6.135), (6.137). If there are CDs based on harmonic oscillators, the essence of the matter is not changed because the observation space is determined by the linear span of q^k, p^k, and 1. When it is necessary to quantize the observation spaces existing for CDs of different types, then such an approach is of interest.

Finally, it is actually possible to quantize the externally and nonpotentially reconfigurable Hamiltonian CAs. But in this chapter we have shown that it is possible to quantize the "completed" transputer-type CAs.

6. Notes and Sources

Physicists studying open quantum systems have provided new methods that open up future possibilities for efficient optimization algorithms. Our control algorithms realize finite and resonance control, which to date have been primarily thought of as theoretical tools.

We discussed mathematical models of classical and quantum systems with focus on Hamiltonian models of classical and quantum-mechanical networks. The observability, controllability, and minimal realizability theorems for the neural networks were formulated. The cellular dynamaton based on quantum oscillators were presented. The proposed models are useful to investigate information processes in the human brain.

Quantum models could make it possible to view the brain as a quantum macroscopic object with new physical properties (Jibu and Yassue, 1995). These relation properties are very much needed in neuroscience for simulation of physical properties in the brain (Gomatam, 1999). However, efficient techniques to solve reconstruction problems of dynamical systems and optimization algorithms are needed. The global optimization algorithms, described in Horst and Pardalos (1985) can be used.

Self-organization and stochastic resonance can occur through the interaction between macroscopic and microscopic levels of a neural dynamaton with intrinsic deterministic noise and an external control forcing. These interactions through mesoscopic dynamics (Freeman, 2000) are the basis for self-organization of brain and behavior. The neural network controlled by noise may induce transitions between the stable states of the neurons, which behaves stochastically and consequently has a continuous power spectrum. The suppression of chaos around the hyperbolic saddle arises only for a critical external field strength and phase. The latter has apparently no role in the stochastic resonance.

The controlled cellular dynamaton can be used for modeling the CA3 region of the hippocampus (a common location of the epileptic focus). This region is the self-organized information flow network of the human brain. This model consists of a hexagonal CD of nodes, each describing a control neural network consisting of a group of prototypical excitatory pyramidal cells and a group of prototypical inhibitory interneurons connected via excitatory and inhibitory synapses. A nonlinear phenomenon in this neural network has been presented.

The brain is a complex controlled dynamical system with distributed parameters that can display behavior which is periodic or chaotic of varying dimensionality (Sackellares et al., 2000). The controlled chaotic behavior in the brain is healthy whereas lower-dimensional chaos or periodicity is an indicator of disease. In the case of epilepsy, the analysis of a bilinear model shows a seizure where many neurons are synchronized. The bilinear lattice has complex controlled dynamics that should correspond to healthier neural tissue. These dynamics are desirable in brain activity because a chaotic state corresponds to an infinite number of unstable periodic orbits that would then be quickly available for neural computation (Freeman, 2000; Haken, 1996).

Chapter 7

MODELING AND GLOBAL OPTIMIZATION IN BIOMOLECULAR SYSTEMS

This chapter proposes an intelligent sensor based on pattern recognition techniques in a functional space of the fluorescence curves and bilinear models for a sensitive element. We consider new nonlinear and bilinear mathematical models of the dynamic self-organization phenomenon in photosynthetic reaction centers and ion channels (Allen et al., 1986; Deisenhofer et al., 1995; Shimazaki and Sugahara, 1980; Gaididei, Kharkyanen, and Chinarov, 1988; Chinarov et al., 1990, 1992; Shaitan et al., 1991; Gushcha et al., 1993, 1994; Mauzerall, Gunner, and Zhang, 1995; Yatsenko, 1996). Section 7.1 presents the results of a theoretical study of the dynamic self-organization phenomenon in a sensitive element which consists of the photosynthetic reaction centers (RC) from purple bacteria that are responsible for the stochastic effect in RC ensembles. The adiabatic approximation is applied to determine the specific role of slow structural (protein/cofactor) modes in the correlated behavior of electronic and structural variables. It is shown that, at certain values of light intensity, the system undergoes bifurcation. The bistability region for the generalized structural variable occurs where the system has two stable states: one characteristic for the dark-adapted sample (i.e., the sample under very low illumination intensity) and the other for the light-adapted sample. The description is based on a solution of the "forward" Kolmogorov equations using the Markov approach. A distribution function describing the probability of finding an electron localized on a particular cofactor with a certain value of the generalized structural variable is used. This modeling is in good agreement with the results of the experimental investigations of transient optical absorbance changes of the isolated RCs from the purple bacterium Rhodobacter sphaeroides. The results indicate that the free energy difference between Q_A^- and Q_B^-

P.M. Pardalos, V. Yatsenko, *Optimization and Control*
of Bilinear Systems, doi: 10.1007/978-0-387-73669-3,
© Springer Science+Business Media, LLC 2008

changes substantially in the different conformational states of the protein/cofactors induced by light over a wide range of illuminating light intensity. Section 7.2 develops an alternative model of a sensitive element using controlled multiple-ion channels. It is established that the conductivity of biological membranes is basically determined by special proteins, which are capable of forming pores that traverse the cell membranes. Ions and small molecules can travel along these microscopic pores. Thus, the cells affect the matter and charge exchange with an intercellular medium. The ionic channels participate in the processes of the formation and transfer of electric pulses along nerve fibers and muscles, and they facilitate the correlation of chemical reactions in cells and in an intercellular medium. Section 7.3 deals with the influence of pollutants on such biological objects as the photosynthesizing systems. It describes an intelligent sensor based on pattern recognition techniques in a functional space of the fluorescence curves.

1.   Control Dynamics and Photosynthetic Centers
1.1   Mathematical Models

The reaction center (RC) in photosynthetic organisms is a characteristic example of a biological electron transfer system. High-resolution X-ray studies of purple bacterial RCs (Deisenhofer et al., 1995; Allen et al., 1986) and molecular dynamics simulations of electron transfer (ET) reactions have stimulated an increased interest in the physical mechanisms involved in the photoinduced ET. The RC is the pigment–protein complex responsible for the primary charge separation in bacterial photosynthesis. The photoexcitation of a bacteriochlorophyll dimer P (Figure 7.1) is followed by an ultrafast (2–4 ps) charge separation, resulting in the

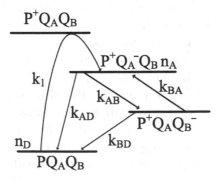

Figure 7.1. Electron transfer (ET) scheme for the RC from purple bacteria *R. sphaeroides.*

formation of the oxidized donor $(P \rightarrow P^+)$, with an efficiency close to 100%. During the next stage ($\sim$150–200 ps), the electron is transfered to the primary quinone acceptor (Q_A) with essentially the same high efficiency. The subsequent slower stage of ET to the secondary quinone acceptor (Q_B), and its possible further recombination with the hole on the pigment pair P or return to the acceptor Q_A, are of particular interest because the rates and efficiencies of these processes depend strongly on the specific configuration of some flexible structural elements of the protein matrix and cofactors. It has been shown that electron localization on each of the quinone acceptors causes structural changes in the RC (Shaitan et al., 1991; Yruela et al., 1994) which strongly influence the efficiency and rates of the ET. This feedback action of the rearranged structure on the charge transfer efficiency may be considered as complex dynamics of electronic and structural variables for the system. In this case, a biological system, from a theoretical point of view, has to be described in terms of stochastic nonlinear dynamic theories (Gardiner, 1985).

In biological systems, the appearance and disappearance of new self-organized states or oscillating regions are possible. It was shown in recent publications (Gushcha et al., 1994; Chinarov et al., 1992) that such correlated behavior of electronic and structural variables leads to the effect of dynamic self-organization (or, better, self-regulation) in macromolecular charge transfer systems. A theoretical description has been developed that describes the situation for a single RC. However, this does not account for the stochastic effects in an ensemble of functioning RCs, which is the situation at hand in a real experiment. The authors describe the processes of RC turnover as a dichotomous noise process. Nevertheless, a more detailed approach to describe the real functioning system is necessary for understanding the dynamics of an ensemble of RCs. The complex nonlinear dependencies and hysteresis effects observed for the RC absorbance changes and the dependence of the primary donor reduction kinetics on various conditions of optical excitation were taken as evidence for dynamic self-regulation. In this work, we present a self-consistent stochastic description of the ensemble of macromolecules undergoing ET, and we demonstrate the importance of nonlinear dynamic effects in the experimentally observed dependencies. We start by developing a theory for the correlated movement of the photoelectron and the RC protein cofactor conformation in a three-level model and then apply the model to experimental data obtained on RCs from the purple bacterium *Rhodobacter sphaeroides (R. sphaeroides)*.

1.2   Kolmogorov Equations and Bilinear Dynamical Systems

Let us consider an ensemble of isolated RCs carrying both primary and secondary quinone acceptors (Q_A and Q_B). The later steps of electron tunneling as well as recombination from both acceptors are of dominant interest for us. In this case, we can neglect the populations of primary charge-separated states; that is, states with an electron localized either on the bacteriochlorophyll monomer or on the bacteriopheophytin molecule, and consider a simpler three-level model for the RCs.

A set of functions $P_i(t, x)$ defining the coupled (in terms of electron i and structural x coordinates) probability of electron localization on the i binding site (i = D, A, B; here, A and B denote the primary or the secondary acceptors, and D denotes the primary electron donor) of the RC with a generalized structural variable x, at time t, is used for the statistical description of the entire RC ensemble. We use the "forward" Kolmogorov equations for determining $P_i(t, x)$ using the Markov approach

$$\frac{\partial P_D(t, x)}{\partial t} = \widehat{L}_D P_D(t, x) - k_1 P_D(t, x)$$
$$+ k_{AD} P_A(t, x) + k_{BD} P_B(t, x),$$
$$\frac{\partial P_A(t, x)}{\partial t} = \widehat{L}_A P_A(t, x) + k_1 P_D(t, x)$$
$$- (k_{AD} + k_{AB}(x)) P_A(t, x) + k_{BA}(x) P_B(t, x),$$
$$\frac{\partial P_B(t, x)}{\partial t} = \widehat{L}_B P_B(t, x) - k_{BA} P_B(t, x)$$
$$+ k_{AB}(x) P_A(t, x) - k_{BD} P_B(t, x), \tag{7.1}$$

where

$$\widehat{L}_i = C_d \frac{\partial}{\partial x} \left[\frac{1}{K_b T} \frac{\partial V_i(x)}{\partial x} + \frac{\partial}{\partial x} \right], \tag{7.2}$$

$V_i(x)$ is the potential of the structural variable x in the case of electron localization on the i binding site, and C_d is the diffusion constant corresponding to the movement of the generalized structural variable x on the potential surface. Several experiments show that photoactivation of RCs leads to reversible structural changes, which relax more slowly than the recombination time of the distribution of the charge-separated states. In this case, the ET between cofactors will be of adiabatic character, and the kinetic constants k_i (Figure 7.1) will depend parametrically upon such slow structural changes. In our treatment, we take into account (see below) the dependence of only two rate constants k_{AB} and

k_{BA} on the single 'slowly varying' structural variable x. Such an approximation should be reasonable for the photosynthetic RC because these rate constants change drastically (as was shown experimentally for the *R.* sphaeroides RC (Malkin et al., 1994)) under the variation of the experimental conditions (temperature, actinic light intensity, etc.).

A number of studies have shown the significance of fast structural dynamics in the description of a functioning RC. With respect to such fast structural changes, the electron transfer can be treated as nonadiabatic. The influence of these fast structural variables is typically included in the expressions for the ET rates in accordance with the well-known Marcus expression for the (nonadiabatic) ET rates.

Our model (equations (7.1) and (7.2)) may be considered as the generalization of the Agmon–Hopfield model for the case of three states of the main system coordinate (population probabilities P_D, P_A, and P_B of the states PQ_AQ_B, $P^+Q_AQ_B$, and $P^+Q_AQ_B^-$, respectively) accounting for all possible transitions between them.

A general solution of the system equations (7.1) and (7.2) cannot be derived without an additional simplification. If we take into account that the relaxation of the RC electronic coordinate occurs much faster than the structural one, the solutions to equations (7.1) and (7.2) can be given in the adiabatic approximation:

$$P_i(t, x) = n_i(t|x)P(t, x), \tag{7.3}$$

$$\frac{\partial P(t, x)}{\partial t} = C_d \frac{\partial}{\partial x}\left[\frac{P(t, x)}{K_b T}\frac{\partial V_{\text{eff}}(x)}{\partial x} + \frac{\partial P(t, x)}{\partial x}\right], \tag{7.4}$$

where $n_i(t|x)$ is the population at the binding site i with a fixed generalized structural variable x, which is averaged over the fluctuation caused by an electronic transition. This structural variable x can be considered as a control mode (Haken, 1978); $P(t, x)$ is the probability density to find the RC with the generalized structural variable x at a particular electron localization site i. $V_{\text{eff}}(x)$ is an effective adiabatic nonequilibrium potential for the control mode x, determined from

$$\frac{\partial V_{\text{eff}}(x)}{\partial x} = \sum_i \frac{\partial V_i(x)}{\partial x} n_i(x), \quad n_i(x) = \lim_{t \to \infty} n_i(t|x). \tag{7.5}$$

The effective potential is of statistical origin (i.e., it depends on the populations n_i) and depends on the intensity I to the actinic light due to the dependence of $n_i(t|x)$ on I. Actinic light is light that produces an identifiable or measurable change when it interacts with a system. The light intensity I (actually, the light-induced turnover rate) in this case stands for the control parameter of the nonequilibrium potential and, therefore, determines the nonlinear dynamic behavior of the system. We now show that bifurcation can arise for the system under consideration.

The equations (7.3) and (7.4), together with equation (7.5) and the balance equations for the population probabilities $n_i(t|x)$ (see below, equation (7.7), for the time-dependent populations of binding sites i; we omitted in these equations the brackets $(t|x)$ near each of the symbols n_i for simplicity), give the correct general description for an ensemble of RCs. This accounts for the effects of the interaction of the photoactivated electron with the adiabatic structural variable x. One should note that equation (7.4) for the RC ensemble corresponds to the stochastic equation for the structural variable x of a single RC, which can be given by

$$\tau_x \frac{dx}{dt} = -\frac{\partial V_{\text{eff}}(x)}{\partial x} + \sqrt{2C_d}\zeta(t), \qquad (7.6)$$

where x is averaged over the fluctuations caused by an electronic transition, as indicated above (i.e., x in both equations (7.4) and (7.6) should be considered as averaged over a time interval longer than the time of charge recombination but shorter than the relaxation time of the adiabatic structural variable x); $\zeta(t)$ describes a δ-correlated random process that models the initial thermal fluctuations of the structural variable of a single RC and $\tau_x = k_b T / C_d$. The theory for a single RC was developed earlier in Gushcha et al., (1995), where the stochastic term $(2C_d)^{1/2}\zeta(t)$ has been neglected. The approach used in the previous work corresponds to the particular case also described here when the RC's distribution function over the structural variable is very narrow. Then, we can introduce the simplified *conformational approach* assuming that, in a particular *conformational state* (which corresponds to a maximum of the distribution function $P(t, x)$), the RCs have almost equal values of the structural variable.

Next, we consider the time evolution of the populations of the different RC levels. This can be described by the following system of balance bilinear equations

$$\frac{dn_D}{dt} = -k_1 n_D + k_{AD} n_A + k_{BD} n_B,$$

$$\frac{dn_A}{dt} = -k_{AD} n_A + k_1 n_D + k_{BA} n_B, \qquad (7.7)$$

$$\frac{dn_B}{dt} = +k_{AB} n_A - k_{BA} n_B - k_{BD} n_B,$$

with normalization conditions

$$n_D + n_A + n_B = 1. \qquad (7.8)$$

In equation (7.7), k_1 is proportional to the intensity of the actinic light I; k_{AD} can be taken as a constant for the sake of simplicity. A dependence of the rate constant k_{AD} on the light intensity for the samples containing only the primary (Q_A) acceptors has been found recently (Schoepp et al., 1992), and it was shown in Goushsa et al. (1995) that this can be due to a small light-induced structural change that has no pronounced influence on other macroscopic parameters of the RC. The k_{AD} value may be determined from experiments; likewise, $k_{AB}/k_{BA} = \exp(\Delta G_{AB}/k_b T)$ can be determined from experiment, where ΔG_{AB} is the energy difference between the $P^+Q_A^-Q_B$ and $P^+Q_A Q_B^-$ states, k_b is the Boltzmann constant, and T is the temperature. We can also set $k_{BD} = 0$, reflecting a negligible probability of a direct pathway for recombination, as compared to the recombination $P^- Q_A Q_B^- \to P Q_A Q_B$ via the primary acceptor Q_A.

Next, we need to find a suitable macroscopic parameter of the RC to describe the correlation between the ET rate constants and the light-induced structural changes. Taking into account the dependence of ΔG_{AB} on the RC structure, we describe, for the sake of simplicity, the RC structural changes in terms of their influence on the free energy difference ΔG_{AB} between the states $P^+Q_A^-Q_B$ and $P^+Q_A Q_B^-$ only. In such a description, we assume that $\Delta G_{AB} = \varphi(x)$, where φ is some function of the structural variable x. The structural changes following electron localization on the quinone acceptors Q_A and Q_B are likely to be very complex.

As an example, we can mention the fast protonation of protein groups that are close to the quinone acceptors, the stronger binding of the singly reduced quinone to the protein pocket as compared to the oxidized one, and the slow conformational rearrangements of the binding pocket itself, which relax on a time scale of minutes. We consider in our theoretical model that the slow conformational rearrangements which influence the value of ΔG_{AB} can be described in terms of a single generalized structural variable x. Such a variable may describe the complex reorganization of the system and can be introduced in a similar way as the generalized "reaction coordinate" (or "perpendicular coordinate") in chemical kinetics. We assume also that ΔG_{AB} is a single-valued function of the structural variable. Hence, for the present consideration, we choose the simplest (i.e., linear) relation between ΔG_{AB} and x without any restriction of generality. We thus take $x = \Delta G_{AB}/k_b T$ as a dimensionless coordinate. Equations (7.1)–(7.6) then describe the dynamics for this structural variable. The stationary populations $n_A(x)$ and $n_B(x)$ (for their definitions see equation (7.5)) can be defined as functions of both actinic light intensity I and structural variable x. Following these lines, the expressions for populations $n_A(I, x)$ and $n_B(I, x)$ can be easily derived from the balance equations taking into account equation (7.8):

$$n_A(I, x) = \frac{I}{I(1 + \exp(x)) + k_{AD}},$$

$$n_B(I, x) = \frac{I \exp(x)}{I(1 + \exp(x)) + k_{AD}}, \qquad (7.9)$$

where we assumed that $k_1 = I$ (both these quantities as well as other rate constants are given in s^{-1}).

Let us consider the changes of the effective potential $V_{\text{eff}}(x)$ for the three-level system under the variable actinic light intensity I. In this case, we can write $V_{\text{eff}}(x) \equiv V_{\text{eff}}(x, I)$ (note that the function $V_{\text{eff}}(x, I)$ is the same function as $V_{\text{eff}}(x)$, with the only difference that the first is written using the direct representation of the dependence on I, whereas the second one has this dependence hidden in x). The potential $V_{\text{eff}}(x, I)$ is determined from equation (7.5) by accounting for the condition

$$n_A(I, x) \ll n_B(I, x), n_p(I, x) \qquad (7.10)$$

which follows directly from both the relationship (7.9) and the inequality $k_{AB}/k_{BA} = \exp(x) \gg 1$. (The inequality (7.10) is always valid except for the case of the saturating intensities of pulsed RC photoactivation when $n_A(I, x) \approx 1$ just after the initial steps of the charge separation. Such a situation does not apply to our situation; thus, the correlation (7.10) is valid for all values of I and x that are of interest.) The low value of the population $n_A(I, x)$ under the conditions assumed here means that the light-induced structural changes in the Q_A binding pocket are much less probable than those in the Q_B binding pocket. We thus restrict the consideration to the structural changes of only the Q_B pocket in view of presently available experimental data (see also the discussion below). This means that the structural variable $x = \Delta G_{AB}/k_b T$ is determined mainly by the local structure around the Q_B site. This may be justified by several arguments. First, the polarity of the Q_B pocket is much larger than that of the Q_A pocket. This favors the conditions where the photoinduced charge separation influences the local structure in the Q_B site by electrostatic interactions more than in the Q_A site. Second, the binding strength of ubiquinone in the Q_B site depends strongly on the Q_B redox state. Third, the lifetime of the $P^+ Q_A Q_B^-$ state is considerably longer than the lifetime of all other charge-separated states in the RC, which allows for a higher effective influence of the electron charge on the surrounding structure. Thus, we can omit for simplicity in equation (7.5) the term $\partial V_A(x) \partial x \, n^A(I, x)$ and, using equations (7.5), (7.8), and (7.10), obtain an expression for the effective potential $V_{\text{eff}}(x, I)$:

$$\frac{\partial V_{\text{eff}}(x, I)}{\partial x} = \frac{\partial V(x)}{\partial x} - f_{\text{B}} n_{\text{B}}(I, x), \tag{7.11}$$

where $V(x) \equiv V_{\text{D}}(x)$ is the initial potential when the electron is localized on the donor and

$$f_{\text{B}} = \left(-\frac{\partial V_{\text{B}}(x)}{\partial(x)} \right) - \left(-\frac{\partial V_{\text{D}}(x)}{\partial x} \right), \tag{7.12}$$

where f_{B} is an additional force acting on the structural variable only in the case of an electron being localized on Q_{B}. In further calculations we use the f_{B} value in units of $k_{\text{b}} T$. This force f_{B} is of stochastic nature, and it describes the perturbation of the local RC structure in the $P^+ Q_{\text{A}} Q_{\text{B}}^-$ state.

For the sake of simplicity but without losing generality, we use the constant force approximation so that f_{B} does not depend on x. An initial effective potential $V(x)$ should be chosen for the determination of the dependence of the structural variable x on the intensity of the actinic light. We take the harmonic potential $V(x) = k(x - x_0)^2/2$, where k is given in energy units and depends on the medium elasticity (k is defined usually as the quantity that essentially depends on the solvent reorganization energy), and x_0 (a dimensionless value) is the equilibrium value of the structural variable in the absence of actinic light (when the electron is on P). Then we obtain from equations (7.11) and (7.12),

$$\frac{\partial V_{\text{eff}}(x, I)}{\partial x} = k(x - x_0) - f_{\text{B}} n_{\text{B}}(I, x), \tag{7.13}$$

After integrating equation (7.13) and taking into account equation (7.9), we get

$$V_{\text{eff}}(x, I) = \frac{k(x - x_0)^2}{2}$$
$$- k(x - x_0) \ln \frac{(1 + \exp(x))I + k_{\text{AD}}}{(1 + \exp(x_0))I + k_{\text{AD}}} + C(I), \tag{7.14}$$

where $(x_{\text{B}} - x_0) = f_{\text{B}}/k$ denotes the maximum shift of the structural variable that could occur under the limiting condition of permanent localization of an electron on Q_{B}; $C(I)$ is an integration constant which depends primarily on the intensity of actinic light. It is easy to show (taking into account initial conditions) that

$$C(I) = \frac{f_{\text{B}}^2}{2k} \left[\frac{I \exp(x_0)}{I(1 + \exp(x_0)) + k_{\text{AD}}} \right]^2. \tag{7.15}$$

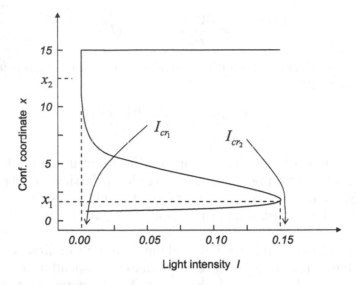

Figure 7.2. System conformational potential obtained for $(x_B - x_0) = 13$ under different levels of stationary actinic light: 1, 0; 2, 10^4 (Icr_1); 3, 1.5×10^{-1} (Icr_2); 4, 1. The curve between 1 and 2 and all the curves between 2 and 3 correspond to intermediate I levels. The second potential minimum appears at $Icr_1 = 10^{-4}$ (the bifurcation point). The value $f = 1$ corresponds to the half-saturating light intensity, ca. 2×10^{15} quanta/(cm^2 s); $x_B = f_B/k + x_0$ is the conformational coordinate value after passing the second bifurcation point (at high actinic light intensity). x_0 is the conformational coordinate value in the absence of light (see text). The curves were obtained for the following values of the RC parameters: $k_{AD} = 13$; $f_B = 7$; $k = 0.5$; $x_0 = 1$ (k_{AD} is in the same units as I, s^{-1}; f_B and k are in the units of $k_b T$; x_0 is dimensionless).

From the analysis of equation (7.14), we obtain, for the case that the condition

$$x_B - x_0 \geq 4 \tag{7.16}$$

is fulfilled, two minima in the effective potential $V_{eff}(x, I)$ (see below).

Figure 7.2 shows the calculated dependencies of $V_{eff}(x, I)$ in the model described above for RCs from R at different values of the actinic light intensity. The minima of the effective potential correspond to the stable states of the system. We denote these states conformational states, where the corresponding values of the structural variable x are termed conformational coordinate values, ξ. We should note here that the initially introduced quantity x in the Kolmogorov equation (7.1) has the meaning of a generalized structural variable for that we have developed the above formalism. The manifold of the states of the electronic-conformational system of the RC can be characterized in general by the manifold of values x. The conformational coordinate ξ, in contrast, describes only the equilibrium configurations of the conformational system. This means that the values x and ξ are identical only for the manifold of equilibrium states when the first derivative of the effective adiabatic potential equals zero (see below). Note that, in this definition, the above-introduced quantities x_0 and x_B belong to the manifold of the conformational coordinate ξ values (nevertheless for these two particular characteristic parameters the structural variable value of the dark-adapted RCs and the maximal shifted structural variable value, we use the previously introduced symbols x_0 and x_B, respectively). We now proceed with the development of the theory using the conformational approach. According to the description given above, the equation for determination of the conformational coordinate values ξ may be obtained from $\partial V_{\text{eff}}(x, I)/\partial x|_{x=\xi=0}$:

$$\xi = x_0 + \frac{f_B}{k} \frac{I \exp(\xi)}{I(1 + \exp(\xi)) + k_{\text{AD}}}. \tag{7.17}$$

From the analysis of the dependence $\xi(I)$, two cases should be distinguished. The first corresponds to the case of weak interaction between the electron localized on Q_B and the local structure in the Q_B site. In this case, the structural changes are not large enough (i.e., the difference $(x_B - x_0)$ would be less than the critical value) to ensure efficient feedback in the electronic-conformational system of the RC. The function $\xi(I)$ in this case grows monotonically and simultaneously with increasing I and approaches asymptotically a constant value at $I \to \infty$ (this rather trivial case is not shown in Figure 7.3). In the second case (i.e., the strong interaction case when the maximal possible structural change described by the difference $(x_B - x_0)$ exceeds the critical value), which is of primary interest for us, the dependence is of S-type in shape (see Figure 7.3). As indicated earlier, such a dependence is typical for the behavior of the reaction coordinate of systems under conditions far from thermodynamic equilibrium. This means that one should expect, in such a situation, the characteristic nonlinear dynamic effects (synergetic effects) in the behavior of certain RC parameters under particular conditions of actinic illumination. In particular, bifurcations should be observable in this case, and they should appear for RCs if $(x_B - x_0) \geq 4$. The values of the conformational coordinate that correspond to the two bifurcation points (where the derivative $\partial I/\partial \xi = 0$) are determined from equation (7.17) as

$$\xi_{1,2} = \frac{x_B + x_0}{2} \pm \sqrt{\left(\frac{x_B - x_0}{2}\right)^2 - (x_B - x_0)}. \tag{7.18}$$

Hence, it follows that, at $(x_B - x_0) < 4$, the bistability region is absent, and the weak interaction case described above is realized. Note the close

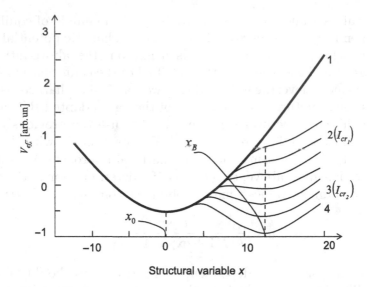

Figure 7.3. Dependence of the conformational values χ on the actinic light intensity for the case $(x_B - x_0) = 13$ Icr_1 and Icr_2 are the intensities of the light corresponding to the bifurcation points χ_1 and χ_2 of the system (see text). Two horizontal arrows indicate the value of the conformational coordinate in the dark (x_0, lower value) and saturation light intensity (x_B, upper value). The curve was obtained for the following values of RC parameters: $k_{AD} = 13$; $f_B = 7$; $k = 0.5$; $x_0 = 1$ (k_{AD} is in the same units as I, s^{-1}; f_B and k are in the units of $k_b T$; x_0 is dimensionless).

proximity of the values ξ_1 and ξ_2, which denote the bifurcation points, to the values x_0 and x_B, which denote the conformational coordinate values in the dark and at saturating actinic light, respectively, for the case $(x_B - x_0) < 4$. This proximity is due to the particular type of adiabatic potential that has been chosen in the present consideration. This means also that an increase of the light intensity above Icr_2 causes only minor further changes of the conformational coordinate value as compared to its value ξ_2 in the upper bifurcation point (see below). Analogously, the decrease of the light intensity below Icr_1 also causes only slight changes of the conformational coordinate value as compared to its value ξ_1 in the low-intensity bifurcation point.

In the case of strong electronic-conformational interaction and feedback in the system (i.e., when $(x_B - x_0) \geq 4$) the RCs are described by an adiabatic potential with only one minimum at $I < 10^{-4}$. The second potential minimum characterizes the new conformational state of the system appearing at the bifurcation point $Icr_1 \approx 10^{-4}$. The latter can be seen in the ξ dependence on actinic light intensity (Figure 7.3). After passing a critical value Icr_1 (the bifurcation point), the system is characterized by three possible stationary states. Two of them are stable and correspond to the potential minima. The third state is unstable and corresponds to the maximum of the potential. After passing the second bifurcation point $I cr_2 \approx 1.5 \times 10^{-1}$, the system remains in the single allowed (light-adapted) state with a considerably displaced conformational coordinate ξ as compared to that in the dark state.

From the above discussion, it follows that the RC conformational coordinate, which represents the control mode for fast electronic transitions, should reveal a nontrivial dependence on actinic light intensity (if $(x_B - x_0) \geq 4$, the validity of such a correlation for purple bacteria RCs is discussed below). The ξ coordinate value increases slowly with an increase of the actinic light intensity from 0 to $I cr_1$. In the region $I cr_1 < I < I cr_2$, two different conformational states coexist with essentially different values of the conformational coordinate ξ. Using a detailed balance of the states connected by the rate constants k_{AB} and k_{BA} (dependent on the variable x), the different values of these rate constants will correspond to different RC conformational states in the bistability region. For the case $I > I cr_2$, the conformational coordinate has a single value and is almost independent of I. Note that, under the conditions of decreasing actinic light intensity back into the bistable region down to the $I cr_1$, the conformational coordinate remains unchanged if the thermally activated transitions between the minima of the adiabatic potential do not succeed in equilibrating thermodynamically the RC conformational states during the variation of I.

From the considerations made above, it can be concluded that the behavior of the electronic-conformational system of the RCs may depend strongly on the prehistory of photoactivation, that is, turnover. Nonlinear dynamic effects should become significant and can be observed experimentally when the relationship (7.16) is fulfilled; that is, when $\Delta G_{AB}(I = \infty) - \Delta G_{AB}(I = 0) > 4 k_b T$. As was shown, this correlation should, indeed, be fulfilled for the purple bacterial RC. One should note, however, that, in accordance with the results of the theory presented here, the measurement of the value $\Delta G_{AB} (I = \infty)$ should be carried out after prolonged illumination of the RCs with light of saturating intensity, such that the relaxation of the conformational coordinate is complete.

1.3 Modeling and Experimental Results

To allow comparison with experiments, we develop the theory further to describe the behavior of a macroscopically accessible parameter of the RC ensemble, that is, the optical absorbance change of P.

Let us consider the change of the RC ensemble distribution over the structural variable in two limiting cases of variation of the actinic light:

(a) a slow adiabatic variation of I, which permits thermodynamic equilibrium to be reached between the conformational states of the system, and (b) a nonadiabatic variation of I, in which the thermodynamic equilibrium is reached near each of two minima of the effective potential $V_{\mathrm{eff}}(x, I)$ independently. In the latter case, we assume that the height of the barrier between the two potential minima is large enough that we may neglect the thermally activated system transitions between conformational states during the period of variation of the actinic light intensity I.

Let N be the total number of RCs. The Boltzmann distribution can be written for the density of states of the structural variable under stationary conditions for the fully thermodynamically equilibrated system:

$$P(x) = Z^{-1} \exp(-V_{\mathrm{eff}}(x, I)/k_b T),$$

$$Z = \int_{\infty}^{\infty} \exp(-V_{\mathrm{eff}}(x, I)/k_b T) dx. \tag{7.19}$$

This case is realized when the rate of light intensity variation is very slow in comparison with the rate of the thermal transitions over the potential maximum. Then the thermally equilibrated system distribution has time to establish itself at each new value of actinic light intensity. The prehistory of the intensity variation does not influence the distribution in that case.

Another extreme case is realized when the thermal transitions over the potential barrier can be neglected (fully nonequilibrium case). This means that the proper quasistationary distribution is reached for each of the two potential minima independently. The potential maximum at $x = x_{\max}$ divides the total RC number N into two parts, $\nu 1 N$ and $\nu 2 N$, where $\nu 1$ and $\nu 2$ are the populations of the RC conformational states. The number of RCs with $x < x_{\max}$ is $\nu 1 N$, whereas the corresponding number with $x > x_{\max}$ is $\nu 2 N$. The boundary conditions for these cases are as follows: at $I = 0$, $\nu 1 = 1$ and $\nu 2 = 0$; at $I > I cr_2$, $\nu 2 = I$ and $\nu 1 = 0$. The resulting distribution function for the nonequilibrium case can be simplified, and we obtain

$$P(x) = \begin{cases} \nu 1 Z_1^{-1} \exp\left(\dfrac{-V_{\mathrm{eff}}(x, I)}{k_b T}\right), \\[2mm] Z_1 = \displaystyle\int_{-\infty}^{x_{\max}} \exp\left(\dfrac{-V_{\mathrm{eff}}(x, I)}{k_b T}\right) dx, \quad x < x_{\max}, \\[4mm] (1 - \nu 1) Z_2^{-1} \exp\left(\dfrac{-V_{\mathrm{eff}}(x, I)}{k_b T}\right), \\[2mm] Z_2 = \displaystyle\int_{x_{\max}}^{\infty} \exp\left(\dfrac{-V_{\mathrm{eff}}(x, I)}{k_b T}\right) dx, \quad x > x_{\max}. \end{cases} \tag{7.20}$$

If I increases from Icr_1 to Icr_2, then $\nu 1$ decreases in accordance with

$$d\nu 1(I) = Z_1^{-1} \exp\left(\frac{-V_{\text{eff}}(x_{\max}, I)}{k_b T}\right) \nu 1 \, dx_{\max} \qquad (7.21)$$

whereas $\nu 2$ increases by the same amount.

If I decreases from Icr_2 to Icr_1, then $\nu 1$ increases,

$$d\nu 1(I) = Z_2^{-1} \exp\left(\frac{-V_{\text{eff}}(x_{\max}, I)}{k_b T}\right) (1 - \nu 1) \, dx_{\max} \qquad (7.22)$$

and $\nu 2$ decreases by the same amount.

The dependence of the population $\nu 1$ of the first conformational state upon the actinic light intensity for the system without thermal transitions was obtained. The dependence is of a characteristic hysteresis type with large width if the activating light intensity in its maximum ($I_{\max}$) is close to or exceeds the value Icr_2. The width of the hysteresis is much smaller if the maximum value of I is smaller than Icr_2. The corresponding dependence for the fully equilibrated system reveals no hysteresis.

Comparison with Experiments. We have carried out some straightforward experiments aimed at revealing the role of nonlinear dynamic effects in isolated photosynthetic RCs from the purple bacterium *R. sphaeroides* (wild-type). The isolation procedure and sample preparation are described elsewhere.

The experimental conditions were pH = 7.5 at room temperature, and the buffer used was 20 mM Tris-HCl with 0.025% lauryldimethylamine N-oxide (LDAO) concentration. The RC suspension, with absorbance $A_{802} = 0.8$, was investigated in a 1-cm path length cuvette. We followed the RC's optical absorbance changes in the maximum of the bacterio-chlorophyll dimer P absorption band ($\lambda = 865$ nm). The testing light intensity was 10^9 quanta/(cm^2 s). Additional continuous wave excitation was provided by an incandescent lamp filtered by an interference filter ($\lambda_m = 850$ nm). First the intensity of continuous wave excitation (actinic light) was increased from zero up to a maximum level $I_{\max}$ of 4×10^{14} quanta/(cm^2 s), and then it was diminished back to the initial low level. The rate of actinic light intensity variation was slow enough to ensure quasistationary experimental conditions.

The experiments showed hysteresis behavior in the optical absorbance ($\Delta A_{865}(I)$) that was proportional to the change of the overall number of photoexcited RCs. The corresponding theoretical value can be defined as

$$\Delta A_{865} \approx \int_{-\infty}^{\infty} dx [n_A(x) + n_B(x)] P(x), \qquad (7.23)$$

where $P(x)$ is determined from equations (7.13) and (7.20)–(7.22); $n_A(I,x)$ and $n_B(I,x)$ are defined by equation (7.9).

The results of model calculations using the theory developed above for different sets of parameters are obtained. The calculations reveal a broad bistability region. The width of the loop is in good agreement with the experimental results and can be explained by the fact that a very small fraction of all RCs are switched into a new conformational state under the experimental conditions used (i.c., the population $\nu 1$ of the first minimum of the conformational potential deviates only slightly from 1).

The good match between the experimental and theoretical plots justifies the approach presented here. The most essential parameters influencing the theoretical curve $\Delta A_{865}/A_{865}$ are k_{AD}, x_0, k, and f_B. The dependencies were obtained for two sets of parameters: (1) $k_{AD} = 13$, $x_0 = 1$, $k = 0.5$, and $f_B = 7$ (the solid line) and (2) $k_{AD} = 6$, $k_0 = 1.8$, and $f_B = 10$ (the dashed line). The corresponding variations of ΔG_{AB} calculated from $x|_{x=\xi} = \Delta G_{AB}/k_B T$ for these two cases (at room temperature) are (1) from 25 to ca. 300 meV and (2) from 25 to ca. 165 meV upon actinic light intensity variation from 0 up to some high, close to saturating, level. The first set of parameters gives a much better match of the theory to the experiment, but it requires a considerably larger variation of ΔG_{AB} as a function of the variation of the activating light intensity.

We should note that the shape of the theoretical hysteresis curve depends strongly on the parameters that determine both the adiabatic potential and the RC distribution function (i.e., k and f_B). This means that the choice of a potential of a different type (Lennard–Jones, Morse, or any other unharmonic potential) could improve the description considerably, even for small variations of the ΔG_{AB} value. This exemplifies the importance of determining the exact shape of the potential experimentally.

We know of no published experiments aimed specifically at studying light-induced ΔG_{AB} changes. The estimates presented in different papers give values from 10 to 100 meV for ΔG_{AB} variation with both k_{AB} and k_{BA} variation and for different external conditions. Such a variation of ΔG_{AB} is more than sufficient to cause the hysteresis behavior in our model, although the shape of the hysteresis loop obtained theoretically for such parameters differs significantly from experiment. We expect that the application of an adiabatic potential of a nonparabolic type in the theory would provide a closer correspondence between the theoretical and the experimentally observed hysteresis curves for the variation of ΔG_{AB} over the range from 10 to 100 meV. To make such an improvement to the theory, however, one should know the exact shape of the conformational potential. This very important information can be obtained, in principle, from a detailed experimental study of RCs, including the study of recombination kinetics under different conditions of illumination. We have obtained preliminary results on the $P^+Q_B^-$ recombination kinetics in RCs subjected to prolonged illumination of high-intensity light (data not shown). These results provide clear evidence for largely different shapes of the conformational potential for RCs in the PQ_AQ_B, $P^+Q_A^-Q_B$, and $P^+Q_AQ_B^-$ states. The principal purpose of the present work has been to develop the theory and to study the role of the slow conformational variable on the RC functioning. The exact parameter values in the model and the proper shape of the potentials will have to be elaborated in a more detailed experimental study.

Our theoretical study reveals the importance of nonlinear dynamic effects (i.e., in our case, the bifurcations of a particular macroscopic parameter) for the function of an ensemble of photosynthetic RCs. We have shown that an interaction between photoseparated charges and cofactor protein conformation may cause pronounced nonlinearity in the dependencies of macroscopic RC parameters (optical absorbance) on the intensity of actinic light. The comparison of the theory with our experimental results is encouraging and reveals good agreement between theoretical and experimental optical absorbance changes. These results have been discussed in terms of variation of free energy difference ΔG_{AB} following the variation of actinic light intensity. The RC in the dark-adapted state was found to be characterized by a low ΔG_{AB} value, whereas higher values of ΔG_{AB} could be obtained after long enough time of illumination of the sample with actinic light of close to saturating intensity. The slow conformational changes caused mainly by the electron localization on Q_B appear to be completed during this period of time. We suggest that the characteristic relaxation time of the structural variable x under consideration is at least on the order of minutes or longer. The existence of such dynamics with relaxation times up to 15–20 min has, for example, been recently detected in photoacoustic studies of RCs from R. sphaeroides.

The two conformational states described by lower and upper branches play an important role in the function of the RC. After prolonged light adaptation (i.e., after the transition to the upper branch of the bistability curve), the conformational state of all the RCs should become almost independent of the variation of the intensity of acting light, thus ensuring a stable region of RC functioning. This region corresponds to a dramatic increase in the ET rate constant $k_{AB}(x)$ and a corresponding decrease of the reverse rate constant $k_{BA}(x)$ as compared with their values in the dark-adapted state, which corresponds to the lower branch of the bistability curve.

2.   Bilinear Models of Biological Membranes

We consider a simple model that simulates the ion transport through a channel with the properties modified by the presence of ions. For channels with a great number of binding states, we derive a set of nonlinear differential equations that generalize the Nernst–Planck equations to the case of channels characterized by the strong interaction of ions with slow conformational degrees of freedom. We study the stationary regime of the functioning of the channel and show that the channel may exhibit dissipative spatially ordered structures whose form depends on the relation between the correlation length of conformational fluctuations and the width of the membrane. Some experimental consequences of this structure are discussed.

It is established that the conductivity of biological membranes is basically determined by special proteins capable of forming pores that traverse the cell membranes. Ions and small molecules can travel along these microscopic pores. Thus, the cells affect the matter and charge exchange with an intercellular medium. The ionic channels participate in the processes of the formation and transfer of electric pulses along nerve fibers and muscles, and they facilitate the correlation of chemical reactions in cells and in an intercellular medium.

The ionic channels are characterized by selectivity (which ions can pass through the channels and how easily they do it) and permeability (which interactions can make the channel close or open). The channels may be divided into two groups (classes) in terms of conductivity control. The first class includes potential-dependent channels, the conductivity of which varies with potential differences applied to the membrane. One representative of this class is the sodium channel, a glycoprotein complex with a molecular weight of about 270 kDa that has atomic groups carrying a constant charge or possessing a large dipole moment. The second class includes the channels that vary their conductivity under the influence of different fields (light, sound, heat, etc.) or because of the interaction with ligand molecules. Thus, the addition of only two acetylcholine molecules to a special membrane protein (the acetylcholine receptor) results in the opening of an ionic channel and several thousand ions pass through it in 10^{-3} s.

Molecular dynamics methods are used to show (e.g., Fisher and Brickman, 1983) that the motion of ions along the channel is similar to the motion of a particle in a potential profile $U(x)$ (x is the position within the pore) representing a set of minima (at x_i) and maxima (at $x_{i,i+1}$, where $x_{i,i+1}$ is the coordinate of the maximum separating the minima at x_i and x_{i+1}). Moreover, the minima correspond to ion-binding sites in the channel, the sites where the ion is in an energetically advantageous environment. The ions surpass the barriers, separating the neighboring binding sites by thermally activated jumps. The rate constant for the jumps from the ith to the $(i + 1)$th binding sites is, according to the theory (Kramers, 1940), denoted

$$W_{i \to i+1} = \frac{D}{2\pi k_B T}[U''(x_{i,i+1}) \mid U''(x_i)]^{1/2} \exp\left[\frac{U(x_{i,i+1}) - U(x_i)}{K_B T}\right],$$

(7.24)

where D is the diffusion constant, k_B is the Boltzmann constant, and T the absolute temperature.

As experiments with X-ray scattering, nuclear magnetic resonance, Messbauer spectra, and fluorescent spectroscopy show (Ringe and Petsko, 1985; Wagner, 1983; Parak and Knapp, 1984), the protein molecules forming the channels can be in different conformational states and, while in heat equilibrium with the environment, can go from one conformational state into another.

When the ion comes to the channel, it interacts with conformational degrees of freedom and can change the conformational state of the protein molecule in the channel.

The existence of such an ion-conformation coupling was demonstrated, for example, in experiments with a gramicidin A channel where two permeable ions, Rb^+ and Na^+, were present in the solution bathing the membrane. It was shown that the potential profile of the Na^+ (Rb^+) ion depends on the species of the ion that had previously passed through the channel. A channel is mostly occupied by one Na^+ ion (Lev, Schagina, and Grinfeldt, 1988). Lev et al. have concluded that the relaxation rates of some conformational degrees of freedom are comparable or smaller than the frequency of ion jumps in the channel. As Ciani (1984) and Läuger (1985) indicated, there is an appreciable difference between fast and slow conformational motions. Läuger observed that the fast conformational degrees of freedom; that is, the motions whose characteristic times are small compared to the time of an ion jump over the barrier, manage to accommodate themselves to the ion and their role amounts to making the ion "heavier" just as the electron becomes heavier in a crystal when it interacts with lattice vibrations (the polaron effect). The situation is different when the ions interact with slow conformational motions whose characteristic correlation times are large as compared to the times of ion transitions between the binding sites. As shown in Ciani (1984) and Läuger (1985), qualitatively new phenomena may be expected here. Thus, Läuger, assuming that the channel can be in two conformational states A and B, introduces four states of the channel: $A°(B°)$ conformation $A(B)$, an empty channel, $A^*(B^*)$ = conformation

$A(B)$, a channel with an ion. Furthermore, considering that the rates of the transition between conformations A and B depend on whether the channel is empty or occupied, Läuger concludes that this model describes the relationship between ion flows even in the case when there is only one ion in the channel at every moment. Similar conclusions were made earlier by Ciani (1984).

In Gaididei et al. (1988) an alternating model was proposed that took into account the interaction of an ion with the slow conformational motions of the channel. As follows from Ciani (1984), there are no two distinct conformational states of the channel. On the contrary, they thought that in the slow conformational degree of freedom with which the ion interacts in the most intensive way, the channel has only one equilibrium position, and as this degree of freedom interacts with an ion flow through the channel, another equilibrium position appears and the channel goes into a bistable mode of functioning. It was shown that because the conductivity of the channel in each of these states is very different, it is reasonable to relate the components of a bistable functioning regime with the experimentally observed open and closed states of a single channel (Chinarov et al., 1990). This is also confirmed by the calculations of the time dependence of an ion current that show a qualitative agreement with experiment.

2.1   Controlled Model of the Channel

When the ion gets into the channel, it interacts with it, attracting certain groups of atoms and repulsing others. As a result, the form of the channel becomes changed, as well as its energy structure and the conditions for ion passage through the channel. This is a very complicated multistage process with many details remaining unclear. Still, its basic properties can be described using our model simulating the operation of the channel with the interaction of ions with slow conformational degrees of freedom taken into account.

Consider an ion channel with two binding sites. Its energy profile corresponds to two ion-binding sites in the channel, and 0 and 3 to the positions of the ions in the electrolyte that flow over the membrane. As mentioned above, the motion of the ions through the channel represents a series of thermally activated jumps over the energy barrier. And the ion transport through a channel with the energy profile is described by the following set of bilinear equations,

$$\frac{d}{dt}N_1 = \Lambda_1 + W_{2\to1}N_2 - (W_{1\to2} + W_{1\to0})N_1,$$

$$\frac{d}{dt}N_2 = \Lambda_2 + W_{1\to2}N_1 + (W_{2\to1} + W_{2\to3})N_2, \qquad (7.25)$$

where $N_i(t)$ ($i = 1, 2$) is the probability of the ion occupying the ith binding site at time t; Λ_1 and Λ_2 are the ion flows from electrolytes to the channel, and $W_{i \to j}$ are the jump rates defined by (7.24).

It is well known that a great number of degrees of freedom participate in any conformational rearrangement. We are not able to describe this process in detail. Therefore, in this chapter a phenomenological approach to the problem is used and it is assumed that the above slow conformational rearrangements of the channel may be related to changes in the orientation of some polar group of atoms. This polar group is situated near ion-binding sites 1 and 2 in the channel. Its participation in the structural rearrangement of the channel will be simulated by its rotation about some fixed axis.

The interaction potential between the ion and the channel protein is assumed to be dependent on the angle by which the polar group is rotated: $U = U(x, \theta)$. We can therefore write at the points x_i and $x_{i,i+1}$ that

$$U(x, \theta) = U(x, 0) + U_\theta(x, 0)\theta + \frac{1}{2} u_{\theta\theta}(x, 0)\theta^2 \qquad (7.26)$$

and the rate of ion jumps may be expressed as

$$W_{i \to j}(\theta) = W_{i \to j}/S_i(\theta), \qquad (7.27)$$

where the function $S_i(\theta)$ has the form

$$S_i(\theta) = (1 + \alpha_i \theta + \beta_i \theta^2)^{-1/2} \exp(\gamma_i \theta + \delta_i \theta^2). \qquad (7.28)$$

Here we have introduced the abbreviations

$$\alpha_i = \frac{\partial}{\partial \theta} \ln \left[U''(x_{i,i+1}, \theta)U''(x_i, \theta) \right]|_{\theta=0},$$

$$\beta_i = \frac{\partial^2/\partial\theta^2 [U''(x_{i,i+1}, \theta)U''(x_i, \theta)]}{U''(x_{i,i+1}, 0)U''(x_i, 0)}\bigg|_{\theta=0},$$

$$\gamma_i = \frac{1}{k_B T} \frac{\partial}{\partial \theta} \left[U(x_{i,i+1}, \theta) - U(x_i, \theta) \right]|_{\theta=0},$$

$$\delta_i = \frac{1}{k_B T} \frac{\partial^2}{\partial \theta^2} \left[U(x_{i,i+1}, \theta) - U(x_i, \theta) \right]|_{\theta=0}.$$

At the same time, the presence of ions in binding sites 1 and 2 affects the orientation of a polar group. It is thus clear that in the situation shown in Figure 7.4b, the charge q_1, situated at point 1, will force the polar group to rotate by the angle $\theta_1 \sim q_1$, and the charge q_2, placed in the symmetrical position 2, will rotate the polar group in the opposite

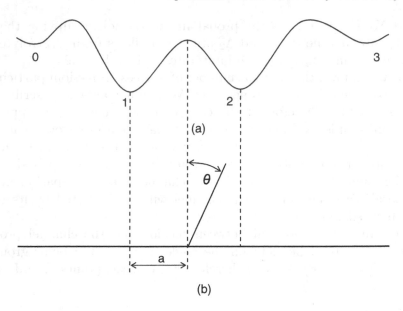

Figure 7.4. (a) Energy profile of a channel with two binding sites 1 and 2. (b) Ideal-
ized scheme of the structural rearrangement of the channel. Polar group with dipole
moment d rotates around the axis which is perpendicular to the plane of the figure.
The polar group rotation changes the rate of ion jumps over the barrier separating
sites 1 and 2.

direction by the angle $\theta_2 \sim q_2$. The resulting rotation angle will be
proportional to the difference $q_1 - q_2$. And if, in the absence of charges
in binding sites 1 and 2, the potential function of the rotational motion
of a polar group is $\frac{1}{2}k\theta^2$, where the quantity k characterizes the elastic
properties of the channel relative to changes in the orientation of a polar
group, then, in the presence of charges q_1 and q_2 in the binding sites, it
transforms to

$$V = \frac{1}{2}k\theta^2 - A(q_1 - q_2)\theta, \qquad (7.29)$$

where the quantity A defines the intensity of the interaction of charges
with a polar group. Thus, if the polar group is modeled by a dipole with
one fixed direction, then $A \simeq d/a^2$ (d is the dipole moment of a polar
group, a is the distance between a polar group and a binding site).

In the present chapter, we study single-ion channels, that is, channels
that contain only one ion at a given instant. In this case, it would seem
that we must set either q_1 or q_2 equal to zero in formula (7.23). However,
if the characteristic frequencies of reorientational motions in the channel
are small or comparable with the frequencies of ion jumps in the channels, these motions are incapable of reacting to every ion separately.

They sense only the average characteristics of an ion flow, and we must replace the instantaneous values of the charges $q_i(t)$ $(i = 1, 2)$ in (7.23) with their average values at a given time $(q_i(t))$.

The brackets $\langle \rangle$ denote an average over the time interval which is large compared to the mean time of ion transitions between binding sites, but small compared to the time of the conformational rearrangement:

$$\langle q_i(t) \rangle = \frac{1}{\Delta t} \int_t^{t+\Delta t} q_i dt' \equiv q\, N_i(t)$$

(q is the ion charge). As a result, the potential function of a polar group is

$$V = \frac{1}{2} k\theta^2 - Aq(N_1 - N_2)\theta. \tag{7.30}$$

The dynamics of the rotational motions of a polar group is described by the Newton equation

$$I \frac{d^2}{dt^2} \theta + \eta \frac{d}{dt} \theta = -\frac{\partial}{\partial \theta} V(\theta),$$

where I is the inertia moment, and the quantity η characterizes the viscosity of a medium in which the motion takes place.

Taking (7.30) into account, this equation may be written as

$$\frac{d^2}{dt^2} \theta + 2\gamma \frac{d}{dt} \theta + \omega^2 \theta = \omega^2 \theta_\infty (N_1 - N_2), \tag{7.31}$$

where $\omega = k^{1/2} I^{-1/2}$ is the frequency of vibrations of a polar group, $\gamma = \eta/(2I)$ is their attenuation, and $\theta_\infty = Aqk^{-1}$ is the limiting angle by which the polar group would rotate if the ion stayed for an infinitely long time at one of the binding sites.

Within the framework of our model (7.25)–(7.28), (7.31), it completely defines the ion transport through the channel. The feedback between the ion flow and the channel structure is also taken into account.

First of all, we study the stationary regimes of the functioning of the channel. We consider the case of a symmetric channel ($W_{0\rightarrow 1} = W_{3\rightarrow 2}$, $W_{1\rightarrow 0} = W_{2\rightarrow 3}$, $S_1(\theta) = S_2(\theta) = S(\theta)$) and assume that an external electrical field is absent ($W_{i\rightarrow j} = W_{j\rightarrow i}$). Thus, the ion transport through the channel is caused by the flux difference $\Lambda_1 - \Lambda_2$. The stationary populations N_i of the binding sites in the channel are denoted

$$\overline{N_1} + \overline{N_2} = (\Lambda_1 + \Lambda_2)/W_{0\rightarrow 1}, \quad \overline{N_1} - \overline{N_2} = \lambda/[1 + \nu S^{-1}(\theta)], \tag{7.32}$$

where the stationary rotation angle θ may be derived from the equation

$$\theta_\infty \lambda = \theta[1 + \nu S^{-1}(\theta)] \equiv F(\theta). \tag{7.33}$$

In (7.32), (7.33), we have introduced the notation

$$\lambda = (\Lambda_1 - \Lambda_2)/W_{0 \to 1}, \quad \nu = 2W_{1 \to 2}/W_{0 \to 1}.$$

When

$$\left(\theta \frac{\mathrm{d}S}{\mathrm{d}\theta} - S\right) S^{-2} \Big/_{\theta = \theta_0} > \nu^{-1}, \tag{7.34}$$

where θ_0 is the root of the equation

$$\theta \frac{\mathrm{d}^2 S}{\mathrm{d}\theta^2} - \frac{2\theta}{S} \left(\frac{\mathrm{d}S}{\mathrm{d}\theta}\right)^2 + 2\frac{\mathrm{d}S}{\mathrm{d}\theta} = 0,$$

the function $F(\theta)$ is nonmonotonic. In the interval

$$F_{\min} < \theta_\infty \lambda < F_{\max}, \tag{7.35}$$

one value of the incoming ion flow λ corresponds to two stationary rotation angles θ_a and θ_c (stationary state with $\theta = \theta_b$ is unstable). Consequently, in the interval (7.35), the ionic channel functions in a bistable region (Figure 7.5). Outside the interval (7.35), the system described by equations (7.25)–(7.28), (7.31) is monostable.

It is worth mentioning that the value of the incoming ion flow λ qualitatively affects the form of the potential function that characterizes the

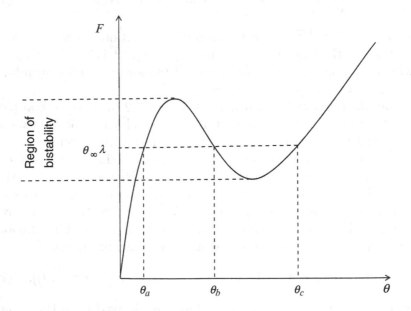

Figure 7.5. Dependence of the steady values of conformational coordinate θ on the incoming ion flow.

motion of a polar group. Really, if we assume that the stationary populations of binding sites in the channel are accomplished faster than the stationary state in the conformational variable θ, then, using Haken's principle of the adiabatic elimination of fast variables, we can derive the following effective equation for the conformational variable θ in equations (7.25), (7.26), and (7.31),

$$I\frac{d^2\theta}{d\theta^2} + \eta\frac{d\theta}{dt} = -\frac{d}{d\theta}V_{\text{eff}}(\theta), \tag{7.36}$$

where the effective potential function $V_{\text{eff}}(\theta)$ has the form

$$V_{\text{eff}} = \frac{1}{2}k\left[\theta^2 - 2\theta_\infty\lambda\int\frac{S(\theta)}{\nu + S(\theta)}d\theta\right]. \tag{7.37}$$

From (7.37), it follows that, with $(\theta_\infty\lambda) \in [F_{\min}, F_{\max}]$, the effective potential function has one minimum and, consequently, the polar group has a single equilibrium position. In the interval (7.35), it has two minima at the points θ_a and θ_c. Accordingly, the polar group has two stable equilibrium positions (Figure 7.6).

It is worth comparing the ion flows passing through the channel

$$J = (N_1 - N_2)W_{1\to2}(\theta)$$

in both equilibrium positions. From (7.27) and (7.32), it is seen that

$$J_a/J_c = (\nu + S(\theta_c))/(\nu + S(\theta_a)). \tag{7.38}$$

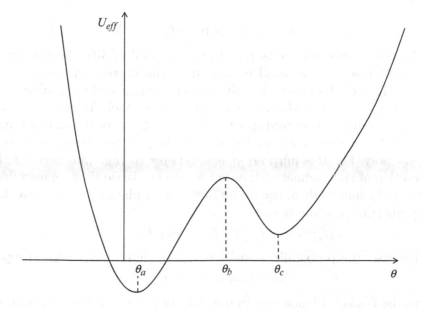

Figure 7.6. Modified conformational potential in the bistable regime.

Let us calculate this ratio for the case when

$$S(\theta) = \exp(\theta^2/\theta_r^{-2}), \tag{7.39}$$

where θ_r is the characteristic rotation angle of a polar group for which the probability of the jumps $W_{1\to 2}$ decreases by a factor of e. With such $S(\theta)$, inequality (7.34), implying the basic possibility of the existence of a bistable regime of functioning, reduces to

$$2\nu > \exp(3/2)$$

or, in dimensional variables, to the ratio between the heights of the energy barriers at the entrance to the channel and inside it

$$U(x_{1,0}) - U(x_{1,2}) > k_B T/10.$$

In other words, at room temperature the bistability may be expected to be exhibited only by the channels for which the difference in the heights of the barriers at the entrance and inside the channel exceeds 2 meV.

Substituting (7.39) into (7.33), we find that, for $\nu \gg 1$, the channel functions are in a bistable regime when

$$\sqrt{\ln \nu} < \lambda\theta_\infty\theta_r^{-1} < \nu/\sqrt{2e}. \tag{7.40}$$

In this case, the equilibrium in the rotation angles of a polar group are defined by

$$\theta_a \simeq \theta_\infty\lambda(1+\nu)^{-1}, \quad \theta_c \simeq \theta_\infty\lambda$$

and

$$J_c/J_a = (1+\nu)\exp(-\theta_\infty^2\lambda^2\theta_r^{-2}).$$

The last expression shows that in the interval (7.40), the current in a state with $\theta = \theta_c$ is small compared to the current in a state with $\theta = \theta_a$. This enables us to identify the components of a bistable state with the open and closed states of a single ion channel. At the same time, the experimentally observed spontaneous switchings of the channel from a closed to an open state and vice versa may be explained in a very realistic way. For this purpose, it is necessary to take into account the interaction of the channel with its environment. To do this, we introduce on the right-hand side of equation (7.36) the random force $f(t)$ that has the properties of white noise:

$$\langle f(t)\rangle = 0, \quad \langle f(t)f(t')\rangle = 2k_B T\eta\delta(t - t').$$

In this case, the probability density of the deviation of a polar group

$$P(\theta, t) = \langle\delta(\theta - \theta(t))\rangle$$

obeys the Fokker–Planck equation which, in the Smolukhovsky approximation, has the form

$$\eta \frac{\partial}{\partial t} P = \frac{\partial}{\partial \theta} \left(\frac{\partial}{\partial \theta} V_{\text{eff}}(\theta) + k_B T \frac{\partial}{\partial \theta} \right) P.$$

The stationary probability distribution has the form

$$P_{st}(\theta) = \exp\left[-\frac{V_{\text{eff}}(\theta)}{k_B T} \right] \bigg/ \int_{-\infty}^{\infty} d\theta \exp\left[-\frac{V_{\text{eff}}(\theta)}{k_B T} \right].$$

In the interval (7.35), this function has two maxima at the points θ_a and θ_c and the minimum at the point θ_b.

Using Kramers's method (Kramers, 1940). Brownian motion, from equation (7.40) we can evaluate the time during which the channel is in the open (τ_a) and the closed (τ_a) states

$$\tau_{a_c} = \pi \left| \frac{V_{\text{eff}}(\theta_b)}{V_{\text{eff}}''(\theta_{a_c})} \right|^{1/2} \exp\left[\frac{V_{\text{eff}}(\theta_b) - V_{\text{eff}}(\theta_{a_c})}{k_B T} \right]. \tag{7.41}$$

Using the explicit expression (7.37) for $V_{\text{eff}}(\theta)$, we conclude from (7.41) that the lifetime of the channel in a discrete conductance state is dependent on the concentration of ions in a solution flowing over the membrane (this concentration is involved in the parameter λ that defines the ion flow coming into the channel) and on the type of ions passing through the channel: the quantity θ_∞ is dependent on the charge of ions and their polarizability.

The dependence of the lifetime of the channel in an open state on the type of ions and their concentration was observed in acetylcholine-activated channels, potassium channels, and calcium channels.

It is to be emphasized that the state of discrete conductance of the channels results from the interaction of an ion flow with slow conformational degrees of freedom that govern the structural properties of the channel. Therefore, single-ion channels also have discrete conduction states that are dependent on the nature and concentration of the ions that pass through the channel.

If the solutions flowing over the membrane have ions of different species, then the presence of a slow conformational degree of freedom in the channel results in an indirect interaction between ion flows. This relationship between ion flows is observed for a number of ionic channels.

2.2 Generalized Equation of Diffusion

We now consider the problem of ion transport through channels with an arbitrary number of binding sites. The ion transport through the channel is, in this case, a series of successive thermally activated jumps over potential barriers, and for channels with M binding sites, it is described by the following set of bilinear equations,

$$\frac{d}{dt}N_i = W_{i+1\to i}N_{i+1} + W_{i-1\to i}N_{i-1} - (W_{i\to i+1} + W_{i\to i-1})N_i,$$

$$i = 1, 2, \ldots M, \tag{7.42}$$

where $N_i(t)$ is the probability for an ion occupying the ith binding site in the channel. The quantities N_0 and N_{M+1} in equation (7.42) are supposed to be constant; they characterize the ion populations in the solutions bathing the membrane near the mouths of the channel. We assume that the ions moving along the channel interact with a slow conformational degree of freedom of the channel. The degree of freedom governs the transport process, affecting the rates of the ion jumps. Moreover, a local change in the form of the channel, for example, a change in the neighborhood of the ith and $(i+1)$th binding sites, determines the rate of the jumps, $W_{i\to i+1}$, between these binding sites. We refer to this local change in the form of the channel as the rotation of the ith polar group and, following Section 7.2.1, conclude that this rotation leads to a decreased jump rate

$$W_{i\to i+1} = WS^{-1}(\theta_i),$$

where the properties of the function $S(\theta)$ are defined by the relation (7.28).

The ions moving along the channel also change its form. Part of these changes is concerned with the conformational degree of freedom $\{\theta_i\}$ which is assumed to be so slow that it is incapable of reacting to every ion individually and, therefore, senses only their average number at the binding site. As in the case of a two-site channel, discussed in Section 7.2.1 (equation (7.3)), we can write an equation for a local change in the channel form such as

$$\omega^{-2}\frac{d^2}{dt^2}\theta_i + 2\gamma\omega^{-2}\frac{d}{dt}\theta_i + \theta_i = \theta_\infty(N_{i+1} - N_i)$$

$$+ \sum_j \varkappa_{ij}(\theta_j - \theta_i), \tag{7.43}$$

where the term $\sum_j \varkappa_{ij}(\theta_j - \theta_i)$ takes into account the presence of the spatial dispersion in structural changes in the channel.

When the number of binding sites in the channel is large enough and the probabilities of the occupation of neighboring wells $N_i(t)$ and $N_{i+1}(t)$, as well as the local changes in the channel form θ_i and θ_{i+1} are not much different, we can go to a continuum limit in equations (7.42) and (2.2). As a result, we have the following equation for the quantities

$N(x,t)$ $\theta(x,t)$ $(\theta_i(t) = \theta(ir,t)$, $N_i(t) \equiv N(ir,t))$, r is the average distance between neighboring binding sites,

$$\frac{\partial}{\partial t}N(x,t) = \frac{\partial}{\partial x}D(\theta)\frac{\partial}{\partial x}N(x,t), \tag{7.44}$$

$$\omega^{-2}\frac{\partial^2}{\partial t^2}\theta(x,t) + 2\gamma\omega^{-2}\frac{\partial}{\partial t}\theta(x,t) + \theta(x,t)$$

$$= l^2\frac{\partial^2}{\partial x^2}\theta(x,t) + r\theta_\infty\frac{\partial}{\partial x}N(x,t), \tag{7.45}$$

where

$$D(\theta) = Wr^2/S(\theta)$$

is the ion diffusion coefficient, and

$$l = \left[\sum_j \varkappa_{ij}(i-j)^2 r^2\right]^{1/2}$$

is the correlation length of the conformational fluctuations.

Equation (7.44) is written for the point inside the channel. It should be supplemented with boundary equations. These are derived as follows. Suppose that the concentrations of transferred particles in solutions on the left and right of the membrane are C_1 and C_2, respectively, and the rate constant of the jump from a solution to an end site of binding ($x = 0$ and $x = L$, $L = Mr$ is the channel length) is $W_+(x)|_{x=0,L}$, and the rate constant of an inverse jump is $W_-(x)|_{x=0,L}$. Then, for the probabilities of occupation of the end sites of binding $N(0,t)$ and $N(L,t)$ we can write

$$\frac{d}{dt}N(0,t) = W_+(0)C_1 - W_-(0)N(0,t) + r^{-1}D(\theta)\frac{\partial N(x,t)}{\partial x}\bigg|_{x=0},$$

$$\frac{d}{dt}N(L,t) = W_+(L)C_2 - W_-(L)N(L,t) - r^{-1}D(\theta)\frac{\partial N(x,t)}{\partial x}\bigg|_{x=1}. \tag{7.46}$$

As a boundary condition for $\theta(x,t)$, we choose

$$\frac{\partial}{\partial x}\theta(x,t)\bigg|_{x=0,L} = 0. \tag{7.47}$$

The set of equations (7.44)–(7.47) generalizes the ordinary diffusion approach to the problem of ion transport through a membrane. It involves the feedback between the ion flow along the channel and the structural changes in the channel generated by this flow.

We note that we do not take into account the effect concerned with a direct ion–ion interaction. This is correct when the ion concentration in the channel is low.

2.3   Structure of a Functioning Channel

We investigate the set of equations (7.44)–(7.47) for the case of boundaries when the solution - channel exchange is much faster than the transport along the channel. Then, at the ends of the channel, we have the sustained occupation probabilities

$$N(0,t) = C_1 W_+(0)/W_-(0) \equiv N_1,$$
$$N(L,t) = C_2 W_+(L)/W_-(L) \equiv N_2. \tag{7.48}$$

As is seen from (7.44) and (7.45), the functions $N(x)$ and $\theta(x)$ that describe the stationary distributions of ions and conformational changes in the channel may be derived from a set such as

$$\frac{\mathrm{d}}{\mathrm{d}x} N = BS(\theta), \tag{7.49}$$

$$l^2 \frac{\mathrm{d}^2\theta}{\mathrm{d}x^2} + r\theta_\infty \frac{\mathrm{d}N}{\mathrm{d}x} - \theta = 0, \tag{7.50}$$

where B is the integration constant defined by the relation

$$B = (N_2 - N_1) \Big/ \int_0^L \mathrm{d}x\, S(\theta). \tag{7.51}$$

The set of equations (7.49), (7.50) has a trivial solution

$$N(x) = N_1 + \frac{N_2 - N_1}{L}x, \quad \theta(x) = r\theta_\infty \frac{N_2 - N_1}{L} \equiv \theta_S. \tag{7.52}$$

By linearizing equations (7.49) and (7.50) near this solution with respect to the fluctuations

$$\delta N \sim \sin\left(n\pi \frac{x}{L}\right) \exp(pt), \delta\theta \sim \cos\left(n\pi \frac{x}{L}\right) \exp(pt), \quad n = 1, 2, \ldots$$

it is easy to conclude that its stability is violated (Re $(p) > 0$) when

$$\theta \frac{\mathrm{d}}{\mathrm{d}\theta} \ln S(\theta)\Big|_{\theta=\theta_S} > 1 + n^2\pi^2 \left(\frac{1}{L}\right)^2. \tag{7.53}$$

Here we have inhomogeneous quasiharmonic structures with a period and an amplitude dependent upon the concentration gradient $(N_2 - N_1)/L$. Thus, at

$$\theta_c^{(1)} < r\theta_\infty (N_2 - N_1)L^{-1} < \theta_c^{(2)},$$

where $\theta_c^{(n)}$ is the root of the equation

$$\theta \frac{d}{d\theta} \ln S(\theta) = 1 + n^2 \pi^2 \left(\frac{1}{L}\right)^2,$$

the channel shows an inhomogeneous structure with

$$\theta(x) = \theta_S \pm \Delta\theta \cos\left(\frac{\pi 2}{L}\right),$$

with the amplitude $\Delta\theta \sim (\theta_S - \theta_c^{(1)})$. In this case, we deal with a super-critical bifurcation.

Such small-amplitude spatial structures are stable only in short enough ion channels. It is shown below that in the channels for which $L \gg l$, spatially inhomogeneous structures with a finite amplitude arise when

$$\theta_S < \theta_c^{(1)}.$$

To show this, we make use of the simplest approximation of the function $S(\theta)$:

$$S(\theta) = 1 + \theta^2 \theta_r^{-2}, \tag{7.54}$$

where θ_r is the value of a conformational change at which the ion diffusion decreases by half. Substituting (7.54) into (7.49) and (7.50), we have

$$l^2 \frac{d^2\theta}{dx^2} + rB\theta_\infty(1 + \theta^2\theta_r^{-2}) - \theta = 0. \tag{7.55}$$

The solutions of this equation that satisfies the boundary conditions $\theta'(x)|_{x=0,L} = 0$ have the form

$$\theta_n(x) = \theta_r[1 - \xi^4 n^4(1 - m + m^2)K^4(m)]^{-1/2}$$
$$\times \left[1 + \xi^2 n^2 K^2(m)\left(1 + m - 3m\,\text{sn}^2\left(\frac{nx}{L}K(m)\mid m\right)\right)\right]. \tag{7.56}$$

Here $\xi = 2l/L$, sn $(z\mid m)$ is the elliptic sine with period $4K$, where

$$K(m) = \int_0^{\pi/2} d\varphi(1 - m\sin^2\varphi)^{-1/2}.$$

is an elliptic integral of the first kind. The modulus of the elliptic integral m $(0 < m < 1)$ is defined from condition (7.51) which, in the present case, is

$$r\theta_\infty(N_2 - N_1)\theta_r^{-1}L^{-1} = \mathcal{F}_n(m),$$
$$\mathcal{F}_n(m) = [1 + \xi^2 n^2(3E(m)K + (m-2)K^2)]$$
$$\times [1 - \xi^4 n^4(1 - m + m^2)K^4)]^{1/2}, \tag{7.57}$$

where

$$E(m) = \int_0^{\pi/2} d\varphi (1 - m \sin^2 \varphi)^{1/2}$$

is an elliptic integral of the second kind.

Analysis shows that when $L < \pi l \sqrt{6}$,

$$\frac{d}{dm} \mathcal{F}_n(m) > 0 \tag{7.58}$$

for all n. In this case, the condition for equation (7.57) to have a solution is

$$r\theta_\infty (N_2 - N_1)\theta_r^{-1}L^{-1} > \mathcal{F}_n(0) \equiv \left(\frac{4 + n^2\pi^2\xi^2}{4 - n^2\pi^2\xi^2}\right)^{1/2}.$$

This condition is the same as (7.53) with $S(\theta)$ denoted by (7.58). Thus, if

$$r\theta_\infty (N_2 - N_1)\theta_r^{-1}L^{-1} > \mathcal{F}_n(0),$$

the stationary spatial structure of the channel is defined by

$$\theta_1(x) = \theta_r \left[\mathcal{F}_1(0) + 6m(16 - \xi^4\pi^4)^{-1/2} \cos\left(\pi\frac{x}{L}\right)\right],$$

where

$$m^2 = 3/23 \frac{(4 - \pi^2\xi^2)^{3/2}(4 + \pi^2\xi^2)^{1/2}}{\pi^2\xi^2(6\pi^2\xi^2 - 4)} \left[\frac{\theta_\infty}{\theta_r}\frac{r}{L}(N_2 - N_1) - \mathcal{F}_1(0)\right].$$

Hence, the channels that satisfy condition (7.58) display a supercritical bifurcation and, in this case, the results from linear perturbation theory are the same as those derived by exactly solving the nonlinear equation (7.56) (Figure 7.7).

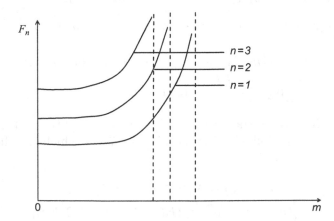

Figure 7.7. Functions $\mathcal{F}_n(m)$ ($n = 1, 2, \ldots$) for short channels (supercritical case).

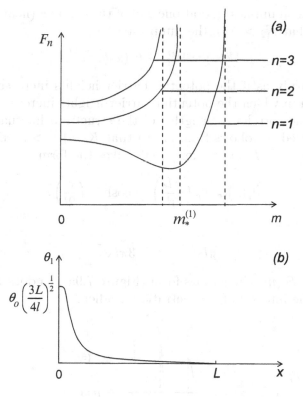

Figure 7.8. (a) Functions $\mathcal{F}_n(m)$ $(n = 1, 2, \ldots)$ for channels whose length exceeds the first critical value (the first subcritical bifurcation). (b) One-barrier dissipating structure in the ionic channel: coordinate dependence of the conformational $\theta(x)$ in the vicinity of the first subcritical bifurcation.

When

$$\pi l \sqrt{6} < L < 2\pi l \sqrt{6}, \qquad (7.59)$$

the function $\mathcal{F}_1(m)$ changes its form: at the point $m = 0$, it has its maximum, whereas it is minimum at the point $m_*(1) \neq 0$ (Figure 7.8a). The character of the other functions $\mathcal{F}_n(m)$ $(n = 2, 3, \ldots)$ remains unchanged. Here we have a subcritical bifurcation, because in the interval

$$\mathcal{F}_1(m_*^{(1)}) < \frac{\theta_\infty}{\theta_r} \frac{r}{L} (N_2 - N_1) < \mathcal{F}_1(0)$$

there occurs a nonharmonic spatially inhomogeneous structure with a finite amplitude (Figure 7.8b)

$$\theta_1(x) \simeq \theta_r \left(\frac{3L}{4l}\right)^{1/2} \mathrm{cn}^2\left(\frac{x}{2l}\bigg|m^{(1)}\right), \qquad K(m^{(1)}) \simeq \frac{L}{2l}. \qquad (7.60)$$

This expression implies that at one end of the channel (near $x = 0$ when we assume that $N_2 > N_1$), the jump rates

$$W(\theta) = W[1 + \theta^2(x)\theta_r^{-2}]^{-1}$$

decrease sharply as if the potential barrier heights increased. The size of the region in which the potential barrier heights increase is primarily defined by the correlation length l of the structural fluctuations in the channel. Indeed, it follows from (7.59) that $K(m)^{(1)} > 3$ and $m^{(1)} \simeq 1$. Therefore, for $x < L$, expression (7.60) takes the form

$$\theta_1(x) = \theta_r \left(\frac{3L}{4l}\right)^{1/2} \cosh^{-2}\left(\frac{x}{2l}\right).$$

When

$$\pi l\sqrt{6} < L < 3\pi l\sqrt{6},$$

the function $\mathcal{F}_2(m)$ changes its form (Figure 7.9a), as does the function $\mathcal{F}_1(m)$. In the interval of concentration gradients,

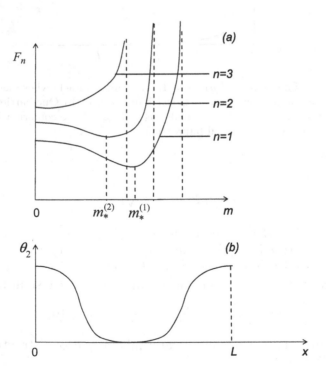

Figure 7.9. (a) Functions $\mathcal{F}_n(m)$ $(n = 1, 2, \ldots)$ after the second critical bifurcation. (b) Two-barrier dissipative structure in the ionic channel whose length exceeds the second threshold value.

$$\mathcal{F}_1(m_*^{(1)}) < \frac{\theta_\infty}{\theta_r} \frac{r}{L}(N_2 - N_1) < \mathcal{F}_2(0),$$

there are, in this case, two stationary spatially inhomogeneous solutions (Figure 7.9b): $\theta_1(x)$ and

$$\theta_2(x) \simeq \theta_r \left(\frac{3L}{8l}\right)^{1/2} \mathrm{cn}^2 \left(\frac{x}{2l}\bigg| m^{(2)}\right), \quad K(m^{(2)}) \simeq \frac{L}{4l}. \qquad (7.61)$$

The conformational changes are then primarily observed at the channel ends. The channel then looks like a single-site one. In the general case when

$$(n+1)\pi l \sqrt{6} > L > n\pi l \sqrt{6},$$

the concentration gradient interval

$$\mathcal{F}(m_*^{(1)}) < \frac{\theta_\infty}{\theta_r} \frac{r}{L}(N_2 - N_1) < \mathcal{F}_{n+1}(0)$$

displays n stationary spatially inhomogeneous solutions

$$\theta_n(x) \simeq \theta_r \left(\frac{3L}{4nl}\right)^{1/2} \mathrm{cn}^2 \left(\frac{x}{2l}\bigg| m^{(n)}\right), \quad K(m^{(n)}) \simeq \frac{L}{2ln}. \qquad (7.62)$$

The channel functioning in the regime that corresponds to the state $\theta_n(x)$ is characterized by the distribution of the ion jump rates $W(\theta) = W[1 + \theta^2(x)\theta_r^{-2}]^{-1}$ similar to that which is typical of the channels with $[n/2]$ ($[x]$ is the integral part of the number x) binding states.

It is reasonable to compare the ion flows

$$J = D(\theta)\frac{\mathrm{d}N(x)}{\mathrm{d}x} \equiv Br^2 W$$

that pass through the channel in each stationary state. Substitution of (7.56) into (7.51) yields

$$J_n = Wr\frac{\theta_r}{2\theta_\infty}\left[1 - \xi^4 n^4(1 - m + m^2)K^4(m)\right]^{1/2}.$$

For the stationary states (7.62), we then have

$$J_n = Wr\left(\frac{\theta_r}{\theta_\infty}\right)^2 \frac{1}{N_2 - N_1}n, \quad n = 1, 2, 3, \ldots .$$

Hence, each of the stationary states $\theta_n(x)$ is associated with a separate ion and, accordingly, we can say that each of the states (7.56) is a state with a different conductivity.

Experiments show that there are channels with a different number of discrete conductance states. Thus, calcium channels of the L-type have two open and one closed state, whereas the ionic channels induced by alameticyne show seven conductivity levels. The nature of these states is unclear. Theoretically, however, discrete conductance states appear to be quite natural and result from the strong nonlinearity of ion transport through biological membrane channels, generated by the interaction of ion flow with conformational degrees of freedom of the channel.

It is interesting to remark that quite recently the dependence of the duration of single current pulses has been determined in a fast K^+-channel of molluscan neuron on a concentration of intracellular $[K^+]$. It was found that, at the current decreases to zero, due to the change $[K^+]$; the rate constant approaches zero irrespective of potential and temperature. It was concluded that the channel closing rate is determined by the value of the current through the channel. Temperature and membrane potential modify the sensitivity of the channel to the current. The lifetimes of a number of conduction states are dependent upon the channel parameters (ratio lL^{-1}, θ_r, etc.), the ion species (W, θ_∞), and the salt concentration in water solutions flowing over the membrane.

3.   Intelligent Biosensors

In this section, we discuss an influence of pollutants on such biological objects as photosynthesizing systems in order to reveal the capabilities and features of their application as the controlling sensor in integral ecological monitoring microsystems. It is proposed to elaborate upon the intelligent sensor on the basis of: (1) neural network technologies; (2) the possibility of separating the characteristics of the substances dissolved in water by means of methods that recognize patterns in a functional space of the fluorescence curves; (3) the results of the chromatographic analysis of standard water samples. This sensor allows us to predict water state and to make optimal decisions for correcting an ecosystem's condition. The efficiency of such a system for water analysis can be improved using the dual measurement principle. This principle suggests a bilinear identification (Yatsenko, 1996) of a biosensor model according to experimental data.

3.1   Ecological Monitoring and Living Objects

Ecological monitoring of the environment is one of the most important tasks of the natural sciences. As a rule, the existing systems used for the analysis and control of an ecological system function according to the principles of discovery of signs characterizing the actions of certain types of pollutants (direct measurements). At the same time, some pollutants may remain unidentified if their actions upon an ecological system are insufficiently explored (i.e., either a physical or a mathematical model may be absent) or if these pollutants cannot be detected with the help of existing monitoring methods. Such imperfections, observed when the ecological monitoring systems are designed, may be overcome if systems oriented to integral pollution sensors are created. In particular, bio-objects in which life processes depend upon the state of the environment can play this role.

Any pollutant exhibits a tendency to be stored in air, water, silt, plants, and animals, and disturbs their life. As for severeness, the pollutants may be classified as lethal, dangerous, harmful, latent, indifferent, or comfortable. The first three classifications are caused by harmful matters in a bio-object at concentrations of levels about one hundred to one thousand times higher than the limit of possible concentration (LPC). Such concentrations are obtainable in a sufficiently simple way by direct measurements. The LPC-level concentrations, especially when many factors are active, cause hidden latent reactions. The direct measurement of such pollutants is difficult and is associated with the use of very expensive equipment. Moreover, during direct measurement, the influence of such pollutants may remain nearly undetected or undetected for a long period of time.

Living objects are sensitive to the influence of low concentrations of pollutants and react upon them when normal physiological processes are disturbed. This feature of bio-objects is proposed as a sensor able to record data about the integral state of an environment.

We describe a method by means of which the substances dissolved in water can be analyzed under dynamic conditions when separate components interact with the sensitive element containing photosynthetic objects. Because these components influence the objects in a different way, there appears the possibility of separating the characteristics of the components by means of the methods recognizing patterns in a functional space of the fluorescence curves. Contrary to classic chromatography, this method may find more applications and, moreover, it is adaptable to measurement conditions. Also, it is possible to use the chromatographic data at the sensor training stage, to perform parallel data processing and scientific analysis. The requirements for sample preparation are low.

3.2   Experimental Results

Materials. Plants, algae, photosynthesizing bacteria, extracted reaction centers (RCs), Langmuir–Blodgett–Shefer films taken from reaction centers of *R. sphaeroides* purple bacteria were used to examine the photosynthetic objects (Gushcha et al.,1995; Yatsenko, 1996). The pure films and the films affected by atrazine solution were investigated experimentally. Also, we examined samples of water taken from different artificial ponds, and such pollutants as heavy metals, chemical toxicants, and herbicides of different concentrations were detected there. The samples were characterized by numbers in the range [0,1] and this interval shows the degree of ecological purity of water.

Methods. The influence exerted by different pollutants on the functional characteristics of the photosynthetic objects in a sensitive element was assessed. The curves of delayed fluorescence and of fluorescence induction were examined for this purpose. The distinctive features of these curves were used to analyze the polluted water. The pattern recognition methods were implemented on a neural chip with probabilistic neurons, and the experimental results obtained for the examined water samples were used to train this chip (the water component composition was known here in advance). The sensitive element characteristic "degradation" was corrected by the method of bilinear identification of the sensor model and the correcting potential on the membrane was changed. All the mentioned methods permit us to provide the sensor with the robust features, to implement the principle of dual measurement and to foresee qualitative non-parametric assessments of low concentrations of pollutants in water ponds. The assessments can be further improved if the chromatographic analysis results are delivered to the chip input.

Apparatus. Figure 7.10 depicts the experimental system, which includes: a film-type biosensor; a neural chip based on probabilistic neurons and located at the sensor output (Section 4); and a bilinear model identifier located in the feedback circuit (Section 3). The signals coming from converter 8 and photo–detector 6 are the input signals for this identifier. The latter generates the required control signal value at the feedback circuit output. The proposed scheme provides dual control principle realization, and here the biosensor operation mode as well as the mode in which the whole monitoring system functions are thus optimized. When the reverse bilinear system is used as signal converter 8, the neural chip may not be applied.

Measurements. At present, many works are known (or have been published) in which the influence of pollutants on object functioning has been studied.

These investigations were aimed at revealing the influence on both the spectral characteristics of such objects (Shimazaki and Sugahara, 1980; Goldfeld and Karapetyan, 1986; Karapetyan and Buhov, 1986: Okamura, Feher, and Nelson, 1982; Feher and Okamura, 1978) and the dynamics of charges in photosynthesizing materials. The results of numerical simulation of photosynthesizing-system dynamics are described in detail in

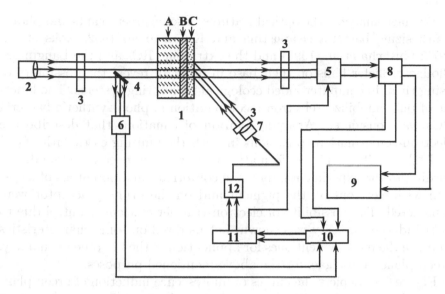

Figure 7.10. Experimental eco-object state estimation setup based on a film-type biodetector. (1) biodetector (A = crystal plate, B = lipid layer, C = monolayers of purple bacteria RC); (2) the source of continuous light; (3) interferometric filters; (4) semitransparent glass plate; (5) photodetector with preamplifier; (6) photodetector with preamplifier and converter; (7) flash lamp; (8) signal converter; (9) neural chip; (10) bilinear model identifier; (11) feedback unit; (12) control signal generator.

Kapustina and Kharkyanen (1992), Noks et al. (1977); Peters, Avouris, and Rentzepis (1978), and Kleinfeld, Okamura, and Feher (1984).

In particular, the essential influence of pollutants on spectral characteristics of radiation of the chlorophyll-bearing materials was discovered. We should also note the results in these papers where influence of pollutants on the kinetics of a charge moving along an electron-transport circuit of photosynthesizing systems are explored. Thus, the existence of different pollutants in a system blocks, at one stage or another, the movement of a photoexcited electron along an electron-transport circuit of photosystems of plants or bacteria and causes, therefore, the disturbance of the kinetics of relaxation of a biological system under its photoexcitation.

Such disturbances are revealed in optical characteristics of photosynthesizing objects in the following ways.

— Relaxation times for a system change under pulsed optical excitation in absorption bands of chlorophyll-bearing pigments.

— The slow fluorescence and the form of the induction fluorescence curve are changed when an object is acted upon by continuous optical excitation.

We have analyzed the optical features of some green plants and photo-synthesizing bacteria (Kapustina and Kharkyanen, 1992; Noks et al., 1977; Gushcha et al., 1993) and the extracting RCs and the Langmuir–Blodgett–Shefer films on their base in order to reveal the possibility of using them in computer-aided ecological monitoring systems. The kinetics of photomobilized electron recombination in photosynthetic bacteria RCs was considered. Analytic solution of equations that describe the electron-conformation transitions in both the limiting cases with "fast" and "slow" diffusion in conformational coordinate space were found. According to the experimental data the conformational potentials of a system with electrons on the pigment and on the primary acceptor were considered. The possibility of electron-transfer efficiency control due to light-induced intensity was shown. Let us dwell on some characteristics that are the most evident ones for application of the bio-objects, namely, green plants and algae, for the above-mentioned purposes.

Figure 7.11 depicts the curves of fluorescence induction of green plant leaves that are in a healthy environment and ones that were acted upon for 2 min by an atrazine–water solution (under the LPC-level concentration: about $10^5 M$). These curves are recorded in no more than 2 min and easily define the features associated with the presence of herbicides in an environment. Such facts make it possible to construct an integral ecological monitoring system on the basis of such a type of biosensor. Note that the induction curves of plant fluorescence are also disturbed when actions of heavy metals and a wide range of toxicants of a chemical nature occur.

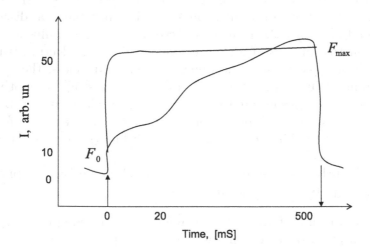

Figure 7.11. Curves of fluorescence induction for green plant leaf. Solid line for plants in unpolluted water; (*) for plants in water solution of atrazine. Arrows indicate: F the moment of turning on the acting light; J, the moment of turning off the light.

Another object of our particular interest is the RC of purple bacteria *R. sphaeroides*. The analysis of their absorption spectra shows essential changes at the wavelength region 750–900 nm under the action of saturating optical pumping. These changes are due to photo-oxidation of the RC and they manifest also in the recovery kinetics of the RC under pulsed optical excitation. It is well established now that the herbicides substitute the secondary quinone acceptors from their localization sites in the RC, causing sufficient changes of RC recovery kinetics. Figure 7.12

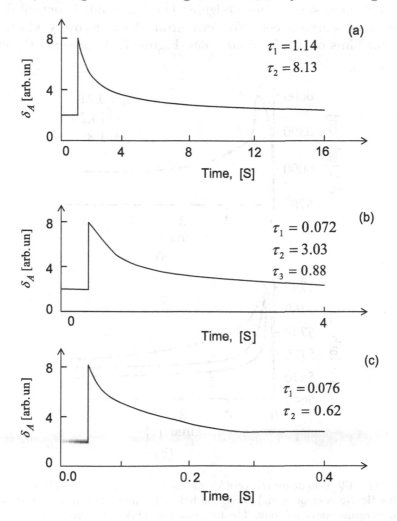

Figure 7.12. Electron donor recovery kinetics of *R. sphaeroides* RCs without addition of inhibitors under pulsed optical excitation: (a) by He–Ne laser ($\lambda=633$ nm, $\tau=1$ s); (b) by Xe-lamp ($\lambda = 450$–650 nm, $\tau = 10$ s). (c) The curve shows the corresponding dependence for a RC suspension with atrazine (10^7 M) under Xe-lamp excitation. The times (in s) of the best exponential approximation of the experimental curves by the calculated ones are given in the figures.

shows the recovery kinetics of isolated *R. sphaeroides* RC suspensions
with (Figure 7.12c) and without (Figure 7.12a, b) addition of herbicides.
One can see the presence of only the short-time component in the curves
in Figure 7.12c. Thus, the herbicides are well detected optically in the
experiments on the recovery kinetics of purple bacteria RC suspensions.
The main result is that the characteristic time of recovery is much shorter
for the samples with herbicides than for those without herbicides.

It is evident that film-type sensors (see this section below) may be
essentially more suitable for designing computer-aided controlled ecological monitoring systems. We have studied the recovery kinetics of
Langmuir films of RC *R. sphaeroides*. Figure 7.13a presents the curve

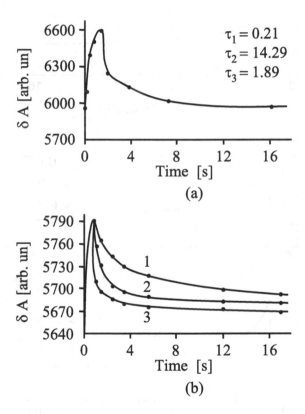

Figure 7.13. Electron donor recovery kinetics of Langmuir-film RCs (20 monolayers)
following He–Ne laser pulse (Is) excitation before (a) and after (b) 5 min processing
with an atrazine water solution. The times and weights of the best exponential approximation of the experimental curve by the calculated one are given in (a). For (b):
curve 1 corresponds to the sample processed with atrazine solution of 10^7 M (the best
approximations with $\tau_1 = 0.21$ (50%), $\tau_2 = 13.0$ (20%), $\tau_3 = 1.89$ (30%)); curve 2
corresponds to 10^{-8} M atrazine solution [$\tau_1 = 0.21$ (70%), $\tau_2 = 12.0$ (20%), $\tau_3 = 1.89$
(10%)]; curve 3 is for 10^{-5} M atrazine solution $\tau_1 = 0.21$ (88%), $\tau_2 = 11.11$ (12%)].

for the samples in the natural unpolluted ecological system (water). This curve is almost the same as for suspensions of RC with unblocked electron transition from the primary to secondary quinone acceptor (see Figure 7.12a). After immersion of the films into the water solution of atrazine (10^5–10^7 M) the electron transfer onto the secondary acceptor was blocked. This was revealed experimentally by the increase of the short-living component weight (Figure 7.13b).

One should point out that the studied Langmuir films were of rather good quality, but improvement of their optical characteristics is necessary for further practical application. These films can serve as the basis of the sensors used in ecological monitoring systems, but the task for the near future consists in making our sensors more perfect.

In conclusion, it may be noted that right now it is possible to model ecological monitoring systems using (as the biosensors) green plants fixing the useful data, in particular, in the form of a disturbed fluorescence induction curve. On the basis of such sensors, integral ecological monitoring systems can be designed both for separate ecological objects and for more extended systems.

3.3   Identification of a Bilinear Sensitive Element

The identification of a sensitive biosensor element (SBE) based on approximation of the input-output map with certain simple models has been considered and investigated.

In the continuous-time case these simple models are bilinear systems (BS), given by:

$$\dot{x}(t) = \left[A_0 + \sum_{i=1}^{p} u_i(t) A_i \right] x(t),$$

$$y(t) = \lambda x(t), \tag{7.63}$$

where the state is $x \in \mathbb{R}^n$ and the (scalar) output is $y(t)$; $u_1(t)$ is the measured signal; $u_2(t), \ldots, u_p(t)$ are the control inputs of the biosensor. The approximation result is as follows.

Let $J \subset [0, \infty)$ and $C \subset [C^*(J)]^p$ be compact sets, where $C^*(J)$ denotes the space of continuous functions $J \to \mathbb{R}$, with the supreme norm. A functional $J \times C \to \mathbb{R}$ is said to be causal if and only if its value in $t \in J$, for $(u_1, \ldots, u_p) \in C$, does not depend on $u_1(r), \ldots, u_p(r)$ for all $t < \tau$; then the following theorem holds.

Theorem 7.1. *Any causal and continuous functional $J \times C \to \mathbb{R}$ can be arbitrarily closely approximated by BSs.*

This result has been independently obtained by Susmann (1976). In the discrete-time case, the former result is not valid, as can be seen with the classic counterexample $y(t) = u^2(t - 1)$. However, a similar result can be shown replacing BSs by state-affiance systems:

$$\dot{x}(t + 1) = \left[A_0 + \sum_{i=1}^{p} f_i(u_i(t), \ldots, u_m(t))A_i\right] x(t),$$

$$y(t) = \lambda x(t), \tag{7.64}$$

where $x \in \mathbb{R}^n$, $y(t)$ is scalar and $f_i(u_1(t), \ldots, u_m(t))$ are monomials in the inputs $u_1(t), \ldots, u_m(t)$. An input-output map is continuous if and only if for every t the output depends continuously on the inputs $u_1(j), \ldots, u_m(j)$, with $0 < j < t - 1$.

Theorem 7.2. *Any continuous SBE map input-output can be arbitrarily closely approximated on a finite time interval and with bounded inputs by state-affine systems.*

At the identification stage it seems convenient, for computational reasons, to use discrete models, as presented by Dang Van Mien and Norman-CyrotHang (1984). They used a least-squares method on a single-input – single-output SBE, verifying the following conditions.

(i) The input-output map can be linearized at several operating points.

(ii) An operating point is described by certain measurable parameters $\theta \in \mathbb{R}^{m-k}$ which, in the state-affine approximation, play the same role as control inputs where $u_{k-1}, \ldots, u_m$ are the parameters θ. In applications, the difficulty lies in the determination of the number of parameters, their physical meaning, and their relation to the dynamic behavior of the system.

In spite of what has been previously said, continuous-time analysis should not be discarded, because it is common in practice when processes and systems described by differential equations are found. This is the motivation for our work on the proposed identification method based on an approximating model given by equation (7.63). The technique used is of the Ritz–Galerkin kind (Mosevich, 1977). The hypotheses i) and ii) mentioned above are not needed, although it is required that the states of the SBE must be observable.

Identification procedure. Let $z(t) \in \mathbb{R}^n$ be the output variables of an SBE when the inputs are $u_1(t), \ldots, u_p(t)$. Both $z(t)$ and $u(t)$ are assumed to be measured. We determine coefficients A_i, $i = 1, \ldots, p$, in such a way that the system in equation (7.63) will produce states $x(t)$ close, in some sense, to $z(t)$, when the inputs are $u_i(t)$ in the time interval $[0, T]$. In order to achieve this, we follow the ideas presented in Mosevich (1977). Let $\{\mu_k(t)\}_{k=0}^{\infty}$ be a basis of $L^2[0, T]$ and define $u_0(t) \equiv 1$. The system in equation (7.63) can be written as

$$\dot{x}(t) = \left[\sum_{j=0}^{p} u_j(t) A_j \right] x(t). \tag{7.65}$$

Observe that the apparently more general

$$\dot{y}(t) = \left[\sum_{j=0}^{p} u_j(t) M_j \right] y(t) + \sum_{j=1}^{p} u_j(t) B_j \tag{7.66}$$

can be written in the form of equation (7.64) by defining

$$x = \begin{bmatrix} y \\ 1 \end{bmatrix}. \quad A_j = [M_j \vdots B_j]. \tag{7.67}$$

Integrating equation (7.65), multiplying by γ_k, and integrating from 0 to T, results in

$$b_k^t \triangleq \int_0^T \gamma_k [x(t) - x(0)] dt = \int_0^T \gamma_k(t) dt \int_0^t \sum_{j=0}^{p} u_j(s) A_j x(s) ds. \tag{7.68}$$

Calling Γ_k the integral of γ_k,

$$b_k^t = \sum_{j=0}^{p} A_j \int_0^T u_j(s) x(s) [\Gamma_k(T) - \Gamma_k(s)] ds. \tag{7.69}$$

The integral on the right-hand side is denoted $(g^{jk})^t$. Equation (7.68) can be repeated in k, until the same number of equations as unknowns is obtained. It can be seen that the number of functions γ_k required is $L = (p + 1)n - 1$. Proceeding in this manner we obtain the system of equations

$$Qa = b, \tag{7.70}$$

where

$$
Q = \begin{bmatrix}
q^{00} & q^{10} & \cdots & q^{L0} \\
q^{01} & q^{11} & \cdots & q^{L1} \\
\cdot & \cdot & \cdots & \cdot \\
\cdot & \cdot & \cdots & \cdot \\
\cdot & \cdot & \cdots & \cdot \\
q^{0L} & q^{1L} & \cdots & q^{LL}
\end{bmatrix},
$$

$$
b = \begin{bmatrix} b_0 \\ b_1 \\ \vdots \\ b_L \end{bmatrix} (L+1) \times n, \quad
a = \begin{bmatrix} A'_0 \\ A'_1 \\ \vdots \\ A^t_p \end{bmatrix} (L+1) \times n.
$$

Solving the linear system of equations (7.69), we calculate the coefficients of the system in equation (7.65). This solution is always possible because of the linear independence of the γ_ks. Furthermore, it must be noticed that we require $u_i(t) \neq u_j(t)$, $0 \leq i, j \leq p$. It is possible to build a rectangular system of the form of equation (7.69) adding equations corresponding to γ_ks with $k > L$; and then solving by least squares. We considered the following equation of the SBE,

$$
\ddot{y} - ay - by - cy^3 = u(y),
$$

or its equivalent

$$
\begin{aligned}
\dot{q}_1 &= q_2, \\
\dot{q}_2 &= aq_2 + bq_1 + cq_1^3 + u(t).
\end{aligned}
\tag{7.71}
$$

The order n of the approximation BS is not known, hence different approximations are made by incorporating additional states. The observed states are (q_1, q_2) and powers and products of q_1 and q_2 are incorporated to augment the dimension of the approximation. This is justified because the functionals of the SBE are analytic, and then we can naturally introduce power series involving those terms. A formal proof and the relationship with Volterra series was given by Brockett (1976). The pollutant characteristics assessment method is presented in Yatsenko and Knopov (1992). The bilinear models are given by

$$
\dot{x} = A_0 x + A_1 x u,
$$

where $x \in \mathbb{R}^n$ and u contains scalar functions. The identification procedure is used to obtain the matrices A_0 and A_1. The independent functions γ_k used in this example were the Laguerre polynomials.

3.4 Separation of Pollutant Characteristics by Neural Chips

Requirements for standard samples of polluted water. The present point suggests that it is necessary to make use of a neural chip based on probabilistic neurons and used to recognize the characteristics of water pollutants. It is supposed that the results of chromatographic analysis can be delivered to the chip input. However, when the neural processing method is applied, it can also compete with the chromatographic methods. In all the cases, when derivation of a mathematical model adequate for an object is not successful at some investigation stage, but there are only experimental data characterizing the behavior of this object under various disturbing actions, pattern recognition training methods can be used (Gushcha et al., 1993; Ackley, Hinton, and Seinowski, 1985).

The main requirements for the pattern recognition training methods are: (a) guarantee of quality and reliability of object state recognition performed by the solution rules yielded by the training process; (b) these solution rules must be easily interpretable and, from the technical point of view, they also must be easily implementable when it is necessary to create a special pattern recognition system; (c) the possibility to operate with object properties of different types.

Let us represent some types of pollutant concentration prognostication problems that can be formulated and solved within the framework of the pattern recognition training problem and let us take the water environment state assessment problems as the example when the assessment is made pursuant to indirect measurements. Assume that the investigation object is the water sample and there exists some measuring facility based on biosensors and an electrical or optical signal is recorded at the facility output. The parameters characterizing such a signal (frequency, phase, amplitude characteristics, etc.) are hereafter referred to as the indirect measurements. Here we follow the series of experiments. A water assay is taken and the direct measurements are performed (chemical analysis, etc.) and according to these results an expert (an expert group) yields the integral water quality assessment made according to the proper assessment scale. For instance: "Water" = ("distilled", "spring", "drinking", "industrial", "domestic"), or: "Water" = ("very pure", "pure", "more likely pure", "more likely polluted", "polluted", "very polluted"). As the result of all such experiments, we have an L-length observation sample V on which the object subsets $V_1, V_2, \ldots, V_m$ ($V = \cup_{j=1}^m V_j$; $V_j \cap V_k = \varnothing$ when $j \neq k$) are determined and they correspond to the water quality classes (patterns) $V_1^*, V_2^*, \ldots, V_m^*$ identified with respect to the assessment scale. Every object $v \in V$ is described by the vector of values $x = (x_1, x_2, \ldots, x_s)$ of the indirect measurements.

The observation sample V is the data required for the training algorithm to operate. Pursuant to the above-mentioned requirements made with respect to recognition systems, state the pattern recognition training problem as follows.

Let V_j be a subset of training sample objects that corresponds to a pattern V_j^* and let $V_{\bar{j}}$ be an object subset corresponding to the rest of the patterns $V_{\bar{j}}^*$.

It is said that when using the training sample V it is required to find sets of signs in the space where every object set V_j can be separated from the set $V_{\bar{j}}$ by the solution rule $F(x)$ that belongs to a solution rule set Φ. And here the quality of recognition of new objects done by the rule $F(x)$ must be guaranteed with reliability $1 - \eta$ and this quality is not lower than the value

$$\epsilon = \frac{\ln N - \ln \eta}{L},$$

as specified in advance. Here N is the extension function of a solution rule set Φ.

To solve the problem in such a statement, recognition training algorithms may be used.

During the training process, the most informative indirect measurements and the ones meeting the pattern sign definition are selected from the indirect measurement set. The final result of the training procedure is the set of solution rules that permit us to prognosticate a water state pursuant to the indirect measurements. And the quality and reliability of the work under the new data are guaranteed for the obtained solution rules.

Separating pollutant characteristics by a stochastic neural chip. The intelligent sensor is the device with the neural chip. We use a neural chip with symmetric recurrent connections where each neural element is stochastic and the firing depends on the weighted sum of inputs.

Let us consider a neural chip consisting of n neurons, and let $w_{ij} = w_{ji}$ be the symmetric connection weight between the ith neuron and the jth neuron. The self-recurrent connection w_i is assumed to be zero. Let h, be the threshold of the ith neuron. The potential of the rth neuron is defined by

$$U_i = \sum_{j=0}^{n} w_{ij} x_j. \tag{7.72}$$

Each neuron changes its state asynchronously depending on U_i, where the new state x_j of the ith neuron is equal to 1 with probability $p(U_i)$ and is equal to 0 with probability $1 - p(U_i)$.

The vector $x = (x_1, \ldots, x_n)$ is called the state of the neural chip. A state transition is mathematically described by a Markov chain with 2^n states x. When all the neurons are connected, they form an ergodic Markov chain, having a unique stationary distribution $\mu(x)$. Every initial state x converges to $f(x)$, and state x appears with relative frequency $f(x)$ over a long course of time.

Let $\nu(x)$ be the probability distribution over x with which an ecological information source emits a signal x. Signals are generated independently subject to $v(x)$ and are presented to a neural chip. This chip is required to modify its connection weights and thresholds so that it will simulate the ecological information source. It is required that the stationary distribution $f(x)$ of the neural chip becomes as close to $v(x)$ as possible. The learning rule is given by

$$\Delta w_{ij} = \Delta w_{ji} = \epsilon(\nu_{ij} - f_{ij}), \tag{7.73}$$

where ϵ is a small constant, ν_{ij} is the relative frequency that both x, and x_j are jointly excited under probability distribution $v(x)$, and f_{ij} is the relative frequency that both x_i, and x_j are jointly excited under $f(x)$, that is, when the neural chip is running freely.

The learning rule is realized by the Hebbian synaptic modification method in two phases (Amari, Kurata, and Nagaoka, 1992). In the first phase, the input learning phase, the connection weight w_{ij} is increased by a small amount whenever both x_i and x_j are excited by an input x; hence, on average the increment of w_{ij} is proportional to ν_{ij}. In the second phase, the free or antilearning phase w_{ij} is decreased by the same small amount whenever both x_i and x_j are excited by the free state transition; hence, the decrement of w_{ij} is proportional to p_{ij} on average.

We consider a situation where the neurons are divided into two parts, namely visible neurons and hidden neurons. Visible neurons are divided further into input and output neurons. In the learning phase, inputs are applied directly to visible neurons. Inputs are represented by a vector $x_\nu = (x_I x_O)$ for visible neurons, where x_I, and x_O correspond to the states on the input and output neurons, respectively, and components on hidden neurons have meaning in this phase. A visible input x_V is generated from the ecological information source. Its joint probability distribution is denoted by $\nu(x_I x_O)$.

In the working or recalling phase, only the input part x_I of the x_V is applied. The stochastic state transitions take place under this condition of fixed x_I, so that the conditional stationary distribution $f(x_H x_O / x_I)$

is realized, where x_H denotes the states on the hidden neurons. The distribution can be calculated from the connection weights thresholds, and the fixed x_I.

In the more general case a Boltzmann machine is required to realize the conditional probability distribution $\nu(x_O/x_I)$ of the ecological state of water as faithfully as possible by learning. The distribution of the state of the hidden neurons is of no concern, but $f(x_O,x_I)$ should be as close to $\nu(x_0,x_1)$ as possible. It can also be shown (Amari, Kurata, and Nagaoka, 1992), that the learning rule gives a stochastic gradient descent of the conditional Kullback information

$$I[\nu; f] = \sum \nu(x_I)\nu(x_O/x_I) \log \frac{\nu(x_I, x_O)}{f(x_I x_O)}, \tag{7.74}$$

where ν_{ij} denotes the relative frequency of $x_i = x_j = 1$ under the condition that x_I and x_O are fixed, and f_{ij} is the relative frequency in the restricted free run when only x_I is fixed.

Another solution of this problem is also possible. Introduce, for instance, the notions "pure water" and "polluted water". According to the direct measurement results, an expert determines the degree of belonging of a given observation object to the "pure water" and "polluted water" notion. The degree of belonging is specified by a number taken from the closed interval [0,1]. For example, if some given observation object is the "pure water" notion standard, then the degree of belonging to it is equal to 1 and, by contrast, its degree of belonging to the "polluted water" notion is equal to 0.

The training is over, and the pattern recognition system yields the prognosis with respect to water quality as the degree of belonging of this observation object to the "pure water" or "polluted water" notion.

Within the framework of the pattern recognition training problem, not only the problem associated with prognostication made with respect to indirect measurements of water quality can be solved, but also the problem of prognostication of concentration of different chemical agents and elements (e.g., herbicides) in this water. In this case, it is guaranteed with reliability $1-\eta$ that the probability of deviation of a prognosticated value from a real value by more than $2g$ does not exceed the value specified in advance.

We can now summarize the results in this chapter:

1. The theoretical and experimental investigations as well as the sensor signal-processing methods and the environment state assessment methods based on neural network technologies are considered here. These investigations can be the basis for improvement of the sensitive sensor elements. The results of studying the neural sensor are based

on the novel principles of information processing performed in physical and biological systems. It is proposed here that the intelligent sensor be elaborated on the basis of:

(a) Neural network technologies;

(b) The possibility to use the chromatographic characteristics of different pollution components in water.

2. Experimental investigations were performed using photosynthetic bacteria to study the influence of herbicides on optical properties of the RC. In the presence of herbicides at the LPC-level, essential changes of the RC's recovery kinetics were observed.

3. Langmuir films of *R. sphaeroides* purple bacteria RCs were determined to be a good detector of water pollution.

4. The influence of herbicides, heavy metals, and some other pollutants (toxicants) on the operation of the green-plant photosynthesis apparatus was analyzed experimentally. The analysis was performed pursuant to the results concerning the influence on kinetic-fluorescent characteristics of plants. The fluorescence induction curves of green plants adequately reflect their response to unfavorable conditions present in the environment.

5. The choice of a nonlinear identification method for biosensor dynamics was substantiated. Algorithms were proposed for identification of the nonlinear model with respect to the experimental input and output data and on the basis of the Ritz–Galerkin method. The possibility of performing the system analysis of biosensor dynamics and construction of an inverse model to classify the pollutants was determined (Yatsenko and Knopov, 1992). Information was also obtained about the character of the processes running in polluted water.

6. Training algorithms that guarantee the quality and reliability of recognizing the water state are proposed and theoretically substantiated. This quality is not lower than the value specified in advance. In this case, it becomes possible to operate with biosensors of different types.

7. Intelligent sensors provide a new microsystem for treating the wide variety of ecological pollutants. Information geometry that originates from the intrinsic properties of a smooth family of probability distributions is also appropriate to the study of the manifold of sensors. The manifold of a simple intelligent sensor with no hidden units is

proved to be l-flat and m-flat, so that it possesses nice properties. The present chapter, together with Amari, Kurata, and Nagaoka (1992), is the first step in constructing a mathematical theory of sensors.

4.   Notes and Sources

The material contained in chapter 7 is based on early papers of Chinarov et al. (1992), Chinarov et al. (1990), Gushcha et al. (1993), and Yatsenko (1992, 1995, 1996). The idea of an intelligent sensor was proposed by Yatsenko (1996). The identification of a sensitive biosensor element is based on the papers of Susmann (1976), and Mosevich (1977).

Chapter 8

MODELING AND ANALYSIS
OF BILINEAR SYSTEMS

The problem of reconstruction of nonlinear and bilinear dynamics has been studied in a number of physical simulations ranging from Whitney's theorem (Loskutov and Mikhailov, 1990; Nicolis and Prigogine, 1977; Nerenberg and Essex, 1986) to state-space representation (Chang, Hübler, and Packard, 1989; Takens, 1981; Grassberger and Procaccia, 1982). This problem is also related to the flow method developed by Cremers and Hübler (1987). The flow method is a procedure for reconstructing a set of coupled maps (CMs) or ordinary differential equations (ODEs) from a trajectory of the system in state space. This chapter presents methods for determining nonlinear dynamical systems and application to small target detection from sea clutter and modeling of nonlinear chaotic dynamics. We show that this methodology may easily be adapted to systems with hidden variables.

In Section 8.1 we present a method for the reconstruction of nonlinear and bilinear equations of motion for systems where all the necessary variables have not been observed. This technique can be applied to systems with one or several such hidden variables, and can be used to reconstruct maps or differential equations.

Section 8.2 outlines the problem of determining sea clutter dynamics and the application of this methodology in detection and classification of small targets. A systematic method of reconstruction of a sea clutter attractor is considered. We explore the use of dynamical system techniques, optimization methods, and statistical methods to estimate the dynamical characteristics of sea clutters. We assume that radar information is in a form of nonlinear time series. Hence we employ a dynamical approach for characterizing a radar signal, based on nonlinear estimation

P.M. Pardalos, V. Yatsenko, *Optimization and Control of Bilinear Systems*, doi: 10.1007/978-0-387-73669-3,
© Springer Science+Business Media, LLC 2008

of dynamical characteristics, by forming a vector of these characteristics, and modeling the evolution of dynamical processes over time.

In Sections 8.3 and 8.4 we discuss the application of global reconstruction techniques to the reconstruction of stochastic models and Fokker–Planck equations for a single-variable distribution function. In order to characterize the nonideal filter, we introduce the effective diffusion coefficient of the filter. The reconstruction of the master equation for one-step processes is considered. We then apply global optimization approaches to reconstruction of dynamical systems related to epileptic seizures.

Section 8.5 discusses whether living things, under the same measuring conditions, can generate signals with different types of dynamics. We also wanted to detect the possible effects of low-intensity microwaves using parameters of deterministic chaos. For this purpose, two sets of electroetinograms were analyzed by methods aimed at recognizing different types of deterministic dynamics. Both sets included time series recorded from objects exposed to low-intensity microwaves and those that were not. The analytical methods are based on nonlinear forecasting and a "surrogate data" technique. Although the experimental conditions were identical for the two sets, we have shown that both have time series with deterministic and stochastic dynamics. We also found that the use of parameters of deterministic dynamics is insufficient to distinguish between the sets.

1.   Global Reconstructing of Models

1.1   Modeling without Hidden Variables

In the flow method as developed by Cremers and Hübler (1987), the dynamics throughout the state space is represented either with a single set of coupled maps (CM), given by

$$y_i(n+1) = f_i(y(n), p), \quad i = 1, \ldots, N, \quad y \in \mathbb{R}^N$$

or a set of ordinary differential equations

$$\dot{y}_i(t) = f_i(y(t), p), \quad i = 1, \ldots, N, \quad y \in \mathbb{R}^N,$$

where p is the vector of free parameters P_i, $i = 1, \ldots, N_c$; N_c is the number of free coefficients. The number of observed variables is assumed sufficient to embed the dynamics. The functions $\{f_i\}$ may be of any form, but are usually taken to be a series expansion. This method has been successfully tested with Taylor- and Fourier-series expansions. In this manner, the modeling is done by finding the best expansion coefficients to reproduce the experimental data. Often the situation arises where the form of the functions $\{f_i\}$ is known, but the coefficients are unknown; for example, this occurs frequently with rate equations for chemical processes. This added information greatly reduces the number of undetermined parameters, thus making the modeling computationally more efficient.

The modeling procedure begins by choosing some trial coefficients. The error in these parameters can be computed by taking each data point $x(t_n)$ as an initial condition for the model equations. The predicted value $y(t_{n+1})$ can then be calculated for CMs as

$$y_i(n+1) = f_i(x(n)), \quad i = 1, \ldots, N,$$

or for ODEs as

$$y_i(t_{n+1}) = x_i(t_n) + \int_{t_n}^{t_{n+1}} f_i(y(t'))dt, \quad i = 1, \ldots, N \qquad (8.1)$$

and compared to the experimentally determined value. Previous work (Cremers and Hübler, 1987) has shown that more stable models can often be obtained by comparing the prediction and the experimental data several time steps into the future. For the present analysis, we predict the value only to the time of the first unused experimental data point. The error in the model is thus obtained by summing these differences

$$\chi_\nu^2 = \frac{1}{N(M-1) - N_c} \sum_{i=1}^{M} \sum_{j=1}^{N} \frac{1}{\sigma_{ij}^2} [y_j(t_i) - x_j(t_i)]^2, \qquad (8.2)$$

where N_c is the number of free coefficients, M is the number of data points, and σ_{ij} is the error in the jth vector component of the ith measurement. The task of finding the optimal model parameters has now been reduced to a χ_ν^2 minimization problem. Thus, the best parameters are determined by

$$\frac{\partial \chi_\nu}{\partial p_i} = 0, \quad \forall\, i.$$

Therefore the ability to determine these coefficients rests upon the strength of the algorithm employed to search through the space of parameters. Because this has been formulated as a standard χ_ν^2 problem, statistical tests can be applied. Typically, $\chi_\nu \simeq 1$ implies that the modeling was successful; however, more sophisticated tests can be applied as well (e.g., F test, etc.). If the experimental uncertainties σ_{ij} are unavailable, this normalization factor can simply be removed from equation (8.2). This means that the χ_ν tests cannot be applied, but the best possible model can still be determined by locating the global minimum of χ_ν in the parameter space.

1.2   Modeling with Hidden Variables

As stated, this method works when all variables can be measured. Unfortunately, this is almost never the case in real experiments. Frequently, one or more of the variables are hidden; that is, cannot be directly measured. This requires that a new method be utilized to reconstruct the equations of motion for the dynamics.

To develop a technique for reconstructing the dynamics of systems with hidden variables, we assume that only one variable w is hidden (i.e., $N_h = 1$), and our experimental data x contain N_0 observables, $x \in \mathbb{R}^{N_0}$. The restriction on N_h is purely for illustration. The model equations are identical to the previous case, with

$$y_i(n+1) = f_i(y(n)), \quad i = 1, \ldots, N \tag{8.3}$$

for a system of maps where $N = N_0 + N_h$, $y \in \mathbb{R}^N$, and $y(n) = x(n) + w(n)$. The predicted values $y(n)$ are calculated from

$$y_i(n+1) = f_i(x(n), w(n)), \quad i = 1, \ldots, N \tag{8.4}$$

thus requiring that we know $w(n)$. Because experimental data are available for the other N_0 variables for all n, we can use those to solve for $w(n)$ as

$$f_i(x(n), w(n)) - x_i(n+1) = 0, \quad i = 1, \ldots, N. \tag{8.5}$$

Having one hidden variable, only one of these equations is needed to solve for $w(n)$. If f_i were a Taylor-series expansion to lth order, then solving for $w(n)$ produces l roots. In practice, because we do not expect our first guess for the model coefficients to be correct, each of the N_0 equations is solved, thus generating lN_0 possible solutions for $w(n)$. To be accepted, these roots are required to be real and to satisfy any known physical bounds upon the value of $w(n)$. Aside from these constraints, there is no a priori method to determine which root is correct. Therefore, each is tried in turn with the best root chosen according to predictive accuracy.

Using one such $w(n)$, we determine $w(n+1)$ from equation (8.4). Now the predicted values $y(n+2)$ can be calculated from $(x(n+1), w(n+1))$ and compared to $x(n+2)$. The error in the model is thus

$$\chi_\nu^2 = \frac{1}{N_0(M-2) - N_c} \sum_{i=2}^{M} \sum_{j=1}^{N_0} \frac{1}{\sigma_{ij}^2} [y_j(i) - x_j(i)]^2. \tag{8.6}$$

The best value of $w(n+1)$ is stored for the next iteration. If at step $n+1$ no acceptable roots can be found, the previous best is used to continue the calculation. Note that we do not use $w(n)$ to immediately calculate $y(n+1)$. This is because the $x(n+1)$ have already been used in equation (8.6) to improperly characterize the accuracy of the model.

This analysis can be extended simply to handle an arbitrary number of hidden variables. If we have at least as many observables as hidden variables $N_0 \geq N_h$, then rather than solving one equation in one unknown, we must solve a system of equations in N_h unknown parameters.

$$f_1(x(n), w(n)) - x_1(n+1) = 0,$$
$$f_2(x(n), w(n)) - x_2(n+1) = 0,$$
$$\vdots$$
$$f_{N_h}(x(n), w(n)) - x_{N_h}(n+1) = 0, \qquad (8.7)$$

where $w \in \mathbb{R}^{N_h}$. Once $w(n)$ has been determined, the analysis proceeds exactly as before.

When $N_h > N_0$, we cannot generate enough equations from the first N_0 model equations using only $x(n)$ and $x(n+1)$ to uniquely determine the N_h initial conditions of w. Therefore we create additional equations using more experimental data points:

$$f_1(x(n), w(n)) - x_1(n+1) = 0,$$
$$\vdots$$
$$f_{N_0}(x(n), w(n)) - x_{N_0}(n+1) = 0,$$
$$f_{N_0+1}(x(n), w(n)) - w_1(n+1) = 0,$$
$$\vdots$$
$$f_N(x(n), w(n)) - w_{N_0}(n+1) = 0,$$
$$f_1(x(n+1), w(n+1)) - x_1(n+0) = 0,$$
$$\vdots$$
$$f_{N_0}(x(n+1), w(n+1)) - x_{N_0}(n+0) = 0. \qquad (8.8)$$

In equations (8.8), the first N_0 equations are the same ones used previously. However, the second N_h equations come from the model equations representing the dynamics of the hidden variables. These are needed to calculate $w(n+1)$ as a function of $w(n)$. The third set of equations determines N_0 more of the $w_i(n)$ from $x(n+2)$, and so on. This continues until enough equations have been generated to determine $w(n)$; $N_h([N_h/N_0]+1)$ equations are needed, where the square brackets indicate the greatest integer.

Although the above method for handling $N_h > N_0$ is straightforward, an alternate approach exists that may be more efficient in some situations. In such cases, there will be $N_u = N_h - N_0$ unknown variables that cannot be reconstructed directly from the N_0 observations at a single time. If the system is not chaotic, we can simply add one unknown parameter for each of the N_u unknown variables representing that variable's value at time t_0. This value can then be iterated at each time step and carried forward to the next step, exactly as is done when no acceptable roots can be found. This makes the optimization of the model coefficients more difficult due to an increase in the number of local minima in the χ_ν landscape. Fortunately, the correct minimum can still be determined by χ_ν satisfying the statistical tests. Knowledge of the proper form of the $\{f_i\}$ will be very useful for reducing the complexity of this search.

However, if the system is chaotic, a single initial condition iterated through the entire data set cannot be expected to remain close to the experimental trajectory even for an accurate model. Thus several initial conditions will be needed for each unknown variable. The time between these initial conditions τ will depend in general upon the rate of information loss; that is, the Lyapunov exponents. An additional term can be added to χ_ν^2 representing the distance between the previous initial condition p_{ij} iterated up to the time of the next initial condition and that initial condition $p_{i,j+1}$ so that

$$\chi_\nu^2 = \frac{1}{N_0(M-2) - N_c} \sum_{i=3}^{M} \sum_{j=1}^{N_0} \frac{1}{\sigma_{ij}^2} [y_j(i) - x_j(i)]^2$$

$$+ \frac{1}{N_u(N_T - 1) - N_T} \sum_{i=1}^{N_u} \sum_{j=1}^{N_T-1} \frac{1}{\sigma_{i,(j+1)}^2} [p_{i,j+1} - f_i^\tau(p_{i,j})]^2, \qquad (8.9)$$

where N_T is the number of parameters needed for each undetermined variable. The added term in equation (8.9) is a simple endpoint-matching condition that helps reduce the number of local minima in χ_ν^2. Because of the persistent complexity of the χ_ν^2 landscape, knowledge of the form of the model equations will be necessary, and they must be limited to only a few free parameters. This variation of our hidden variables reconstruction is generally the least robust of the options described, but it may be useful in some special situations.

Although the discussion of hidden variables has thus far focused upon maps, it can easily be extended to reconstructing hidden variables in continuous systems (ODEs). For this, equation (8.5) becomes

$$f_i(x(t_n), w(t_n)) - \dot{x}_i(t_n) = 0; \qquad (8.10)$$

so we must calculate the first derivatives of the observables $\dot{x}$. The modeling process then proceeds exactly as in equations (8.3)–(8.6) except that a modification of equation (8.1) replaces equation (8.4). The extension to multiple hidden variables is equally straightforward by simple modifications of equations (8.5)–(8.9) and with the same restrictions as before. The only caveat to this process comes from the introduction of additional noise that occurs during the process of computing the derivatives of the experimental datasets $\dot{x}(t_n)$ and the integration of the model equations necessary to calculate $w(t_{n+1})$ and $y(t_{n+2})$. To compensate for these errors, the σ_{ij} should be adjusted appropriately, thereby making it possible for the value of χ_ν to satisfy the statistical tests. This computational noise is discussed further in the examples of the Rössler, Lorenz, and bilinear systems

$$\dot{x} = Ax + Bxu, \tag{8.11}$$
$$y = Cx. \tag{8.12}$$

Specific examples of the reconstruction of hidden variables is now given for simulated data. This is an effective method of testing because the correct model parameters will already be known. The first example is for two coupled bilinear maps,

$$x(n+1) = \lambda_1 x(n)[1 - x(n)] + dy(n),$$
$$y(n+1) = \lambda_2 y(n)[1 - y(n)] + dx(n). \tag{8.13}$$

The experimental data were generated using $\lambda_1 = 2.0$, $\lambda_2 = 3.5$, and $d = 0.2$. For the test using equations (8.13), we took y to be the hidden variable. We begin by assuming that the form of the equations of motion is known, and λ_2 and d are the only free parameters. We show the χ_ν landscape for these two parameters. This landscape has two minima: $\lambda_2 = 3.5$, $d = 0.2$; and $\lambda_2 = 3.875$, $d = 0.275$. According to the value of χ_ν, the first minimum qualifies as a solution, whereas the second minimum fails badly. In this case, the χ_ν landscape was sufficiently simple that even an unaided gradient search would have a high probability of locating the correct minimum.

For the previous test, the noise in the data was nothing more than roundoff error in the last significant digit. To better study the effects of noise, we have added bandlimited noise in the range $-\varepsilon$ to ε to the experimental data. We use the same model as above, so we still have λ_2 and d as our free parameters. We investigate the effects of dynamical noise. This means that the noise was added to the experimental system at each step in the mapping. This would be like making perfect measurements of a dynamical system that was being constantly perturbed. The error in the model is thus

$$s^2 = \sum_{i=1}^{M} \sum_{j=1}^{N_0} [y_j(i) - x_j(i)]^2,$$

as in equation (8.6) but not normalized with σ_{ij} versus the maximum noise amplitude ε, for the two local minima. We see that for very small noise levels, the error of the first minimum s_1 is $s_1 \simeq \varepsilon$, corresponding to $\chi_\nu \simeq 1$, and that the error of the second minimum s_2 is approximately constant and consistently fails any χ_ν tests. Because the experimental data no longer represent a trajectory of the true system with complete accuracy, the model coefficients will not be precise and will vary from one dataset to another. The average difference between the model parameters and the correct values is estimated as η versus ε, where $\eta^2 = (\lambda_1^m - \lambda_1)^2 + (d^m - d)^2$ with the superscript m indicating the coefficients obtained from the model.

When the noise reaches approximately 2%, the trends just noted begin to break down and it becomes increasingly difficult to distinguish the two minima. For approximately 5% noise, both minima now satisfy our statistical tests, and it is no longer possible to determine which set of parameters is correct. The point at which this occurs is not generic, but rather depends specifically upon the system being studied. Clearly, the added noise has the effect of smoothing the s landscape, thus the first minimum is affected most drastically. For the current case, to be able to determine which minimum is the correct solution for up to 5% noise is a significant achievement. Note that in all these simulations, equation (8.4) is used only to predict one time step forward. When the predictions are carried several time steps into the future, the false minima become more shallow, providing greater tolerance to noise.

Next we consider data obtained from a set of ODEs. The first such example is the bilinear system,

$$\dot{x} = -y - z,$$
$$\dot{y} = x + ay,$$
$$\dot{z} = b + (x - c)u, \qquad (8.14)$$

where we have used $a = 0.343$, $b = 1.83$, $u = kz$, $k = 1$, and $c = 9.75$ to generate the data. For this modeling, we use the method for reconstructing ODEs and assume a model of the same form as equations (8.14) with b and c unknown. We have used the experimental data with $z(t)$ hidden. We have again calculated the χ_ν landscape for the free parameters. In this case, there is only one minimum, and it occurs at the proper parameter values. In this case, however, the minimum is not as sharp as the previous example. This smoothing of the landscape comes from errors

introduced through the calculation of the derivatives and integrals. The
greatest error comes from the differentiation which was done simply as

$$\dot{x}(t_n) = \frac{x(t_{n+1}) - x(t_{n-1})}{t_{n+1} - t_{n-1}}.$$

This error was not incorporated into χ_ν so as to illustrate the cumulative
effect of these computational errors. The most remarkable aspect of this
example is that the data only contain two excursions along the axis of
the hidden variable, even though the full attractor has a fully developed
funnel, and yet the hidden variables were still reconstructed effectively.
When a spline fit or other more accurate differentiation scheme is employed, the minimum becomes much more distinct.
    We also consider the system

$$\dot{x} = \sigma(y - x),$$
$$\dot{y} = rx - y - xz,$$
$$\dot{z} = -bz + xy.$$

The data were generated with $\sigma = 50$, $r = 50$, and $b = 8/3$. For this
case, the time between data points is much larger than the previous
example. This tends to amplify differentiation errors, thus causing the
χ_ν landscape to be even more strongly smoothed. Again, we have not
incorporated these errors into χ_ν so as to illustrate the magnitude of the
effect. In fact, the minimum is still at the correct parameter values; but,
if the differentiation errors are not symmetrically distributed, the minimum may wander some from the true values. The modeling in this case
is remarkably successful, considering the coarseness of the experimental
data.
    In both of the examples where the models were ODEs only one mini-
mum was found in the landscape. Because of the integration needed to
predict the value of the observables at the next time, equation (8.1),
we are no longer simply doing a one-step prediction as was done for
the CM. Thus stable solutions are preferred and fewer local minima occur in the χ_ν landscape. This implies that it may be easier to generate
models based on ODEs. We have, of course, chosen systems that could
be modeled with a finite number of polynomial terms. If this fails, one
may simply need to choose a different set of expansion functions. When
appropriate expansion functions were selected, the only real limitations
came from failings in the search algorithm employed to minimize χ_ν
or from noise present in the data or introduced through the modeling
technique.

1.3   Controlling Chaos

Much progress has been made in recent years in the area of controlling nonlinear systems without the continuous feedback required by traditional methods (Jackson, 1989). The success of this nonlinear control theory hinges upon the creation of a good model for the system. To control a system, we consider the following equations:

$$\dot{\mathbf{x}} = \mathbf{f}(\mathbf{x}) + \mathbf{F}(t) \quad \text{(experimental dynamics)},$$
$$\dot{\mathbf{q}} = \mathbf{h}(\mathbf{q}) \quad \text{(model dynamics)},$$
$$\dot{\mathbf{u}} = \mathbf{g}(\mathbf{u}) \quad \text{(goal dynamics)}.$$

The equation for the experimental dynamics represents the actual system to be controlled. The goal dynamics represents the system to which we wish to drive the original system. The driving force needed to control the system is

$$\mathbf{F}(t) = \mathbf{g}(\mathbf{u}) - \mathbf{h}(\mathbf{u}).$$

Previous work has shown that $\mathbf{g}(\mathbf{u})$ cannot be chosen arbitrarily, but is subject to certain stability constraints and restrictions on the initial conditions when the driving force is applied. These issues have been detailed elsewhere (Jackson, 1989), and we assume that all these conditions are met.

For effective entrainment of the experimental system to the goal dynamics, a model must be constructed for which $\mathbf{h} \simeq \mathbf{f}$. Chang, Hübler, and Packard (1989) have shown that when the model is not perfect the distance between the experimental system and the goal dynamics scales linearly with the error in the model coefficients. This indicates that the control is stable despite possible small modeling errors produced during the hidden variable reconstruction.

One case that can arise is a system in which some of the variables are hidden from observation, thus hidden during modeling, but all the variables may be controlled. For numerical experiments, we consider

$$\dot{\mathbf{x}} = \mathbf{f}(\mathbf{x}) + \mathbf{F}(t), \quad \mathbf{f}(\mathbf{x}) = \begin{cases} \sigma(y - x), \\ rx - y - xu_1, \\ -bz + xu_2; \end{cases}$$

$$\dot{\mathbf{u}} = \mathbf{g}(\mathbf{u}), \quad \mathbf{g}(\mathbf{u}) = \begin{cases} \sigma'(v - u), \\ r'u - v - uv_1, \\ -b'w + uv_2. \end{cases}$$

Here $\mathbf{x}(t)$ is the experimental system; $u_1 = z$, $u_2 = y$, $v_1 = w$, $v_2 = v$; and $\mathbf{u}(t)$ is the goal system. We have considered the specific case where

$$\sigma = 10.0, \quad \sigma' = \sigma;$$
$$r = 50.0, \quad r' = 125.0;$$
$$b = \frac{8}{3}, \quad b' = b.$$

During the modeling process, we assumed that $y(t)$ was hidden, but the original parameters were reconstructed without difficulty. Thus this case is not significantly different from previous studies (Cremers and Hübler, 1987). For the goal system, we have chosen simply to drive the system to a higher Reynolds number.

This example was presented because there may be important situations where some variables are hidden during the modeling process, but can still be controlled. One such case might be in chemical systems where the reactants are known, but only a subset of these can be continuously monitored. By using the techniques of Section 8.1.2, we can reconstruct the rate equations for the chemical reactions. To control the system may then be simple because one might easily be able to add reactants to the system even though the concentrations of those reactants are hidden from observation. Thus the combination of hidden variable reconstruction and nonlinear control may have important applications in chemical systems.

The second example we wish to consider differs from the previous case in that we wish to control the system even when some of the variables are not controllable; that is, a driving force cannot be applied. Consider the following example,

$$\dot{\mathbf{x}} = \mathbf{f}(\mathbf{x}) + \mathbf{F}(t), \quad \mathbf{f}(\mathbf{x}) = \begin{cases} \sigma(y - x), \\ rx - y - xu_1, \\ -bz + xu_2; \end{cases}$$

$$\dot{\mathbf{u}} = \mathbf{g}(\mathbf{u}), \quad \mathbf{g}(\mathbf{u}) = \begin{cases} \sigma'(v - u), \\ r'u - v - uv_1, \\ f_z(\mathbf{u}); \end{cases}$$

$$\mathbf{F}(t) = \mathbf{g}(\mathbf{u}) - \mathbf{f}(\mathbf{u}).$$

In this case, we again took $y(t)$ to be the hidden variable, but we also chose to make $z(t)$ an uncontrollable variable, so $\mathbf{F}$ applied only to the x and y components. (There is no reason to assume that the hidden variables are also uncontrollable variables, or vice versa.) We have an uncontrollable variable, therefore we can no longer choose to drive the system to an arbitrary goal. Instead, we must restrict our goals just to altering the controllable variables. Because the driving there could be applied only in the plane of the Poincaré map, the goal was also required to be a Poincaré map. For our test, we again chose to drive the system to a different r value:

$$\sigma = 10.0, \quad \sigma' = \sigma;$$
$$r = 50.0, \quad r' = 125.0;$$
$$b = \frac{8}{3}.$$

It is important to understand the differences between the driving force here and the fully controlled case. When all the variables were controllable, even though our goal did not attempt to alter the dynamics of either $x(t)$ or $z(t)$, we could have done so had we so chosen. In the present case, we cannot drive $z(t)$ directly, so we must restrict our goals to those systems that leave the z dynamics unchanged. Obviously then, not all systems can be entrained by this method, but it will be possible in many cases. Well-chosen goal dynamics is important for the entrainment.

The problem of having uncontrollable variables is an interesting one that is still under investigation. Current results indicate that it may be possible to pick specific goals such that the uncontrollable variable can be entrained indirectly to a new dynamics. The types of entrainment that are possible in systems with one or many uncontrolled variables is an area of continuing research.

2.   Nonlinear Dynamics of Sea Clutter and Detection of Small Targets

2.1   Non-Gaussian Signals and Backscattering Process

At present, algorithms for detection of small-size objects and small targets are mainly based on statistical models (Pardalos, Murphey, and Pitsoulis, 1998; Grundel et al., 2007). They make it possible to compute or predict detection characteristics with an accuracy defined by the degree of accounting for nonlinear behavior of a signal and noise. However, the assumption of Gaussian noise limits potentialities of the statistical approach to detecting objects, especially small-size ones. For example, the efficient Swerling type models (Swerling, 1957) that take into account only Gaussian noise are not applicable for detection systems with high resolution and small illuminated surface areas.

To describe signals of non-Gaussian type, the rules of combinational resonant scattering caused by nonlinear hydrodynamic waves at the boundary of two optically different media should be taken into account. In this chapter, it is assumed that nonlinear processes of sea clutter can be adequately described by means of nonlinear finite-dimensional dynamic models with control (Cherevko and Yatsenko, 1992).

The following quantities may serve as control in such models: a signal modulation within the impulse, frequency spectrum, interpulse frequency spectrum, interpulse frequency returning, parameters of synthesized aperture, and others. A state vector of the dynamical model has a correlation between noise and object signals.

The equation of the identifying model has the form:

$$\dot{x} = f(x, \Theta, u), \quad y = g(x, v), \tag{8.15}$$

where x is a vector of the state of a dynamic signal model; u is a random process; f is a nonlinear function of C^m class; g is a nonlinear function of C^1 class; v is a random noise generated by a radar system. The solution of such a system can be characterized by the K-distribution which fairly well takes into account the correlation properties of a real seaclutter signal. In this case, the target model can be represented by the Rice-squared distribution in which the power parameters are modulated by dynamical multipath reflection from the sea surface. However, due to the limitation mentioned above, an attractor model is suggested as an alternative approach to the description of reflection processes. Within the framework of the Hamiltonian formalism, we can construct an oscillatory model of the sea clutter model with chaotic behavior. The given assumption was tested (Cherevko and Yatsenko, 1992) with experimental data of radar signals for a station with high resolution. The obtained results demonstrate to a certain degree that it is not useful to use common statistical models to describe reflection processes.

This requires revision of the existing estimation methods. To this end, it is possible to pass from the Hamiltonian model of a signal with noise of dynamical chaos type to the quantum one which in turn is provided with the ability for noise suppression. This section discusses the given questions.

2.2 Sea Clutter Attractor

We consider the problem of simulating the process of sea clutter according to experimental data represented by time series. The main goal of the simulation was the determination of the minimum number of degrees of freedom of a dynamic system required for an adequate description of scattering processes (minimal realization). For simulation of sea clutter we have used a composite K-distribution as the target clutter model; it describes coherent and incoherent signals reflected from the sea surface

$$Q_{H-K}(G)$$

$$= \frac{2b^\nu G}{\Gamma(\nu)} \int_0^\infty dx\, x\, x^{\nu-1} e^{-bx} \frac{1}{x} \exp\left(-\frac{G^2 + A^2}{x}\right) I_0\left(\frac{2GA}{x}\right), \qquad (8.16)$$

where b is a scale parameter of order; ν is a shape parameter; Γ is the gamma function; G is a module $|G|$ of a sum of coherent terms A with the noise terms s.

Statistical approach to simulation dates back to Boltzmann who supposed that chaotic properties of physical systems result from the interaction of the great number of their degrees of freedom. But the advances in the theory of dynamic systems compel us to revise such an approach. With this purpose, a numerical experiment was conducted in which the data of radar scattering represented by time series were used. Each time series $\{y_0(t), y_0(t+T), \ldots, y_0[(t+(n-1)T]\}$, where T is a quantization time, represented the evolution of a sea clutter signal at a fixed slant radar range. The number of time series members varied from 1500 to 10,000. The method described by Takens (1981) was used to analyze the dynamic chaos. The experiment was based on a numerical procedure for computing a number of the attractor characteristics. We describe the algorithm for computing only one characteristic. Consider a set of points N on the attractor embedded into the n-dimensional phase space. Let $y_i = \{y_0(t_i), \ldots, y_0[(t_i - (n-1)T]\}$. If point y_i is chosen we can calculate its distance $y_i - y_j$ to the remaining $N-1$ points. This permits calculating data points that fall within the closed r-radius sphere with the midpoint y_i.

By repeating the process for all i, we obtain the quantity

$$C(r) = \frac{1}{N^2} \sum_{i,j}^{N} \Theta(r - |x_i - x_j|), \qquad (8.17)$$

where Θ is the Heaviside function. The relation between correlation dimensionality and a correlation function is based upon the power law

$$C(r) \approx r^M, \qquad (8.18)$$

where M is the correlation dimension. It is seen from equation (8.18) that the correlation dimension can be found from plotting $C(x)$ on a log – log graph. A domain in which the power law (8.18) obeys the linear dependence calculated numerically and the slope of the corresponding line was defined on the basis of using the numerical differentiation algorithm. Because the sequence of estimates defined from the family of plots parameterized by n converges we obtained a consistent estimate of the attractor dimension.

The numerical simulation has shown that slope values tend to some limiting value. This value defines the correlation dimensionality being within the limits of 6.6 to 6.9 (our estimate equals 6.75). Numerical values of such parameters as the Lyapunov exponents, Kolmogorov entropy, and the like, were defined as well.

As a whole, the simulation has shown that the processes of signal scattering on the sea surface for radar with high resolution should be considered as a result of deterministic dynamics incorporating a limited number of variables. The attractor dimensionality equal to 6.75 explains a random character of properties of reflection processes. Its value says that for modeling of time series no more than seven independent variables are required (three oscillators of two variables each and a one-dimensional equation of first order). A signal reflected from a small-size target is described by an additive sum of the chaotic process and a signal from the target itself. To describe the latter, we developed an identification method based on the self-organization approach.

2.3 Mathematical Model of Sea Clutter

The following system of differential equations was used as a finite-dimensional model describing the returned signal (Dmitriev and Kislov, 1989),

$$T_1 \dot{X}_1 + X_1 = F(Z_k), \quad T_2 \dot{X}_2 + X_2 = X_1, \ldots, T_N \dot{X}_N + X_N = X_{N_1},$$
$$\ddot{Z}_1 + \alpha_1 \dot{Z}_1 + \beta_1^2 Z_1 = \beta_1^2 X_N, \quad \ddot{Z}_2 + \alpha_2 \dot{Z}_2 + \beta_2^2 Z_2 = \beta_2^2 Z_1, \ldots,$$
$$\ddot{Z}_k + \alpha_k \dot{Z}_k + \beta_k^2 Z_k = \beta_k^2 Z_{k-1}, \tag{8.19}$$

where T_i are time constants, α_i and β_i are dissipation coefficients and resonance frequencies, and $F(Z)$ is a characteristic of nonlinear element. Realization of the required chaotic oscillation can be expected in the system with $k = 3$. We performed numerical simulation of the system (8.19) with parameters ($\beta_1 = 1.0$; $\beta_2 = 1.7$; $T = 2.0$; $\alpha_1 = 0.1$; $\alpha_2 = 0.17$), $F(Z) = MX \exp(-Z^2)$. We have shown that with an increase of M there arise self-oscillations; then a bifrequency mode is born and quasiperiodic oscillations are established. The further growth of the gain coefficient results in synchronization of bifrequency oscillation which results in the resonance torus with $\rho = 3/5$. For $M > 18$ the failure of the two-dimensional torus and transition to stochasticity occur. In case of large M when β_3 changes, the zones of synchronism and chaos alternate as well as zones of increase and decrease of the amplitude of oscillations in connection with resonance frequency relations. Both regular and statistical resonances arise in the system. For the values of M in the neighborhood of 15 there are zones of synchronism, quasiperiodicity, and chaos. Stochastic modes are realized on the basis of both bifrequency and three-frequency oscillations.

Simulation of the required chaotic mode is provided by the optimal choice of system parameters by a criterion of proximity of principal characteristics of the attractor to values typical for reflected radar signals.

The adequate description of sea clutter signals for radar stations with high resolution is provided by a finite-dimensional stochastic equation with control. However, the universally adopted statistical approach to simulating the processes of scattering of high-resolution radar station signals should be used with caution. Before designing the estimation algorithm the numerical simulation for checking the character of dynamical behavior should be performed. If, nevertheless, there is a dynamic chaos it can be adequately described by a difference equation of small dimensionality.

3. Global Reconstruction and Biomedical Applications

3.1 Nonparametric Models for Epilepsy Data

In this section the Nadaraya–Watson (NW) and the local linear polynomial regression (LLPR) methods are briefly introduced. They can be used for estimation in the stochastic models for EEG spike and wave (SW) activity.

Nadaraya–Watson kernel estimator. The NW kernel estimator is a well-known method for nonparametric function fitting (Robinson, 1983; Miwakeichi et al., 2001). According to this method, the estimate of f in a nonlinear stochastic model

$$x_t = f(x_{t-1}, x_{t-2}, \ldots, x_{t-m}) + e_t, \qquad (8.20)$$

at a point $(z_{t-1}, z_{t-2}, \ldots, z_{t-m})$ of the state space is obtained as a weighted average of all the data $(x_1, x_2, \ldots, x_N)$. Here $f : \mathbb{R}^m \to \mathbb{R}$ is a smooth map, e_t represents dynamical noise, and m is a positive integer (embedding dimension). Specifically,

$$\widehat{f}(z_{t-1}, z_{t-2}, \ldots, z_{t-m}) = \sum_{i=m-1}^{N} \frac{\prod_{j=1}^{m} K(|z_{i-j} - x_{i-j}|, h)}{\sum_{i=m+1}^{N} \prod_{j=1}^{m} K(|z_{i-j} - x_{i-j}|, h)}. \qquad (8.21)$$

Here, $K(|z - x|, h)$ is a kernel function, and the tuning parameter h is some positive real number (width of the kernel).

Local linear polynomial regression. LLPR is a particular case of local polynomial regression (LPR; Cleveland, 1979). In general, the LPR method approximates the function f in (8.20), in a neighborhood of each point x_0 of the state space, as a local (multivariate) polynomial; that is,

$$f(x) = p^n(x - x_0, \theta(x_0)), \tag{8.22}$$

where $\theta = \theta(x)$ is the vector of coefficients of the n-degree polynomial p.

To estimate the coefficients, the following least-squares problem is solved,

$$\widehat{\theta} = \arg \min_{\theta \in \mathbb{R}^{p+1}} \left\{ \sum_{i=1}^{n} [y_i - f(x_0) - j(x_0)(x - x_0)]^2 K\left(\frac{x_i - x_0}{h}\right) \right\}, \tag{8.23}$$

where the y_is are dependent observed real values, the x_is are independent observed values lying in the state space, and K is some kernel function. The value of f at x_0 is estimated by the first component of the coefficient θ; that is, $\widehat{f}(x_0) = \widehat{\theta}(x_0) \equiv p^k(0, \widehat{\theta})$.

In the case $n = 1$ it is called local polynomial regression (LLPR), and (8.23) becomes

$$\{\widehat{f}(x_0), \widehat{j}(x_0)\} = \arg \min_{f(x_0),\, j(x_0)} \left\{ \sum_{i=1}^{n} [y_i - f(x_0) - \right.$$
$$\left. - j(x_0)(x_i - x_0)]^2 K\left(\frac{x_i - x_0}{h}\right) \right\}. \tag{8.24}$$

Here $\widehat{j}(x_0)$ provides an estimate of the vector of first derivatives (gradient) at the point x_0.

3.2 Reconstruction of the Parameter Spaces of the Human Brain

Let

$$X(t, a) \tag{8.25}$$

be a single time series of one characteristic parameter X at a fixed value of the set of rule parameters a. If the observation time τ_{obs} is long enough, we can then construct the corresponding stationary distribution function

$$f(X, a), \quad \int f(X, a) dx = 1. \tag{8.26}$$

We assume that the process (8.25) is a Markovian one and therefore, for continuous processes for the statistical description of the time evolution, the following Fokker–Planck equation is used

$$\partial_t f(X, at) = \partial X[D(X, a), \partial_X f] + \partial_X[A(X, a)f]. \qquad (8.27)$$

We must remark that the kinetic form of the Fokker–Planck equation is used here (see Klimontovich, 1990), where the comparative analysis of the Ito, Stratonovich, and kinetic forms of stochastic equations was carried out.

In equation (8.27) there are two unknown functions, $D(X, a)$ and $A(X, a)$, which define, correspondingly, the diffusion and friction in any point X at a given value of the rule parameter a.

In order to obtain information about the structure of these functions, we can use the stationary solution of equation (8.27). It is possible to represent this solution as follows,

$$f(X, a) = f(X_0, a) \exp\left(-\int_{X_0}^{X} \frac{A(X', a)}{D(X', a)} dX'\right). \qquad (8.28)$$

We also use the consequence of this equation:

$$A(X, a) = -D(X, a)\partial_x \ln f(X, a). \qquad (8.29)$$

The left side of equation (8.28) can be found on the basis of experimental data (see equations (8.25) and (8.26)). Thus, equation (8.28) represents one equation for two unknown functions $A(X, a)$, $D(X, a)$. In order to obtain additional information about the structure of these functions, it is useful to eliminate (exclude), in reality of course only partially, the natural and external statistical noise by a special method of filtration. As a result we obtain the new time series

$$X_{\text{filt}}(t, a) \qquad (8.30)$$

and the new corresponding stationary one-variable distribution function

$$f_{\text{filt}}(X, a), \quad \int f_{\text{filt}}(X, a)dX = 1. \qquad (8.31)$$

It is natural to assume that this distribution function is governed by the Fokker–Planck equation with some constant diffusion coefficient

$$\partial_t f_{\text{filt}}(X, a, t) = D_{\text{filt}}\partial_{XX} f_{\text{filt}} + \partial_X(A(X, a)f_{\text{filt}}). \qquad (8.32)$$

The constant diffusion coefficient D_{filt} characterizes the degree of non-ideality of the filter and the roundoff error in the last significant digit of the computer. The value of D_{filt} can be determined by the special control numerical experiment (see Conclusions).

The stationary solution of equation (8.32) has the form

$$f_{\text{filt}}(X, a) = f_{\text{filt}}(X_0, a) \exp\left(-\frac{1}{D_{\text{filt}}} \int_{X_0}^{X} A(X', a)dX'\right). \qquad (8.33)$$

The left side of this equation is known from the experimental time series (see equations (8.25), (8.29), and (8.30)); therefore equation (8.33) is the equation for the one unknown function $A(X, a)$. Thus, this function is defined by the expression

$$A(X, a) = -D_{\text{filt}}\partial_x \ln f_{\text{filt}}(X, a). \qquad (8.34)$$

This function characterizes the structure of the dynamic equation.

It is convenient in some cases (see below) to use the "effective energy", the "effective Hamilton function". By the definition (Klimontovich, 1995, 1999) for the distribution functions (8.26) and (8.30), the effective Hamiltonian functions are determined by the expression

$$H_{\text{eff}}(X, a) = -\ln f(X, a),$$
$$H_{\text{eff}}^{\text{filt}}(X, a) = -\ln f_{\text{filt}}(X, a). \qquad (8.35)$$

It is possible now to rewrite equation (8.34) in the form

$$A(X, a) = D_{\text{filt}}\partial_x H_{\text{eff}}^{\text{filt}}(X, a). \qquad (8.36)$$

Thus, the function $A(X, a)$ is defined by the effective Hamilton function for the time series after filtration.

We can now use the effective Hamilton function $H_{\text{eff}}(X, a)$ to represent the relation (8.29) in the form

$$D(X, a) = A(X, a)/\partial_X H_{\text{eff}}. \qquad (8.37)$$

This equation is an example of the fluctuation-dissipation relation (FDR). It is established here on the basis of the experimental data.

Thus, two equations, (8.36) and (8.37), allow us to find two nonlinear functions $D(X, a)$ and $A(X, a)$ that define the structure of the Fokker–Planck equation (8.27). With the help of this equation the time evolution of the one-variable distribution function can be described at any value of the rule (control) parameter a.

It is evident, that, in general, the structure of functions $D(X)$, $A(X)$ is very complicated because the one-variable time series contains information about the complex motion of the multidimensional system. Only in the simplest cases is it possible to find the approximate analytical expressions for the functions $D(X, a)$, $A(X, a)$.

The reconstruction of the master equation on the basis of experimental data. In the paper by Klimontovich (1990) two different forms of the master equations for the Markov processes which are called 'birth-death processes' or 'generation-recombination processes' were considered. We use the more convenient name of 'one-step processes'.

If, for example, the range of the variable X is a discrete set of states with labels n, then the statistical process is characterized by two functions ("recombination" and "generation")

$$r_n(a), \quad g_n(a), \tag{8.38}$$

or by the corresponding diffusion and dissipative functions

$$D_n(a) = \frac{1}{2}(r_n + g_n), \quad A_n(a) = r_n - g_n. \tag{8.39}$$

Here, as above, a is a set of rule parameters.

Two time series, (8.25) and (8.30), are replaced now by the following time series

$$n(t, a), \quad n_{\text{filt}}(t, a), \tag{8.40}$$

for the discrete set of states in time at any given value of the rule parameters. The corresponding distribution functions are represented in the following forms,

$$f_n(a) = \exp(-H_{\text{eff}}(n, a)),$$
$$f_n^{\text{filt}}(a) = \exp(-H_{\text{eff}}^{\text{filt}}(n, a)). \tag{8.41}$$

Here we have introduced, as above, the effective Hamiltonian functions.

From the master equation for the stationary state, there follows the relation (Klimontovich, 1990)

$$D_n(a) = \frac{1}{2}A_n(a)\frac{f_{n-1}(a) + f_n(a)}{f_{n-1}(a) - f_n(a)}, \tag{8.42}$$

which in the continuous limit ($1/n \leq 1$), is transformed to a simpler relation

$$D_n(a) = -A_n(a)/\partial_n \ln f_n(a), \tag{8.43}$$

corresponding to the expression (8.29). Equations (8.30) and (8.31) also give the examples of the fluctuation–dissipation relations for the discrete and continuous variables, respectively.

In order to obtain the second equation for the definition of the two unknown functions $D_n(a)$ and $A_n(a)$, we again use the filtration procedure. As a result, we obtain two characteristics

$$n_{\text{filt}}(t, a), \quad f_n^{\text{filt}}(a). \tag{8.44}$$

If D_{filt} is, as above, the characteristic diffusion coefficient, then we have the following equation for the definition of the $A_n(a)$ function

$$A_n(a) = 2D_{\text{filt}} \frac{f_{n-1}(a) - f_n(a)}{f_{n-1}(a) + f_n(a)}. \tag{8.45}$$

In the continuous limit, it follows from this equation that

$$A_n(a) = D_{\text{filt}} \partial_n \ln f_n^{\text{filt}}(a), \tag{8.46}$$

which corresponds to expression (8.34).

Thus, it is possible to reconstruct from experimental data not only the Fokker–Planck equation for the one variable distribution function, but also the corresponding master equation for one-step processes.

The reconstruction of the Fokker–Planck equation for the multivariable distribution function from experimental data. The method for reconstructing the Fokker–Planck and master equations has been based until now on the information from the one-variable time series (8.25). We saw that the principal possibility exists to reconstruct the kinetic equations for the one-variable distribution functions $f(X, t)$ and $f_n(t)$. The question now arises: how do we reconstruct the kinetic equations for multivariable distribution functions? In order to answer this question it is necessary, at first, to use the notion of system dimension.

The time series (8.25) describes, in general, complicated motion with a very high number of degrees of freedom. The dimension of this motion is much higher than the effective dimension of the 'dynamic process' which follows from the time series (8.25) after filtration. Such a process is described by the time series (8.30). We assume that the process after adequate filtration is an almost deterministic complicated motion, almost 'deterministic chaos.'

We now have a new 'simpler' problem: determining the minimal dimension, the minimal number of macroscopic variables required for the description of the 'deterministic chaos.' At present there exists a method for the solution of this problem. It is based on Takens' theorem (Takens, 1981) and on the methods that have been used for a long time for pattern recognition and medical diagnostics, and in recent years in the synergetic approach for the theory of neural nets and in the theory of cognition (Haken, 1988, 1991).

Let us suppose, for example, that the minimal number of the independent macroscopic variables for the time series $X_{\text{filt}}(t)$ equals three and that the physical interpretation of these variables is also known. We can use, for this function, the following notations,

$$X_{\text{filt}}(t, a), \quad Y_{\text{filt}}(t, a), \quad Z_{\text{filt}}(t, a). \tag{8.47}$$

If the time observation is long enough, we can then construct the corresponding distribution function of the variables X, Y, Z:

$$f_{\text{filt}}(X, Y, Z, a) \equiv f_{\text{filt}}(X, a), \quad X = (X_1, X_2, X_3);$$

$$\int f_{\text{filt}}(X, a)dx = 1. \tag{8.48}$$

Let us again characterize the fluctuation of the filter by the constant diffusion coefficient D_{filt}. Then the Fokker–Planck equation for the distribution function $f_{\text{filt}}(X, a)$ can be written in the form

$$\partial_t f = D_{\text{filt}} \partial_{X_i X_i} f_{\text{filt}} + \partial_{X_i}(A_i(X, t) f_{\text{filt}}(X, t)). \tag{8.49}$$

The unknown functions $A_i(X, a)$ may be expressed via the derivative of the stationary distribution (8.48):

$$A_i(X, a) = -D_{\text{filt}} \partial_{X_i} \ln f_{\text{filt}}(X, a). \tag{8.50}$$

For the one-variable distribution function, this equation coincides with equation (8.34). The right side of this equation is defined from the experimental data, therefore the expression (8.50) serves as the definition functions $A_i(X, a)$ from the experimental time series (8.47).

In order to obtain the corresponding diffusion function, we use, from experimental data, the time series

$$X_i(t, a), \quad i = 1, 2, 3. \tag{8.51}$$

Here we again use the information about the physical nature of the macroscopic variables $X_i(t, a)$ and assume that their measurement is possible. We have, as a result, experimental information about the structure of the static distribution

$$f(X, a), \quad \int f(X, a)dX = 1. \tag{8.52}$$

The time evolution is described by the corresponding Fokker–Planck equation

$$\partial_t f(X, t) = \partial_{X_i} D_i(X, a) \partial_{X_i} f + \partial_{X_i}(A(X, a)f). \tag{8.53}$$

If the function (8.52) is the stationary solution of this equation, we then have a second equation for the definition of the diffusion functions $D_j(X, a)$ from the experimental data

$$D_i(X, a) = -A_i(X, a)/\partial_{X_i} \ln f(X, a), \quad i = 1, 2, 3. \tag{8.54}$$

We can again, as in the case of the one-variable distribution function, introduce the effective Hamiltonian function

$$f(X, a) = \exp(-H_{\text{eff}}(X, a)), \tag{8.55}$$

and represent equation (8.54) in the form

$$D_i(X, a) = A_i(X, a)/\partial_{X_i} H_{\text{eff}}(X, a), \quad i = 1, 2, 3. \tag{8.56}$$

In concluding this section, it is useful to make the following remark. In accordance with the definitions (8.36) and (8.50), the functions $A_i(X, a)$ have for constant diffusion coefficient D_{filt} the gradient form. The relation (8.56) is more general. Now only the ratio $A_i(X, a)/D_i(X, a)$ obeys the potential condition.

We have shown that by using two time series, $X(t, a)$ and $X_{\text{filt}}(t, a)$, it is possible to reconstruct the Fokker–Planck and master equations. But the problem of how to extract practically the dynamic noise from the observed time series arises.

In order to solve this problem, it is necessary to use some special numerical procedure for the filtration of natural and external noise from the observed time series $X(t, a)$. Obviously, it is impossible to eliminate completely the natural and external noise by some numerical procedure. Only the 'almost dynamic time series' after some special numerical filtration can be obtained. We characterized the statistical properties of such an 'almost dynamic time series' $X_{\text{filt}}(t, a)$ by some constant diffusion coefficient D_{filt}.

It is necessary, however, to keep in mind that a wide variety of different filtration procedures exists, and that different methods of filtration can lead to different results.

Indeed, in the work of Chennaoui et al.(1990), for example, a low-pass filter was used for the dynamic time series, particularly, from a logistic equation. In this case, the fractal dimension of the filtered time series of course increases. In order to decide if the time series is a pure dynamic or filtered motion, it is necessary to take into account the following (Chennaoui et al., 1990).

For dynamic motion, the attractor can be reconstructed from each of two time series

$$X(t, a) \quad \text{and} \quad \frac{dX(t, a)}{dt}, \tag{8.57}$$

and in both cases the fractal dimension is the same. But for a low-pass filtered time series

$$X_{\text{l-p filt}}(t, a) \quad \text{and} \quad \left(\frac{dX(t, a)}{dt}\right)_{\text{l-p filt}}, \tag{8.58}$$

the fractal dimensions are different. The fractal dimension for the derivative is smaller. So it is possible to decide, by comparing the fractal dimensions of the time series (8.57) and (8.58), whether the time series $X(t, a)$ is filtered. It is shown in Chennaoui et al. (1990) that for the low-pass filtered Rössler time series, the fractal dimension for the derivative decreases remarkably.

This procedure has been applied in Chennaoui et al. (1990) to the experimental data for the rotational Taylor–Couette flow. In the experimental series, the 'almost dynamic motion' was filtered by an RC lowpass filter, and the filtered time series was later reconstructed. In Chennaoui et al.(1990) the following results were obtained. The information dimension $D = 2.17$ for the experimental 'almost dynamic time series', $D = 2.50$ for the RC lowpass filtered time series, and $D = 2.18$ for the reconstructed time series.

We see that, indeed, the influence of lowpass filtration on the value of the dimension is remarkable. Thus the method of reconstruction of the "almost dynamic time series" may be effective.

We can also distinguish the difference between the time series (8.57) and (8.58) by the method presented in the previous sections. Namely, we can consider the change of the time series by the lowpass filter as 'evolution in the space of rule parameters' and use the S-theorem to compare the relative degree of order (or chaos) of the processes (8.57) and (8.58). One can expect that the degree of chaos of the time series after the lowpass filtration will be increased.

We can also reconstruct the Fokker–Planck equations and the master equations for the processes as considered in Chennaoui et al. (1990).

The choice of the filtering method, of course, is a problem. We now want to discuss the alternative methods of filtration based on the Takens theory (Takens, 1981).

This method of filtration uses the transformation to the orthonormal basis (Haken, 1988, 1991) in the Takens phase space and is based on the difference in the physical nature of the time series with statistical noise and that of the 'almost dynamic time series'. As before, this difference is expressed, particularly, in the difference of dimensions.

The problem may be solved in the following way. We replace the time series (8.25) by a corresponding series in discrete time

$$X(t, a) \rightarrow X_k(a), \quad t = k\tau, \quad k = 1, \ldots, N, \quad N = \tau_{\text{obs}}/\tau, \tag{8.59}$$

and choose, by some numerical experiment, the characteristic time interval τ. After this we introduce the n-dimensional vector with sufficiently large n,

$$Y_k(a) = (X_k, X_{k+1}, \ldots, X_{k+n-1}). \tag{8.60}$$

The number n depends of course upon the character of the statistical noise for the time series (8.25).

The process y_k is described approximately by the process $Y_k^{(l)}$ in the space with dimension $l < n$ with some orthonormal basis. The basis vectors should be chosen so that the approximation error ε_l is minimal. The algorithm for the choice of the basis vectors is based on the Karhunen–Loeve theorem. It can also be determined by the Neymark algorithm.

The value of ε_l decreases with the growth of l and tends to some constant value defined by the level of the statistical noise. The characteristic dimension $m < n$ (the 'embedding dimension' in the Takens theory) can be found from the beginning of the plateau on the plot of error ε_l versus l.

For example, for the Rössler equation, the dimension $m = 3$. Now we can consider the problem of filtration of the time series. We assume that the series in discrete time can be represented in the form

$$Y_k = Y_k^{\text{filt}} + y_k, \quad Y_k = (X_k, X_{k+1}, \ldots, X_{k+n-1}). \tag{8.61}$$

Here Y_k^{filt} is the filtered series in discrete time for the 'almost pure dynamic time series' with the dimension m, and y_k is the statistical noise.

We can again expand the function y_k in an orthonormal system of vectors, but the number of vectors now equals $m > n$. If the statistical noise distributes on all variables n, then the dispersion

$$\langle (Y_k^{\langle m \rangle} - Y_k^{\text{filt}})^2 \rangle \tag{8.62}$$

will be approximately m/n times less than the dimension of the statistical noise $\langle (y_k)^2 \rangle$ in equation (8.61).

Thus, the time series $Y_k^{\langle m \rangle}$ gives the first approach to the pure filtered time series Y_k^{filt}. Such filtering may be repeated. After q iterations, the dispersion equation (8.62), in an ideal situation, will be approximately proportional to $(m/n)^q$.

It is obvious that the full elimination of statistical noise is impossible and we can only obtain an "almost dynamic time series." The incompleteness of the filtration is characterized in the previous sections by the constant diffusion coefficient of the filtration D_{filt}.

We see that the method of reconstruction of the kinetic equation considered previously can indeed be realized. Moreover, we have criteria of the relative degree of order of different nonequilibrium states of open systems for the control of the filtration processes in different time series.

The methods considered above can be used not only for time series, but also for space series and corresponding time–space spectra.

We considered here the possibility of reconstructing the Fokker–Planck equation and equation for one-step processes. Obviously, it is possible to enlarge the class of processes and to reconstruct the equations for the description of different kinds of turbulent motion.

4. Global Optimization Approaches to Reconstruction of Dynamical Systems Related to Epileptic Seizures

4.1 Nonlinear Dynamics and Epilepsy

The existence of a complex chaotic, unstable, noisy, and nonlinear dynamics in the brain requires another approach to identification and simulation of brain activity. These approaches are opposite to the universally accepted stochastic simulation of random processes with given distribution. Next we discuss the possibility of using a global optimization approach to reconstruction of brain dynamics under the assumption that the diagnosis information comes in the form of a nonlinear time series. We consider a method for global reconstruction of nonlinear models for systems where all the necessary variables have not been observed. This technique can be applied to systems with one or several such hidden variables, and can be used to reconstruct maps or differential equations of brain dynamics. The quadratic programming approach to reconstruction of dynamical process is considered. We propose the possibility of the global reconstruction of the Fokker–Planck equation for the multivariable distribution function which reflects the complexity of the motion of the considered brain. Finally, an application is given of the reconstructing technique to solve different problems, such as reconstruction of the epileptic brain dynamics from time series of the EEG with noise.

Prior to the 1980s, researchers had always assumed that to study the dynamics of the brain with many degrees of freedom, time-series measurements of all the variables, or derivatives thereof, were necessary to generate state-space representations of the dynamics. For the brain, derivatives are particularly difficult to employ due to the complexity and the noise problems. In 1989, Chang, Hübler, and Packard and Ruelle (Loskutov and Mikhailov, 1990) noted that a state-space representation of the dynamics could be reconstructed from a single time series through the use of delay coordinates. This delay-coordinate reconstruction would then be topologically equivalent to the dynamics of the true system. Whitney had shown much earlier (Whitney, 1936) that any compact manifold with dimension m can be embedded in $\mathbb{R}^{2m+1}$.

Takens extended this (Takens, 1981) by proving that an embedding can be obtained for any system from only a single time series by using $2m+1$ delay coordinates. Although this combination of ideas thus far has been extremely useful in studying nonlinear systems, several difficulties arise in their application that we hope to address with an alternative method for reconstructing these hidden variables.

In order to create a modeling technique in which existing information can be incorporated, the resulting model can be interpreted neurodynamically and robust to noise, we base our technique upon the global optimization approach (Horst and Pardalos, 1995; Du, Pardalos, and Wang, 2000; Pardalos and Romeijn, 2002; Pardalos and Resende, 2002; Pardalos et al., 2004), kinetic approach (Klimontovich, 1990), dynamical approach (Sackellares et al., 2000) and flow method (Cremers and Hübler, 1987). The flow method is a procedure for reconstructing a set of coupled maps (CMs) or ordinary differential equations (ODEs) from a trajectory of the system in state space. We show that this may be easily adapted to the presence of hidden variables.

4.2   Reconstructing Equations of the Epileptic Brain from Experimental Data

The brain dynamics throughout the state space is represented either with a single set of coupled maps

$$y_i(n + 1) = f_i[y(n), \gamma, a], \quad i = 1, \dots, N, \quad y \in \mathbb{R}^N$$

or a set of ordinary differential equations

$$\dot{y}_i(t) = f_i[y(t), \gamma, a], \quad i = 1, \dots, N, \quad y \in \mathbb{R}^N,$$

where $a = a_1, \dots, a_{N_c}$ are unknown parameters. The number of observed variables is assumed sufficient to embed the dynamics. The functions $\{f_i\}$ may be of any form, but are usually taken to be a series expansion. This method has been successfully tested with Taylor- and Fourier-series expansions. In this manner, the modeling is done by finding the best expansion coefficients to reproduce the experimental data. Often the situation arises where the form of the functions $\{f_i\}$ is known, but the coefficients are unknown; for example, this occurs frequently with rate equations for epileptic processes. This added information greatly reduces the number of undetermined parameters, thus making the modeling computationally more efficient.

The modeling procedure begins by choosing some trial coefficients. The error in these parameters can be computed by taking each data point $x(t_n)$ as an initial condition for the model equations. The predicted value $y(t_{n+1})$ can then be calculated for CMs as

$$y_i(n+1) = f_i[x(n), a], \quad i = 1, \ldots, N,$$

or for ODEs as

$$y_i(t_{n+1}) = x_i(t_n) + \int_{t_n}^{t_{n+1}} f_i[y(t'), a]dt', \quad i = 1, \ldots, N \qquad (8.63)$$

and compared to the experimentally determined value. It is well known that more stable models can often be obtained by comparing the prediction and the experimental data several time steps into the future. For the present analysis, we predict the value only to the time of the first unused experimental data point. The error in the model is thus obtained by summing these differences

$$F = \frac{1}{N(M-1) - N_c} \sum_{i=1}^{M} \sum_{j=1}^{N} \frac{1}{\sigma_{ij}^2} [y_j(t_i) - s_j(t_i)]^2, \qquad (8.64)$$

where N_c is the number of free coefficients a_i, M is the number of data points, and σ_{ij} is the error in the jth vector component of the ith measurement. The task of finding the optimal model parameters has now been reduced to a minimization problem. Thus, the best parameters are determined by

$$\min_a F(a, y),$$
$$\alpha_i^{\min} \leq \alpha_i \leq \alpha_i^{\max}, \quad i = 1, \ldots, r, \qquad (8.65)$$

where α_i are the system characteristics of the epileptic brain (Haken, 1996) (fractal dimension, pointwise dimension, information dimension, generalized dimension, embedding dimension, Lyapunov dimension, Lyapunov exponents, metric entropy, etc.).

Let N be a minimal embedding dimension and a be a slow parameter. As a result, we should solve the *constrained optimization problem* (8.65). The constrained optimization algorithm was implemented as a function in MATLAB 7 running on a UNIX computer.

Therefore the ability to determine these coefficients rests upon the strength of the algorithm employed to search through the space of parameters. This has been formulated as a standard F_ν^2 identification problem, thus the normal statistical tests can be applied. Typically, $F_\nu \simeq 1$ implies that the modeling was successful; however, more sophisticated tests can be applied as well, for example, the F test. If the experimental uncertainties σ_{ij} are unavailable, this normalization factor can simply be removed from equation (8.64). This means that the F_{ν} tests cannot be applied, but the best possible model can still be determined by locating the global minimum of F_{ν} in the parameter space.

4.3 Quadratic Programming Problem

In this section we discuss a possible reconstruction of a dynamical model of the epileptic brain system (DS) analyzing the spatiotemporal dynamical changes in the EEG. Our research is based on temporal dynamical analysis (Iasemidis et al., 2001). We used 28 electrodes for subdural and depth EEG recording. We have typically analyzed continuous EEG signals for at least 1 hour before to 1 hour after seizure, sampled with a sampling frequency of 200 Hz and lowpass filtered at 70 Hz.

Next we introduce the definitions of the T-index (Iasemidis et al., 2001) as a measure of distance between the mean values of pairs of $STL_{\max}$ profiles over time.

Definition 8.1. *By the T-index (or T-signal) at time t between electrode sites i and j, we mean the variable*

$$T_{ij} = \sqrt{N} \times E(STL_{\max,i} - STL_{\max,j}/\sigma_{ij})(t), \qquad (8.66)$$

where $E(\cdot)$ is the sample average difference for the $(STL_{\max,i} - STL_{\max,j})$ estimated over moving window $\theta_t(\lambda)$ defined as

$$\theta_t(\lambda) = \begin{cases} 1 & \text{if } \lambda \in [t - N - 1, t], \\ 0 & \text{if } \lambda \notin [t - N - 1, t], \end{cases} \qquad (8.67)$$

where N is the length of the moving window; $\sigma_{ij}(t)$ is the simple standard deviation of the $STL_{\max}$ differences between electrode site i and j within the moving window $\theta_t(\lambda)$; STL (Short Time Lyapunov) is the operator of numerical estimation of $L_{\max}$; $L_{\max}$ is the maximum Lyapunov exponent (the Kolmogorov–Sinai entropy).

Let us consider a measurement system (MS) registering T signals. To each T-signal (TS) corresponds a real sensor signal of certain form, which is transmitted to the sensors and appears in their input possibly distorted, and corrupted by random noise.

The input $T(t)$ is a stochastic process, described in terms of probability density functions. Let the TS be denoted $T_1, T_2, \ldots, T_N$. A nonlinear stochastic model is given as

$$T_t = f(T_{t-1}, T_{t-2}, \ldots, T_{t-m}) + \epsilon_t, \qquad (8.68)$$

where $f : \mathbb{R}^m \longrightarrow \mathbb{R}$ is a smooth map, which has been designed as the "skeleton," ϵ_i represents dynamical noise; and m is a positive integer, the so-called embedding dimension.

The data for solving the reconstruction problem is represented as $\{T, y\} = \{T_i, y_i\}_1^n$, where $n = N - m$, $T_i \in \mathbb{R}^m$ are the rows of $\mathbf{T}$, and $y_i \in \mathbb{R}$. Here we present a nonlinear method for the reconstruction of the function f given by (8.68) using the approach of Horst and Pardalos (1995).

Let

$$\hat{f} = \arg \min_{y \in H} R[g] = C \sum_{i=1}^{n} |y_i - g(T_i)|_\epsilon + \frac{1}{2}\|g\|_H, \qquad (8.69)$$

where

$$|T|_\epsilon = \begin{cases} 0 & \text{if } |T| \leq \epsilon, \\ |T| - \epsilon & \text{otherwise} \end{cases} \qquad (8.70)$$

is the so-called ϵ-insensitive loss function, C is a positive number, and H is a reproducing kernel Hilbert space with norm $\| \ \|_H$ (Vapnik, 1995, 1998). It can be proved that the function $\hat{f}$ can be expressed by $\hat{f}(T) = \sum_{i=1}^{\infty} \omega_i \gamma_i(T) + b$, where the set of functions $\{\gamma_i(T)\}_{i=1}^{\infty}$ is a basis of H.

Consider the following optimization problem,

$$\left\{ \begin{aligned} \min_{x,x^s} & \frac{1}{2} \sum_{i=1}^{n} \sum_{j=1}^{n} (x_i^s - x_i)(x_i^s - x_j)K(T_i, T_j) \\ & - \sum_{i=1}^{n} x_i^s(y_i - \epsilon) + \sum_{i=1}^{n} x_i(y_i + \epsilon) \end{aligned} \right\} \qquad (8.71)$$

with the constraints

$$0 \leq x_i, \quad x_i^s \leq C, \quad i = 1, \dots, n,$$

$$\sum_{i=1}^{n} (x_i^s - x_i) = 0,$$

where $K(T_i, T_j)$ is the reproducing kernel of H. The function $f(T)$ can be expressed as

$$f(T) = \sum_{i=1}^{n} (x_i^s - x_i)K(T, T_i) + B.$$

It can be demonstrated that the solution of this problem leads to several coefficients $\beta_i = (x_i^s - x_i)$ equal to zero, and so the data points associated with them are not involved in the last expression. The remaining data points are called support vectors and they contain all the information needed for the approximation of the function f. Notice that if the number of support vectors is small, the calculation time of the method will be reduced.

5. Stochastic and Deterministic Dynamics in Electroetinograms

5.1 Experimental Data

It is well known that various characteristics of biological and physical processes can be extracted from recorded signals. These characteristics may be necessary for a better understanding and adequate prognosis of the processes. Two principal classes of characteristics referring to different data representations are now used: statistical (or stochastic) (Klimontovich, 1990; Casdagli, 1989) and chaotic (or strange attractors; Hao, 1984). Statistical parameters that can be mentioned include variance, mean, coefficients of regression models, and spectral magnitudes. Chaotic phenomena can be characterized, in particular, by the Kolmogorov entropy and by the fractal, correlation, and embedding dimensions.

Note that it is necessary to estimate the proper statistical characteristics for a better prediction of stochastic processes and of the strange attractor parameters for chaotic cases as well. This phenomenon can be explained by the difference between the physical natures of the random stochastic and deterministic chaotic processes. Such a difference can be found, in particular, in the variances of the prediction quality depending on the prediction time (Jimenez et al., 1992). Therefore, up to now, either statistical or chaotic parameters have been applied for the analysis of experimental data depending on the adopted hypothesis concerning the nature of the process considered.

On the other hand, the measured time series can, in principle, demonstrate both stochastic and chaotic behavior. Appropriate numerical procedures for distinguishing between these two types of behavior have recently been proposed (Jimenez, et al., 1992; Kaplan and Glass, 1992; Wayland et al., 1993). In the following, we apply procedures for the differentiation of stochastic and chaotic signals of electroetinograms (EEG) recorded from the retinas of living Rana temporaria. Note that we use the word "chaotic" to mean low-dimensional dynamics (Kostelich and Schreiber, 1993).

The aims of this section are: (1) to clarify the question of the possible presence of different types of dynamics in biosignals measured under identical experimental conditions; (2) to study whether the effects of low-intensity microwaves (MWE) can be detected by deterministic parameters (i.e., those of low-dimensional chaos) only.

After the measurements and the first data processing step, the experimental data consisted of 10 scalar time series for retinas with MWE and 10 scalar time series for those without.

Each time series was recorded from the native retinas as electrorctinograms (EEG), using a conventional experimental procedure. In this case every EEG corresponds to a single retina. The number of samples for each time series was fixed at 450.

In a first step the data were preprocessed so as to receive final data sets with zero mean and variance for subsequent analysis. Note that analysis of EEG shape and variance for the two sets of retinas did not reveal any difference between them.

Typical examples of time series (the first 300 points) for retinas with and without MWE are presented in Figure 8.1 and Figure 8.2, respectively. As can be seen, the behavior of the time series seems to be random, without obvious singularities that could be used for differentiating between these series.

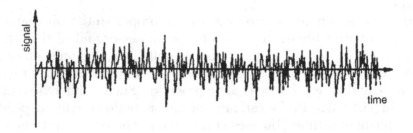

Figure 8.1. Example of experimental data obtained from retinas exposed to low-intensity microwaves (MWE, relative units).

Figure 8.2. Example of experimental data obtained from retinas without MWE (relative units).

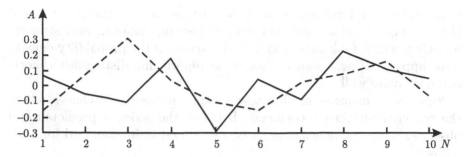

Figure 8.3. The autocorrelation values for A versus the time series number N. The solid line shows data from retinas without MWE, the dashed line data from retinas with MWE. The autocorrelation time equals 1.

Various methods were applied to establish the features of the time series that were useful for distinguishing between the sets of retinas with and without MWE. In particular, regression models were constructed based on the available data, but the coefficients of these models do not reflect the physical aspects clearly enough. On the other hand, estimations of autocorrelation values and power spectra for time series were made to compare the above-mentioned sets.

The autocorrelation values for the analyzed data are shown in Figure 8.3. A statistically significant difference between the two sets could not be found as regards these values. At the same time, for every time series a broadband spectrum without obvious peculiarities was observed. A description of the latter is omitted here for the sake of brevity.

5.2 Methods for the Analysis of Time Series

As mentioned above, the data considered have broadband spectra. At the same time, some chaotic systems with similar spectra can show very similar behavior (Abarbanel, Brown, and Kadtkeet, 1990; Kostelich and Schreiber, 1993). Therefore, for a correct analysis it is necessary to determine the closeness of every time series to the stochastic or chaotic class of signals.

A few approaches have already been described for the clarification of the nature of time series. The "surrogate data" method (Kennel and Isabelle, 1992; Mitschke and Dammig, 1993) is widely employed for this purpose. However, the surrogate data technique fails in certain situations. Therefore, what we use here is a nonlinear forecasting approach together with the surrogate data method.

Because only 450 time points were available for each time series, we tested these algorithms (briefly described below) at 450 time points using both chaotic and random time series. In particular, the logistic and Henon maps, random numbers with Gaussian, uniform, and strongly non-Gaussian K (Jakeman, 1980) distributions of the probability density were applied. It was shown that these algorithms distinguish chaotic behavior quite well.

First, as a measure of closeness to the stochastic or chaotic class, the centered correlation coefficient between the series of predicted and observed values (Jimenez et al., 1992) was used in its standard form, as follows,

$$C = \frac{\langle (X_{\text{pred}} - \overline{X}_{\text{pred}})(X_{\text{obs}} - \overline{X}_{\text{obs}}) \rangle}{[\langle (X_{\text{pred}} - \overline{X}_{\text{pred}})^2 \rangle \langle (X_{\text{obs}} - \overline{X}_{\text{obs}})^2 \rangle]^{1/2}}. \tag{8.72}$$

The designations in (8.72) are analogous to those presented by Jimenez et al. Thus, predicted and observed values obtained using the well-known Takens' procedure for the time series analyzed are denoted X_{pred} and X_{obs}, respectively.

Various predictors for distinguishing chaos from noise have been proposed previously (Farmer and Sidorowich, 1987; Casdagli, 1989; Abarbanel et al., 1990; Kostelich and Schreiber, 1993). Here we use the local optimal linear reconstruction predictor described by Jimenez et al. (1992), although, in our opinion, other constructions can also be used successfully. The brief description of the predictor used is given below.

Let $Y_1, Y_2, \ldots, Y_N$ be a measured scalar time series. To obtain the ensemble of n-dimensional vectors (where n is the embedding dimension) the Taken's procedure was performed. The vectors from this ensemble are designated $X_1, X_2, \ldots, X_l$.

For each member X_p of the ensemble, the set of nearest vectors (neighbors) denoted as $X_1, X_2, \ldots, X_k$ is found. Then, a new set of vectors $S_1, S_2, \ldots, S_l$ is defined, each vector being the difference between a neighbor and the average of all neighbors

$$S_p = X_p - \frac{1}{k} \sum_{i=1}^{k} X_i. \tag{8.73}$$

Here k is the number of neighbors. Let S_i be the vector corresponding to the neighbor X_i. Then, the set of coefficients D_j is computed to obtain S_p by means of the orthonormal Gram–Schmidt basis from the set $S_1, S_2, \ldots, S_l$. Thus,

$$S_p = \sum_{i=1}^{k} D_i S_i. \tag{8.74}$$

The essential feature of the set of coefficients D_j is that they remain constant after a short delay time d. Let $\widetilde{X}_p$ be a vector which is received from X_p after the delay time. Then, analogously to (8.73), $\widetilde{S}_p$ is obtained as follows

$$\widetilde{S}_p = \widetilde{X}_p - \frac{1}{k}\sum_{i=1}^{k}\widetilde{X}_i \tag{8.75}$$

and $\widetilde{S}_p$ can be expressed similarly to (8.74)

$$\widetilde{S}_p = \sum_{i=1}^{k} D_i \widetilde{S}_i. \tag{8.76}$$

For (8.76), it is necessary to satisfy certain conditions (Jimenez et al., 1992). In particular, the approximation (8.76) is true for low-dimensional dynamics.

Thus, it is not difficult to obtain the predicted vectors from (8.76) for the vectors of the initial ensemble $X_1, X_2, \ldots, X_k$ and to calculate the centered correlation (8.72). Choosing various time delay parameters and the embedding dimension, we have obtained the best values for the centered correlation in all time series measured. These values have been used to differentiate between stochastic and chaotic time series.

At the same time, delay time and embedding dimension have been obtained for every time series as corresponding to the best value for the centered correlation. A similar approach has been employed, for example, by Mees et al.(1992), although other methods for estimating the time delay and embedding dimension also exist (Abarbanel, Brown, and Kadtke, 1990).

The following steps were undertaken for recognition of the nature of the time series in the situation considered here. First of all, two thresholds were selected for the magnitudes of the centered correlations. If the magnitude of the centered correlation is less than the threshold denoted by L_1, then we consider the corresponding time series to be stochastic. However, if the centered correlation value is higher than the second threshold designated as L_2, then the respective time series is considered chaotic.

Here, the threshold values $L_1 = 0.3$ and $L_2 = 0.6$ were chosen, taking into account the results for the different time series described previously (Jimenez et al., 1992). It is significant that not only centered correlation values were considered in the differentiation process, but also the behavior of these values depending on the prediction time. If the centered correlation (i.e. the quality of the prediction) decreases noticeably as the prediction interval increases, it indicates that the time series is chaotic. In contrast, if the quality of the prediction does not decrease the data are assumed to be of a stochastic nature.

For comparison with the procedure described above, the surrogate data technique (Kennel and Isabelle, 1992) was also applied and a so-called z statistic was computed for every time series. This statistic serves to discriminate the set of prediction errors for the real data from those for faked (purely random) data (Kennel and Isabelle, 1992). Denote as A and B the set of prediction errors for the real and faked data, respectively. Then, we calculate the Mann–Whitney rank-sum statistic

$$ U = \sum_{i=1}^{N_2} \sum_{j=1}^{N_3} \theta(A_i - B_j), \tag{8.77} $$

where the number of elements in sets A and B is designated N_2 and N_3, respectively, and θ is the Heaviside step function. The z statistic is derived from U as follows,

$$ z = \frac{U - \frac{N_2 N_3}{2}}{\left(\frac{N_2 N_3 (N_2 + N_3 + 1)}{12} \right)^{1/2}}. \tag{8.78} $$

If the z statistic is less than -2.33, the conventional null hypothesis (that the two sets have the same distribution) is rejected and the conclusion can be made at a 0.01 confidence level that the given time series has a chaotic character.

5.3   Numerical Results

The centered correlation coefficients versus the number of time series considered for retinas with MWE are presented in Figure 8.4, and the analogous plot of data from retinas without MWE is shown in Figure 8.5. Here, for simplicity, the numbers of the time series with a large centered correlation are situated at the end of the abscissa. Thus, according to the chosen threshold L, three and two time series have a chaotic nature for retinas with and without MWE, respectively.

The number of time series having a stochastic nature is five for retinas with MWE and six for those without MWE, taking into account the threshold L. At the same time, some time series (numbers 6 and 7 in Figure 8.4, 7 and 8 in Figure 8.5) fall into an intermediate range of values for the centered correlation. However, in all these time series the quality of the prediction decreases distinctly as the prediction interval increases. Therefore these time series were recognized as having a chaotic nature.

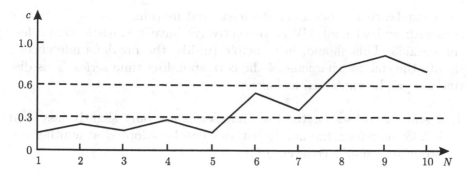

Figure 8.4. The centered correlation C versus the time series number N for retinas with MWE.

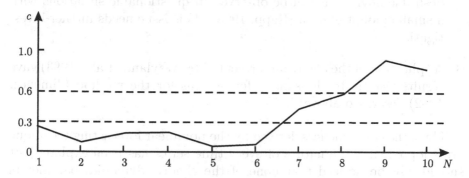

Figure 8.5. The centered correlation C versus the time series number N for retinas without MWE.

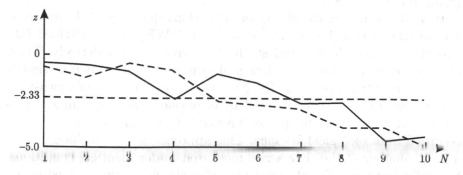

Figure 8.6. z statistic versus the time series number N. The solid line shows data for retinas without MWE, the dashed line data for retinas with MWE.

The results of the calculation of the values of the z statistic are presented in Figure 8.6. It can be clearly seen from Figure 8.6 that the overwhelming majority of z values confirm the inferences derived by the nonlinear forecasting method described above. Only two time series recognized earlier as being of a stochastic nature (numbers 5 and 4 for the sets with and without MWE, respectively) have z statistic values less than -2.33. This should, in principle (unlike the previous inferences), signify the chaotic character of the corresponding time series. This distinction can be explained as follows.

1. z statistic values for the two time series are only slightly less than -2.33; therefore, the null hypothesis can be adopted at another appropriate confidence level, 0.1.

2. The results obtained using the surrogate data technique may depend greatly on the concrete parameters of the algorithm. For example, undesirable deviations can be observed in questionable situations with a small amount of data (Rapp, 1993). This issue needs further investigation.

3. Application of the algorithm presented by Wayland et al., (1993) gave results leading to the same inferences as for the method (Jimenez, 1992) described above.

Thus, the consequences derived by the nonlinear forecasting technique are adopted (i.e., the nature of every time series has been explored). It should also be noticed that none of the chaotic dynamics parameters described can discriminate effectively between the EEG signals of the groups with and without MWE. This inference can be deduced by analyzing Figures 8.4–8.6.

It can be seen from the above considerations that each of the analyzed sets of time series (i.e., with and without MWE) can be divided into two classes with chaotic and stochastic dynamics, respectively. Thus, objects under the same experimental conditions can generate signals with different types of dynamics. Note that previously signals measured experimentally were characterized by only one kind of dynamics. This means that a more careful approach is needed when analyzing time series recorded from biological systems. This approach should include, firstly, a large number of the time series measured under identical conditions. Secondly, the dynamics of each time series should be determined using, for example, the methods described above. Thirdly, each time series has to be characterized by parameters appropriate for a certain type of dynamics.

It is also necessary to note that no effect of low-intensity microwaves was revealed using the chaotic dynamics parameters considered here. However, a more detailed analysis is needed because the stochastic time series should be characterized by statistical parameters.

6. Notes and Sources

In this chapter we showed that EEG activity can be described through nonlinear and bilinear dynamical systems reconstructed by the global optimization method and through a stochastic model estimated by the support vector machine. We developed a method for reconstructing nonlinear equations of motion for the brain where all the necessary variables have not been observed. This technique can be applied to biological systems with one or several such hidden variables, and can be used to reconstruct maps or differential equations. The effects of experimental noise were discussed through specific examples.

In this chapter the performance of global optimization, and support vector machine methods were considered. A global optimization method was confirmed as a useful method for determining the underlying dynamics of the epileptic brain. We have interpreted epilepsy as the output of a random nonlinear dynamical system with noise. It was shown that the standard deviation of noise could be estimated by means of correlation dimension.

We considered the simplest possibility of the reconstruction of the Fokker–Planck equation for a one-variable distribution function of the T-index which reflects, of course, the complexity of the motion of the brain considered. In order to characterize the nonideal filter we introduced the effective diffusion coefficient of the filter. The reconstruction of the master equation for one-step processes was considered. We discussed some aspects of the problem of reconstruction of the Fokker–Planck equation for the many-variable distribution function.

The material gathered in Chapter 8 is a synthesis of the papers of Du, Pardalos, and Wang (2000), Horst and Pardalos (1995), Pardalos et al. (2001), Sackellares et al. (2000), Yatsenko (1986), Cremers and Hübler (1987), and Klimontovich (1990). The results presented in Section 8.5 are inspired by the paper of Iasemidis et al. (2001). The idea of a support vector machine first appeared in a book by Vapnik (1995, 1998).

References

Abarbanel, H., Brown, R., and Kadtke, J. (1990). Prediction in chaotic nonlinear systems: Methods for time series with broadband Fourier spectra. *Physical Review A*, 41(4):1782–1807.

Abraham, R. and Marsden, J. (1978). *Foundation of Mechanics*. Benjamin-Cummings. Reading, MA.

Ackley, D.H., Hinton, G.E., and Seinowski, T.J. (1985). A learning algorithm for Boltzmann machines. *Cognitive Sciences*, 9:147–169.

Ado, I. (1947). Representing Lie algebras by matrices. *Uspekhi Mat. Nauk*, 22(6): 159–173.

Aganović, Z., Gajic, Z. (1995). Linear Optimal Control of Bilinear Systems. Springer, Berlin–New York.

Agrachev, A. and Sachkov, Yu. (2004). Control Theory from the Geometrical Viewpoint. Series: Encyclopedia of Mathematical Sciences , Series Volume 87. Springer, Berlin–Heidelberg–New York–Hong Kong–London–Milan–Paris–Tokyo.

Allen, J.P., Feher, G., Yeates, T.O., Rees, D.C., Deisenhofer, J., Michel, H., and Huber, R. (1986). Structural homology of reaction centers from Phodopseudomonas sphaeroides and Rhodopseudomonas viridis as determined by X-ray diffraction. *Proc. Natl. Acad. Sci. USA.*, 83:8589–8593.

Allen, L. and Eberly, J. (1975). *Optical Resonance and Two Level Atoms*. Wiley, New York.

Amari, S. (1982). Differential geometry of curved exponential families – curvatures and information loss. *The Annals of Statistics*, 10:357–385.

Amari, S. (1985). *Differential-Geometrical Methods in Statistics*. Lecture Notes in Statistics 28. Springer–Verlag, New York.

Amari, S. and Nagaoka, N. (1993). *Methods of Information Geometry 191*. Oxford University Press, Providence, RI.

Amari, S., Kurata K., and Nagaoka, H. (1992). *IEEE Trans. Neural Networks*, 3: 260–271.

Anderson, B. and Moore, J. (1979). *Optimal Filtering*. Prentice-Hall, Englewood Cliffs, NJ.

Andreev, N. (1982). Differential-geometric methods in control theory. *Automatic and Remote Control*, 10:5–46.

Arnold, V. (1968.) Singularities of smooth mappings *Uspekhi Mat. Nauk.* 23(1):3–44.

Arnold, V. (1983). *Mathematical Methods of Classical Mechanics*. Springer–Verlag, New York–Berlin–London–Tokyo.

Bacciotti, A. (2004). Stabilization by means of state space depending switching rules. *Syst. & Control Lett.*, 53(3-4):195–201.

Bailleul, J. (1978). Geometric methods for nonlinear optimal control problems. *J. Optim. Theory Appl.*, 25(4):519–548.

Baillieul, J. (1998). The geometry of controlled mechanical systems. In Baillieul, J. and Willems, J., eds., *Mathematical Control Theory*, pp. 322–354. Springer-Verlag, New York.

Bandurin, V., Zinovyev, A., and Kozorez, V. (1979). On stability of equilibrium of a free physical pendulum in potential well. *Dokl. AN USSR, Ser. A*, 6:478–482 (in Russian).

Barone, A. and Paterno G. (1982). *Physics and Applications of the Josephson Effect*. Wiley Interscience, New York.

Belavkin, V., Hirota, O., and Hudson, R. (1995). *Quantum Communications and Measurement*, Plenum Press, New York.

Bendjaballah, C., Hirota, O., and Reynaud S. (1991). *Quantum Aspect of Optical Communications*, Lecture Note in Physics, Springer, Berlin.

Benes, V.E. (1981). Exact finite dimensional filters for certain diffusions with nonlinear drift. *Stochastics*, 5:65–92.

Bloch, A. (1998). Optimal control, optimization, and analytical mechanics. In Baillieul, J. and Willems, J., eds., *Mathematical Control Theory*, pp. 268–321. Springer–Verlag, New York.

Bloch, A. and Crouch, P. (1996). Optimal control and geodesic flows. *Syst. and Control Lett.*, 28(2):65–72.

Boothby, W. (1975). *An Introduction to Differentiable Manifolds and Riemannian Geometry*. Academic Press, London.

Boothby, W. and Wilson, W. (1979). Determination of the transitivity of bilinear system. *SIAM J. Contr. Optimiz.*, 17(2):212–221.

Bose, T. and Chen, M. (1995). Conjugate gradient method in adaptive bilinear filtering. *IEEE Trans. Signal Processing*, 43:349–355.

Braginskii, V. (1970). *Physical Experiments with Test Bodies*. Nauka, Moskow (in Russian).

Brandt, E. (1989). Levitation in physics. *Science*, 243:349–355.

Brandt, E. (1990). La Lévitation. *La Recherche*, 224:998–1005.

Braunbeck, W. (1939). Freishwebende Körper in electrischen and magnetischen Feld. *Z. Phys.*, 112(11–12):753–763.

Brockett, R. (1972). System theory of group manifolds and coset spaces. *SIAM J. Contr.*, 10:265–284.

Brockett, R. (1973). Lie theory and control systems defined on spheres. *SIAM J. Appl. Math.*, 25(2):213–225.

Brockett, R. (1975). Volterra series and geometric control theory. In *Proc. 1975 IFAC Congress*. ISA, Philadelphia.

Brockett, R. (1976). Finite- and infinite-dimensional bilinear systems. *J. of the Franklin Institute*, 301:509–320.

Brockett, R. (1979). Classification and equivalence in estimation theory. In *Proc. of the 18th IEEE Conf. on Decision and Control*, pp. 172–175.

Brockett, R. (1981). Nonlinear systems and nonlinear estimation theory, In Hazewinkel, M. and Willems, J.C., eds., *Stochastic Systems: The Mathematics of Filtering and Identification and Applications*. Reidel, Dordrecht.

Brockett, R. (1985). On the reachable set for bilinear systems, variable structure systems with application to economy and biology, In Ruberti, A. and Mohler, R.,

Lecture Notes in Economics and Mathematical Systems. New York, Springer, pp. 54–63.

Brockett, R.W. (1976). Volterra series and geometric control theory. *Automatica*, 12:167–176.

Brockett, R.W. (1979). Lie algebras and Lie groups in control theory. In Novikov S.P., ed., *Mathematical Methods in System Theory*. Nauka, Moscow (in Russian).

Brockett, R.W. (1980). Estimation theory and the representation of Lie algebras, In *Proc. 19th IEEE Conf. on Decision and Control*. Albuquerque.

Bruni, C. (1971). On the mathematical models of bilinear systems. *Ricerche di Automatica*, 2(1):11–26.

Bucy, R. and Joseph, P. (1968). *Filtering for Stochastic Processes with Application to guidance.*, Wiley Interscience, New York.

Butenko, S., Murphey, R., and Pardalos, P., eds. (2003). *Cooperative Control: Models, Applications and Algorithms*. Kluwer Academic, Boston.

Butkovskiy, A. (1991). *Phase Portraits of Control Dynamical Systems*. Kluwer Academic, The Netherlands.

Butkovskiy, A. and Samoilenko, Yu. (1990). *Control of Quantum-Mechanical Processes and Systems*. Kluwer Academic Publishers, The Netherlands.

Carrol, C. and Hioe, F. (1988). Three-state systems driven by resonant optical pulses. *J. Opt. Soc. Am. B.*, 5(6):1335–1340.

Casdagli, M. (1989). Nonlinear prediction of chaotic time series. *Physica D*, 35: 335–356.

Chang, K., Hübler, A., and Packard, N. (1989). *Quantitative Measures of Complex Dynamical Systems*. Plenum, New York.

Chen, C., Leung, C., and Yau, S. (1996). Finite dimensional filters with nonlinear drift IV: Classification of finite dimensional estimation algebras of maximal rank with state space dimension 3. *SIAM J. Control Optim.*, 34(3):179–198.

Chennaoui, A., Pawelzik, K., Liebert, W., Schuster, H., and Pfister, G. (1990). Attractor reconstruction from filtered chaotic-series. *Phys. Rev. A*, 41(8):4151–4159.

Chentsov, N. (1982). *Statistical Decision Rules and Optimal Inference*. 53. American Mathematical Society, Providence, RI.

Cherevko, V. and Yatsenko, V. (1992). Control systems and modelling of signals for back scattering from sea surface. *Cyber. Comput. Technol., Ergatic Control Syst.*, 96:107–113.

Chikte, D. and Lo, J.T. (1981). Optimal filters for bilinear systems with nilpotent Lie algebras. *IEEE Trans. Autom. Control*, 24(6):948–953.

Childers, D. and Durling, A. (1975). *Digital Filtering and Signal Processing*, West, St. Paul, MN.

Chinarov, V., Gaididei, Yu., Kharkyanen, V., and Pivovarova, N. (1990). Selforganization of a membrane ion channel as a basis of neuronal computing, In Proc. Intern. Symp. Neural Network and Neural Computing, Prague, pp. 65–67.

Chinarov, V., Gaididei, Yu., Kharkyanen, V., and Sit'ko, S. (1992). Ion pores in biological membranes as self-organization bistable systems. *Phys. Rev. A*, 46(8): 5232–5241.

Chiou, W. and Yau, S. (1994). Finite dimensional filters with nonlinear drift II: Brockett's problem on classification of finite dimensional estimation algebras. *SIAM J. Contr. and Optimiz.*, 32(1):297–310.

Ciani, S. (1984). Coupling between fuxes in one-particle pores with fluctuating energy profiles. *Biophys. J.*, 46:249–252.

Cleveland, W. (1979). Robust locally weighted regression and smoothing scatterplots. *J. Am. Statist. Assoc.*, 74:829–836.

Cremers, J. and Hübler, Z. (1987). Construction of differential equations from experimental data. *Z. Naturforsch*, 42a:797–802.

Crouch, P. (1984). Solvable approximations to control systems. *SIAM J. Contr. Optimiz.*, 32(1):40–54.

Crouch, P. (1985). Variation characterization of Hamiltonian systems. *IMA J. Math. Contr. and Inf.*, 3:123–130.

D'Alessandro, P. (2000). Topological properties of reachable sets and the control of quantum bits. *Syst. & Control Lett.*, 41(3):213–221.

D'Alessandro, P., Isidory, A. and Ruberti, A. (1974). Realization and structure theory of bilinear dynamical systems. *SIAM J. Contr. Optimiz.*, 12:517–535.

Dang Van Mien, H. and Norman-Cyrot, D. (1984). Nonlinear state affine identification methods; application to electrical power plants. *Automatica*, 20(2):175–188.

Daniel, M. and Viallet, C. (1980). The geometrical setting of gauge theories of the Yang–Mills type. *Rev. Mod. Phys.*, 52:175–197.

Danilin, Yu., and Piyavsky, S. (1967). On one algorithm of searching for an absolute optimum, *Theory Optimal Decision*, No. 2:25–37 (in Russian).

Davis, M. and Marcus, S. (1981). An introduction to nonlinear filtering. In Hazewinkel, M. and Willems, J.C., editors, *The Mathematics of Filtering and Identification and Applications*, pp. 53–75. Reidel, Dordrecht.

Dayawansa, W. (1998). Recent advances in the stabilization problem for low dimensional systems. In Jakubczyk, B. and Respondek, W., eds., *Geometry of Feedback and Optimal Control*, pp. 165–203. Marcel Dekker, New York.

Deisenhofer, J., Epp, O., Sinning, I., and Michel, H. (1995). Crystallographic refinement at 2.3 Å resolution and refined model of the photosynthetic reaction centre from Rhodopseudomonas viridis. *J. Mol. Biol.*, 246(3):429–457.

Derese, I. and Noldus, E. (1981). Existence of bilinear state observers for bilinear system. *IEEE Trans. Autom. Control.*, AC-26(2):590–592.

Devyatkov, N. (1978). An effect of EHF band-EMI on biological objects. *Uspekhi Fiz. Nauk.*, 10(3):453–454.

Diniz, P. (1997). *Adaptive Filtering: Algorithms and Practical Implementation*. Kluwer Academic, Boston.

Dirac, P. (1958). *The Principles of Quantum Mechanics*. Clarendon Press, Oxford.

Dmitriev, A. and Kislov, V. (1989). *Stochastic Oscillations in Radiophysics and Electronics*. Nauka, Moscow.

Dong, R., Tam, L., Wong, W., and Yau S. (1991). Structure and classification theorems of finite-dimensional exact estimation algebras. *SIAM J. Contr. and Optimiz.*, 29:866–877.

Du, P.-Zh., Pardalos, P., and Wang, J. (Editors). (2000). Discrete Mathematical Problems with Medical Applications. In *DIMACS Series*, Vol. 55. American Mathematical Society.

Dubrovin, B., Novikov, S., and Fomenko, A. (1984). *Modern Geometry – Methods and Applications, Part. 1*, Graduate text in mathematics, Vol. 93. Springer–Verlag, New York.

Elkin, V. (1999). *Reduction of Nonlinear Control Systems: A Differential Geometric Approach*. Kluwer Academic, Dordrecht.

Ermoliev, Y. and Wets, R. (1984). *Numerical Techniques for Stochastic Optimization Problems*. Int. Inst. Applied System Analysis, Luxemburg, Austria.

Espana, M. and Landau, I. (1978). Reduced order bilinear models for distillation columns. *Automatica*, 14(3): 345–355.

Faibusovich, L. (1988a). Collective Hamiltonian method in optimal control problems. *Cybern. Syst. Anal.*, 25(2): 230–237.

Faibusovich, L. (1988b). Explicitly solvable non-linear optimal control problems. *Int. J. Control*, 48: 2507–2526.

Faibusovich, L. (1991). Hamiltonian structure of dynamical systems which solve linear programming problems. *Phys. D*, 53(2-4):217–232.

Fante, R. (1988). *Signal Analysis and Estimation*. Wiley, New York.

Farmer, J. and Sidorowich, J. (1987). Predicting chaotic time series. *Phys. Rev. Lett.*, 59(8):845–848.

Feher, G. and Okamura, M.Y. (1978). Chemical composition and properties of reaction centers. In Clayton, R.K. and Systrom W.R., eds., *The Photosynthetic Bacteria*, pp. 349–386. Academic Press, New York.

Feynman, R. (1982). Simulating physics with computers. *Int. J. Theoret. Phys.*, 21 (6–7):467–488.

Feynman, R. (1985). Quantum mechanical computers. *Optics News*, 11:11–20.

Fisher, W. and Brickman, J. (1983). Ion specific diffusion rates through trans-membrane protein channels: a molecular dynamics study. *Biophys. Chem.*, 18: 323–337.

Fliegner, T., Kotta, Ju., and Nijmeijer, H. (1996). Solvability and right-inversion of implicit nonlinear discrete-time systems. *SIAM J. Contr. Optimiz.*, 34(6): 2092–2115.

Fliess, M. (1975). Un outil algebrique: Les series formelles non commutatives. In *Proc. CNR-CISM Symp. on Algebraic System Theory*, Udine, Italy, June 16–27.

Fliess, M., Livine, J., Martin, and Rouchon, P. (1999). A Lie-Bäcklund approach to equivalence and flatness of nonlinear systems. *IEEE Trans. Aut. Contr.*, 44(5): 928–937.

Fnaiech, F., Ljung, L., and Fliess, M. (1987). Recursive Identification of Bilinear Systems. *Int. J. Contr.*, 45(2):453–470.

Fraser, A. and Swinney, H. (1986). Independent coordinates in strange attractor from mutual information. *Phys. Rev. A*, 33:1134–1140.

Freeman, W. (2000). *Neurodynamics: An Exploration of Mesoscopic Brain Dynamics*. Springer-Verlag, London.

Freeman, W. and Skarda, C. (1985). Spatial EEG patterns, non-linear dynamics and perception: the neo-Sherringtonian view. *Brain Res. Rev.*, 10:147–175.

Gafka, D. and Tani, J. (1992). Chaotic behavior, strange attractors and bifurcation in magnetic levitation systems. In *International Symposium on Nonlinear Phenomena in Electromagnetic Fields*, January 1992, ISEM-Nagoya, Japan.

Gaididei, Yu., Kharkyanen, V., and Chinarov, V. (1988). Voltage-dependent ion channel in a biological membrane as a self-organizing nonequilibrium system. Preprint ITP-88-77E, Kiev.

Gardiner, C. (1985). *Handbook of Stochastic Methods*. Springer-Verlag, Berlin.

Gauthier, J. and Bornard, G. (1982). Controlabilite des systemes bilineares. *SIAM J. Contr. Optimiz.*, 20(3):377–384.

Gauthier, J., Kupka, I., and Sallet, G. (1984). Controllability of right invariant systems on real simple Lie groups. *Syst. Contr. Lett.*, 5(3):187–190.

Giannakis, G. (1987). Cumulants: A powerful tool in signal processing. *Proc. IEEE*. 75(9):1333–1334.

Goka, T., Tarn, T., and Zaborszky, J. (1973). On the controllability of a class of discrete bilinear systems. *Automatica*, 9:615–622.

Goldfeld, N.G. and Karapetyan, N.V. (1986). Photosynthesis and Herbicides. *J. Soviet Union Chem. Soc.*, 31(6): 567–576.

Gomatam, R. (1999). Quantum theory and the observation problem. In Núñez, R. and Freeman, W., eds., *Reclaiming Cognition. The Primacy of Action, Intention and Emotion*, pp. 173–190. Imprint Academic, UK.

Gough, J., Belavkin, V., and Smolyanov, O. (2005). Hamilton–Jacobi–Bellman equations for quantum optimal feedback control. *Journal of Optics B: Quantum and Semiclassical Optics*, 7: S237–S244.

Gounaridis, C. and Kalouptsidis, K. (1986). Stability of discrete-time bilinear systems with constant-inputs. *Int. J. Control*, 43:663–669.

Gushcha, A., Kapoustina, M., Kharkyanen, V. et al. (1995). A new approach to experimental investigation of dynamical self-organization in reaction centers of purple bacteria. *J. Biol. Phys.*, 21(4):265–272.

Gradshteyn, I. S. and Ryzhik, I. M. (2000). *Tables of Integrals, Series, and Products*. 6th ed. Academic Press, San Diego.

Grassberger, P. and Procaccia, I. (1982). Characterization of strange attractors. *Phys. Rev. Lett.*, 50:346–349.

Grasselli, O. and Isidori, A. (1981). An existence theorem for observers of bilinear systems. *IEEE Trans. Aut. Contr.*, AC-26:1299–1300.

Griffiths, P. (1983). *Exterior Differential Systems and the Calculus of Variations*. Birkhuser, Boston.

Gröbler, T., Barna, G. and Ërdi, P. (1998). Statistical model of the hippocampal CA3 region. I. The single-cell module; bursting model of the pyramidal cell. *Biol. Cybern.*, 79:301–308.

Grover, L. (1996). A fast quantum mechanical algorithm for database search. Proc. of 28th Annual ACM Symposium on the Theory of Computing (STOC), pp. 212–219.

Grundel, D., Murphey, R., Pardalos, P., and Prokopyev, O., eds. (2007). *Cooperative Systems, Control and Optimization, Lecture Notes in Economics and Mathematical Systems*, Vol. 588. Springer, New York.

Gushcha, A., Dobrovolskii, A., Kapoustina, M, Privalko, A., and Kharkyanen, V. (1994). New physical phenomenon of dynamical self-organization in molecular electron-transfer systems. *Phys. Lett. A*, 191(5-6):393–397.

Gushcha, A.J., Dobrovolsky, A.A., Kharkyanen, V.N., and Yatsenko, V.A. (1993). In *Euroanalysis 8, European Conference on Analytical Chemistry*, Edinburgh, 5-11 September 1993, Book of Abstracts, pp. 112.

Gutman, P. (1981). Stabilizing controllers for bilinear systems. *IEEE Trans. Aut. Contr.*, 25:917–922.

Hageronn, P. (1988). *Non-Linear Oscillations, 2nd edition*. Oxford University Press, New York.

Haken, H. (1978). *Synergetics*. Springer-Verlag, New York.

Haken, H. (1988). *Information and Self-Organisation*. Springer, Berlin.

Haken, H. (1991). *Synergetic Computers and Cognition*. Springer, Berlin.

Haken, H. (1996). *Principles of Brain Functioning: a Synergetic Approach to Brain Activity, Behaviour, and Cognition*. Springer-Verlag, New York.

Hammouri, H. and Gauthier, J. (1988). Bilinearization up to output injection. *Int. J. Contr.*, 46:455–472.

Hanba, S. and Yoshihiko, M. (2001). Output feedback stabilization of bilinear systems using dead-beat observers. *Automatica.*, 37:915–920.

Hao, B. (1984). *Chaos*. World Scientific, Singapore.

Haynes, G. and Hermes, H. (1970). Nonlinear controllability via Lie theory. *SIAM J. Contr.*, 8:450–460.

Hazewinkel, M. (1982). Control and filtering of a class of nonlinear but 'homogeneous' systems. *Lect. Notes Inform. Sci. Contr.*, 19:123–146.

Hazewinkel, M. (1986). Lie algebraic method in filtering and identification. In Albeverio, S., Blanchard, P., and Hazevinkel, M., eds, *Stochastic Processes in Physics and Engineering*, pp. 159–176. D. Reidel, Dordrecht.

Hazewinkel, M. (1995). The linear systems Lie algebra, the Segal–Shale–Weil representation and all Kalman–Bucy filters. *Syst. Sci. Math. Sci.*, 5:94–106.

Hazewinkel, M. and Marcus, S. (1982). On Lie algebras and finite-dimensional filtering. *Stochastics*, 7:29–62.

Helstrom, C. (1976). *Quantum Detection and Estimation Theory*, Academic Press, New York.

Helstrom, C. (1995). *Element of Signal Detection and Estimation*. Prentice Hall, Englewood Cliffs, NJ.

Hendricson, A. (1995). The molecular problem: Exploiting structure in global optimization. *SIAM J. Optim.*, 5:835–857.

Hermann, R. (1973). *Constrained Mechanics and Lie Theory*, Second edition, Interdisciplinary Mathematics, Vol. XVII. Math. Sci. Press, New York.

Hiebeler, D. and Tater, R. (1997). Cellular automata and discrete physics. In Lui Lam, ed., *Introduction to Nonlinear Physics*, pp. 143–166. Springer, New York–Berlin.

Hirota, O. and Ikehara, S. (1982). Minimax strategy in the quantum detection theory and its application to optical. communication. *Trans. IECE Japan*, E65:627–642.

Hirschorn, R. (1977). Invertibility of control systems on Lie groups. *SIAM J. Contr.*, 15:1034–1049.

Hodgkin, A. and Hukley, A. (1952). The components of membrane conductance in the giant axon of Loligo. *J. Physiol.*, 116:473–496.

Hopcroft, J. and Ullman, J. (1979). *An Introduction to Automata Theory, Languages and Computation*. Addison-Wesley, Reading, MA.

Hopfield, J. (1994). Neurons, dynamics and computation. *Phys. Today*, 47(2):40–46.

Hopfield, J. and Tank, D. (1985). "Neural" computation of decision in optimization problems. *Biol. Cybern.*, 52:141–152.

Horst, R. and Pardalos, P. (1995). *Handbook of Global Optimization*. Kluwer Academic, Dordrecht.

Hull, J. (2004). Levitation applications of high-temperature superconductors. In A. V. Narlikar, ed., *High Temperature Superconductivity 2: Engineering Applications*, pages 91–142. Springer-Verlag, New York.

Hunt, L. (1972). Controllability of general nonlinear systems. *Math. Syst. Theory*, 12:361–370.

Hunt, L., Su. R, and Meyer, G.(1983). Global transformations of nonlinear systems. *IEEE Trans. Autom. Contr.*, AC-28:24–31.

Husmoller, D. (1975). *Fiber Bundles*. Springer-Verlag, New York.

Iasemidis, L.D., Pardalos, P.M., Shiau, D.-S., and Sackellares, J.C. (2001). Quadratic binary programming and dynamic system approach to determine the predictability of epileptic seizures. *J. Combinat. Optim.*, 5(1):9–26.

Ionescu, L. and Monopoli, R. (1975). On the stabilization of bilinear systems via hyperstabilty. *IEEE Trans. Automat. Contr.*, 20:280–284.

Isidory, A. (1995). *Nonlinear Control Systems* Springer, New York.

Isidory, A. and Krener, A. (1984). On the synthesis of linear input-output responses for nonlinear systems. *Syst. Contr. Lett.*, 4(1):17–22.

Jackson, E. (1989). *Perspectives of Nonlinear Dynamics*, Cambridge University Press, New York.

Jakeman, E. (1980). On the statistics of K-distributed noise. *J. Phys. A*, 13:31–48.

Jeans, J. (1925). *The Mathematical Theory of Electricity and Magnetism*. Cambridge University Press, New York.

Jibu, M. and Yassue, K. (1995). *Quantum Brain Dynamics and Consciousness: An introduction*. John Benjamin, Amsterdam.

Jimenez, J., Moreno, A., Ruggeri, G., and Marcano, A. (1992). Detecting chaos with local associative memories *Phys. Lett. A*, 169:25–30.

Jurdjevic, V. (1997). *Geometric Control Theory*, Fulton, W. et al., eds. Cambridge University Press, New York.

Jurdjevic, V. (1998). Optimal control, geometry, and mechanics. In Baillieul, J. and Willems, J., eds, *Mathematical Control Theory*, pp. 227–321. Springer, New York.

Jurdjevic, V. and Sallet, G. (1984). Controllability properties of affine fields. *SIAM Contr. Optim.*, 22(3):501–508.

Jurdjevic, V. and Susmann, H. (1972). Control system on Lie group. *J. Diff. Eqs.*, 12(2):313–329.

Kalman, R. and Bucy, R. (1961). New results in linear filtering and prediction theory. *J. Basic. Eng., Trans. ASME*, 83D:95–107.

Kalman, R., Falb, S., and Arbib, M. (1969). *Topics in Mathematical System Theory*. Mc Graw–Hill, New York.

Kandel, E., Schwartz, J., and Jessell, T. (1991). *Principles of Neural Science*. Elsevier, New York.

Kaneko, K. (1993). The coupled map lattice. In Kanepo, K., ed. *Theory and Application of Coupled Map Lattices*, pages 1–50. John Wiley and Sons, Chichester.

Kaneko, K. (2001). *Complex System: Chaos and Beyond: A Constructive Approach with Applications in Life Sciences*. Springer-Verlag, Berlin.

Kaplan, D. and Glass, L. (1992). Direct test for determinism in a time series. *Phys. Rev. Lett.*, 68:427–430.

Kapustina, M.T. and Kharkyanen, V.N. (1992). Investigation of photomobilized electron recombination kinetics in bacteria reaction centers as a method for electron-conformation interaction studies. *Preprint*, Institute of Theoretical Physics, Kiev, Ukraine, ITF-92-3R.

Karapetyan, N.V. and Buhov, N.G. (1986). Variable chlorophyll fluorescence as an indicator of the physiological state of plants. *Physiol. Plants*, 33(33):1013–1028.

Kennel, M. and Isabelle, S. (1992). Method to distinguish possible chaos from colored noise and to determine embedding parameters. *Phys. Rev. A*, 46:3111–3118.

Khapalov, A. and Mohler, R. (1996). Reachable sets and controllability of bilinear time-invariant systems: A qualitative approach. *IEEE Trans. Aut. Contr.*, 41(9):1342–1346.

Kinnaert, M. (1999). Robust fault detection based on observers for bilinear systems. *Automatica*, 35:1829–1999.

Kleinfeld, D., Okamura, M.Y., and Feher, G. (1984). Electron-transfer kinetics in photosynthetic reaction centers cooled to cryogenic temperatures in the charge-separated state: Evidence for light-induced structural changes. *Biochemistry*, 23:5780–5786.

Klimontovich, Yu. (1990). Ito, Stratonovich and kinetic forms of stochastic equations. *Physica A*, 163(2):515–532.

Klimontovich, Yu. (1995). *Statistical Theory of Open Systems*. Kluwer Academic, Dordrecht.

Klimontovich, Yu. (1999). Entropy and information of open systems. *Phys-Usp.*, 42(4):375–384.

Knopov, P. (1984). Estimation of unknown parameters of an almost periodic function in the presence of noise. *Kibernetika*, 6:83–87 (in Russian).

Kobayashi, S. and Nomizu, K. (1963). *Foundations of Differential Geometry*, Vol. I. Interscience, New York.

Koditschek, D. and Narendra, K. (1985). The controllability of planar bilinear systems. *IEEE Trans. Aut. Contr.*, AC-30:87–89.

Kostelich, E. and Schreiber, T. (1993). Noise reduction in chaotic time-series data: a survey of common methods. *Phys. Rev. E*, 48:1752–1763.

Kotta, Ju. (1983). On an inverse system of a special class of bilinear systems with many inputs and outputs. *Izv. Acad. of Sci. of the Estonian SSR*, 32(3):323–326.

Kozorez, V. and Cheborin O.G. (1977). On stability of equilibrium in a system of two ideal current rings. *Dokl. Akad. Nauk UkrSSR, Ser. A*, 4:80–81.

Kozorez, V. and Yatsenko, V. (1985). Differential-geometrical methods of analysis of nonlinear controlled circuits with a Josephson junction. In *Int. Conf Theor. Electrical Engineering*, pages 87–88. Nauka, Moscow (in Russian).

Kozorez, V., Kolodeev, I., and Kryukov, M. (1975). On potential well of magnetic interaction of ideal current loops. *Dokl. AN USSR, Ser. A*, 3:248–249.

Kozorez, V., Malitskiy, R., Pardalos, P., Negriyko, A., Udovitskaya, E., Khodakovskiy, V., Ismaili, K., Cheremnykh, O., Yatsenko, V., and Yatsenko, L. (2006). Cryogenic-optical sensor for high-sensitive gravitational measurements. *J. Aut. Inform. Scienc*, 38(4):54–68.

Kramers, H.A. (1940). Brownian motion in a field of force and the diffusion model of chemical reactions. *Physica*, 4:284–304.

Krasovskiy, A. (1990). Problems of physical theory of control. *Autom. Remote Contr.*, 1:3–28.

Krener, A. (1973). On the equivalence of control systems and the linearization of nonlinear systems. *SIAM J. Contr.*, 11:670–676.

Krener, A. (1975). Bilinear and nonlinear realizations of input-output maps. *SIAM J. Contr.*, 13(4):827–834.

Krener, A. (1977). The high order maximal principle and its application to singular extremals. *SIAM J. Contr.*, 15:256–293.

Krener, A. (1998). Feedback linearization. In Baillieul, J. and Willems, J., eds, *Mathematical Control Theory*, pp. 86–98. Springer-Verlag, New York.

Krishnaprasad, P. and Marcus, S. (1982). On the Lie algebra of the identification problem. In *IFAC Symposium on Digital Control*. New Delhi.

Kučera, J. (1966). Solution in large of control problem: $\dot{x} = (A(1-u) + Bu)x$. *Czech. Math. J.*, 91:600–623.

Kuchtenko, A.I. (1963). *The Invariance Problem in Automation*. Tekhnika, Kiev (in Russian).

Kulhavý, R. (1996). *Recursive Nonlinear Estimation: a Geometric Approach*. Springer, London.

Kuroda, M., Tanada, N., and Kikushima. Y. (1992). Chaotic vibration of a body levitated by electromagnetic force. In *Int.Symp. on Nonlinear Phenomena in Electromagnetic Fields*, January 1992, ISEM-Nagoya, Japan.

Kushner, H. (1967). Dynamical equation for optimal nonlinear filtering. *J. Diff. Eq.*, 3:179–185.

Landau, L. and Lifshitz, E. (1976). *Mechanics*. Pergamon Press, Oxford, New York.

Larner, R., Speelman, B., and Worth, R. (1997). A coupled ordinary differential equation lattice model for the simulation of epileptic seizures. *Chaos*, 9:795–804.

Läuger, P. (1985). Tonic channels with conformational substrates. *Biophys. J.*, 47: 585–595.

Lebedev, D.V. (1978). Controlling the motion of a rigid body under incomplete information about its angular velocity. *Autom. Remote Contr.*, No.12, 5–11.

Lee, H. and Marcus, S. (1980). The Lie algebraic structure of a class of finite dimensional filter. In Byrnes, C.I. and Martin C.F., eds., *Algebraic and Geometric Methods on Linear System Theory, Lectures in Applied Math. 18*, pp. 277–297. American Mathematical Society, Providence, RI.

Lev, A., Schagina, L., and Grinfeldt, A.. (1988). Changes of the energy profile of gramicidin A ionic channel dependent on the ratio of cations of different species in the flux passing through the channel. *Gen. Physiol. Biophys.*, 7:547–559.

Likharev, K. and Ul'rikh, B. (1978). *Systems with Josephson Junctions. Fundamentals of the Theory.* Moscow State Univ. (in Russian).

Lloyd, S. (1993). A potentially realizable quantum computer. *Science*, 261:1569–1571.

Lo, J.T. (1973a). Finite-dimensional sensor orbits and optimal nonlinear filtering. *IEEE Trans. Inform. Theory*, IT-18(5):583–588.

Lo, J.T. (1973b). Signal detection on Lie groups. In Mayne, D.O. and Brockett, R.W., eds., *Geometric Methods in System Theory*, pp. 295–303, MA: Reidel, Boston.

Lo, I. (1975). Global bilinearization of systems with control appearing linearly. *SIAM J. Contr.*, 13:879–884.

Longchamp, R. (1980). Stable feedback control of bilinear systems. *IEEE Trans. Automat. Contr.*, AC-25:302–306.

Loskutov, A. and Mikhailov, A. (1990). *Introduction to Synergetics.* Nauka, Moscow (in Russian).

Mahler, G. and Weberruss, V. A. (1995). *Quantum Networks: Dynamics of Open Nanostructures.* Springer, Berlin.

Malkin, S., Churio, M.S., Shochat, S., and Braslavskvy, S.E. (1994). Photochemical energy storage and volume changes in the microsecond time range in bacterial photosynthesis-a laser induced optoacoustic study. *J. Photochem. Photobiol.*, 23:79–85.

Manakov, S. (1976). A note on integration of Euler equations as for dynamics of n-dimensional solid body. *Funct. Anal. Appl.*, 10(4):328–329.

Marcus, L. (1973). General theory of global dynamics. In Mayne, D.Q. and Brockett, R.W., eds, *Geometric Methods in System Theory*, pages 150–158. D. Reidel, Dordrecht.

Marcus, S. and Willsky, A. (1976). Algebraic structure and finite dimensional nonlinear estimation. *Lecture Notes in Economics and Mathematical Systems, Mathematical System Theory* 131, pp. 301–311 Springer, New York.

Mauzerall, D.C., Gunner, M.R., and Zhang, J.M. (1995). Volume contraction on photoexcitation of the reaction center from Phodobacter sphacrodes R-26: Internal probe of dielectrics. *Biophys. J.*, 68:275–280.

Mees, A., Aihara, K., Adachi, M., Judd, K., Ikeguchi, T., and Matsumoto, G. (1992). Deterministic prediction and chaos in squid axon response. *Phys. Lett. A*, 160:41–45.

Melnikov, V. (1963). On the stability of the center for time-periodic perturbations. *Trans. Moscow. Math. Soc.*, 12(11):1–56.

Menskii, M.B. (1983). *The Group of Paths: Measurements, Fields, Particles*, Nauka, Moscow (in Russian).

Martinez Guerra, R. (1996). Observer based tracking of bilinear systems: a differential algebraic approach. *Appl. Math. Lett.*, 9(5):51–57.

Mishchenko, A. (1970). Integral of geodesic flows on Lie groups. *Funct. Anal. Appl.*, 4(3):73–74.

Mitschke, F., Dammig, M. (1993). Chaos versus noise in experimental data. *Int. J. Bifurcation Chaos*, 3:693–702.

Mitter, P. (1980). Geometry of the space of gauge orbits and the Yang–Mills dynamical system. In Hoof G.T. et al., eds, *Recent Developments in Gauge Theories*, (Cargese School, Corsica, August 26 – September 8, 1979). Plenum Press, New York.

Mitter, S. (1979). On the analogy between mathematical problems of nonlinear filtering and quantum physics. *Richerche di Automatica*, 10:163–216.

Miwakeichi, F., Raminez-Pardon, R., Valdes-Sosa, P. and Ozahi, T. (2001). A comparison of non-linear non-parametric models for epilepsy data. *Comput. Biol. Med.*, 31:41–57.

Mohler, R. (1973). *Bilinear Control Processes*. Academic Press, New York.

Moon, F. (1988). Chaotic vibrations of a magnet near a superconductor. *Phys. Lett. A*, 132(5):249–251.

Moon, F. (1992). *Chaotic and Fractal Dynamics*. John Wiley & Sons, New York.

Morris, C. and Lecar, H. (1981). Voltage oscillations in the barnacle giant muscle fiber. *Biophys. J.*, 35:193–213.

Mosevich, J.W. (1977). Identifying differential equations by Galerkin's method. *Math. Comput.*, 31(137):139–147.

Murphey, R. and Pardalos, P., eds. (2002). *Cooperative Control and Optimization*. Kluwer Academic, Boston.

Nerenberg, M. and Essex, C. (1986). Correlation dimension and systematic geometric effects. *Phys. Rev. A*, 42:7065–7072.

Nicolis, G. and Prigogine, I. (1977). *Self Organization in Nonequilibrium Systems*. Wiley, New York.

Nijmeijer, H. and Respondek, W. (1988). Dynamic input-output decoupling of nonlinear control systems. *IEEE. Trans. Aut. Contr.*, AC-33:1065–1070.

Nijmeijer, H. and Van der Schaft, A. (1985). Controlled invariance for nonlinear systems. *IEEE. Trans. Aut. Contr.*, AC-27(4):904–914.

Nijmeijer, H. and Van der Schaft, A. (1990). *Nonlinear Dynamical Control Systems*. Springer-Verlag, New York.

Noks, P.P., Lukashev, E.P., Kononenko, A.A., Venediktov P.S., and Rubin, A.B. (1977). Possible role of macromolecular components in the functioning of photosynthetic reaction centers of purple bacteria. *Mol. Biol.*, 11(5):1090–1099.

Okamura, M.Y, Feher, G., and Nelson N. (1982). Reaction centers. In Govindjee, N.Y, editor, *Photosynthesis. Energy Conversion by Plants and Bacteria*, pp. 195–272. Academic Press, New York.

Ozava, M. (1984). Quantum measuring processes of continuous observables. *J. Math. Phys.*, 25(1):79–87.

Parak, F. and Knapp, E.W. (1984). A consistent picture of protein dynamics, *Proc. Natl. Acad. Sci.*, 481:7088–7092.

Pardalos, P. (1988). Applications of global optimization in molecular biology. In Carlsson, C. and Eriksson, I., eds., *Global Multiple Criteria Optimization and Information Systems Quality*, pp. 91–102. Abo Akademis Tryckeri, Finland.

Pardalos, P. (1998). In Hager, W. and Pardalos, P., eds., *Optimal Control: Theory, Algorithms, and Applications*. Kluwer Academic, Boston.

Pardalos, P. and Guisewite, G. (1993). Parallel computing in nonconvex programming. *Ann. Oper. Res.*, 43:87–107.

Pardalos, P. and Li, X. (1990). Parallel branch and bound algorithms for combinatorial optimization. *Supercomputer*, 39:23–30.

Pardalos, P. and Principe, J., eds. (2002). *Biocomputing*, co-editors: P.M. Pardalos and J. Principe. Kluwer Academic, Boston.

Pardalos, P. and Resende, M., eds. (2002) *Handbook of Applied Optimization*, Oxford, UK.

Pardalos, P. and Romeijn, E., eds. (2002). *Handbook of Global Optimization: Heuristic Approaches,* Vol. 2. Kluwer Academic, Boston.

Pardalos, P., Floudas, C., and Klepeis, J. (1999). Global optimization approaches in protein folding and peptide docking. In *Math. Sup. for Molec. Biol.*, DIMACS Series, 47. pp. 141–171. Amer. Math. Soc.

Pardalos, P., Gu, J. and Du, B. (1994). Multispace Search for Protein Folding, In Biegler L.T. et al., editors, *Large Scale Optimization with Applications, Part III: Molecular Structure and Optimization*, The IMA Volumes in Mathematics and its Applications, 94, pp. 47–67, Springer-Verlag, New York.

Pardalos, P., Knopov, P., Urysev, S., and Yatsenko, V. (2001). Optimal estimation of signal parameters using bilinear observation. In Rubinov, A. and Glover, B., eds., *Optimization and Related Topics*. Kluwer Academic, Boston.

Pardalos, P., Liu, X., and Xue, G. (1997). Protein conformation of a lattice model using tabu search. *J. of Global Optimiz.*, 11(1):55–68.

Pardalos, P., Murphey, R., and Pitsoulis, L. (1998), A GRASP for the multitarget multisensor tracking problem, In *DIMACS Series*, Amer. Math. Soc., 40:277–302.

Pardalos, P., Pitsoulis, T., Mavridou T., and Resende, M. (1995). Parallel search for combinatorial optimization: Genetic algorithms, simulated annealing, tabu search and GRASP. In Ferreira, A. and Rolim, J., editors, *Parallel Algorithms for Irregularly Structured Problems*, Proceedings of the Second International Workshop, IRREGULAR'95, Lyon, France, Sept. 1995. Lecture Notes in Computer Science, 980: pp. 317–331, Springer-Verlag, New York.

Pardalos, P., Sackellares, J.C., and Yatsenko, V. (2002). Classical and Quantum Controlled Lattices: Self-Organization, Optimization and Biomedical Applications. In P.M. Pardalos and J. Principe, editors, *Biocomputing*, pages 199-224. Kluwer Academic, Dordrecht.

Pardalos, P., Sackellares, P. Carney, L., and Iasemidis, L., eds. (2004). *Quantitative Neuroscience*. Kluwer Academic, Boston.

Pardalos, P., Sackellares, J.C., Yatsenko, V., and Butenko, S. (2003). Nonlinear Dynamical Systems and Adaptive Filters in Biomedicine. *Ann. Oper. Res.*, 119:119–142.

Pardalos, P., Tseveendorj, I., and Enkhbat, R., eds. (2003). *Optimiz. Optim. Contr.*, World Scientific.

Pardalos, P. and Xue, G. (1999). Algorithms for a class of isotonic regression problems. *Algorithmica*, 23(3):211–222.

Du, D., Pardalos, P., and Wang, J., eds. (2000). *Discrete Mathematical Problems with Medical Applications*, DIMACS Series, Vol. 55, American Mathematical Society.

Pavlovskii, Y. and Elkin, V. (1988). Decomposition of models of control processes. *J. Math. Sciences*, 88(5):723–761.

Peterka, V. (1981). Bayesian approach to system identification. In Eyichoff, P., ed., *Trends and Progress in System Identification*, pp. 142–147. Pergamon Press, Oxford, UK.

Peters, K., Avouris, Ph., and Rentzepis, P.M. (1978). Picosecond dynamics of primary electron-transfer processes in bacterial photosynthesis. *Biophys. J.*, 23:207–217.

Petrov, B., Ulanov, G., Goldenblat, I., and Ulyanov, S. (1982). *Problems of Control for Relativistic and Quantum Dynamical Systems*. Nauka, Moscow (in Russian).

Prants, S. (1990). Lie algebraic solution of Bloch equations with time-dependent coefficients. *Phys. Lett. A*, 144(4/5):225–228.

Prants, S. (1991). Quantum dynamics of atoms in modulated laser fields. *J. Sov. Laser Res.*, 12(2):165–195 (in Russian).

Pugachev, V. and Sinitsin, I. (1987). *Stochastic Differential Systems. Analysis and Filtering*. John Wiley & Sons, New York.

Quinn, J. (1980). Stabilization of bilinear systems by quadratic feedback controls. *J. Math. Anal. Applic.*, 75:66–80.

Ray, D. (1992). Inhibition of chaos in bistable Hamiltonian systems by a critical external resonances. *Phys. Rew. A*, 46(10):5975–5977.

Ringe, D. and Petsko, C.A. (1985). Mapping protein dynamics by x-ray diffraction. *Prog. Biophys. Molec. Biol.*, 45:197–235.

Rink, R. and Mohler, R. (1968). Completely controllable bilinear systems. *SIAM J. Contr. Optim.*, 6:477–486.

Rink, R. and Mohler, R. (1971). Reachable zones for equicontinuous bilinear control process. *Int. J. Contr.*, 14:331–339.

Robinson, P. (1983). Non-parametric estimation for time series models. *J. Time Ser. Anal.*, 4:185–208.

Ross Cressman, R. (2003). *Evolutionary Dynamics and Extensive Form Games.* The MIT Press, Cambridge, MA.

Ryan, E. and Buckingham, N. (1983). On asymptotically stabilizing feedback control of bilinear systems. *IEEE Trans. Aut. Contr.*, 28(8), 863–864.

Sackellares, C., Iasemidis, L., Shiau, D., Gilmore, R., and Roper, S. (2000). Epilepsy — when chaos fails. In Lehnertz, K., Arnold, J, Grassberger, P. and Elger, C., eds., *Chaos in Brain?*, pp. 112–133. World Scientific, Singapore.

Sagle, A. and Walde, R. (1972). *Introduction to Lie Groups and Lie Algebras.* Academic Press, New York.

Sakurai, J. (1994). *Modern Quantum Mechanics.* Addison-Wesley, Reading, MA.

Samoilenko, Yu. and Yatsenko, V. (1991a). Quantum mechanical approach to optimization problems. In *Proc. Intern. Conf. Optim.*, Vladivostoc, IPU, Moscow.

Samoilenko, Yu. and Yatsenko, V. (1991b). Adaptive estimate the signal acting on macroscopic body in a controlled potential well, *Report of National Academy of Science of Ukraine*, 3:81–86.

Savage, L.J. (1954). *The Foundations of Statistics.* John Wiley, New York.

Schättler, H. (1998). Time-optimal feedback control for nonlinear systems. In Jakubczyk, B. and Respondek, W., eds., *Geometry of Feedback and Optimal Control*, pp., 383–421. Marcel Dekker, New York.

Schoepp, B., Parot, P., Lavorel, J., and Vermeglio, A. (1992). Charges recombination kinetics in bacterial photosynthetic reaction centers: Conformational states in equilibrium pre-exist in the dark. In Breton, J. and Vermeglio, A., eds., *The Photosynthetic Bacterial Reaction Center II*, NATO ASI Ser., 237:331–339. Plenum Press, New York.

Semenov, V. and Yatsenko, V. (1981). Dynamical equivalence and digital simulation. *Cybern. Comput. Technol. Complex Control Syst.*, 96:107–113.

Seraji, H. (1974). On pole shifting using output feedback. *Int. J. Contr.*, 20(5):721–726.

Shaitan, K., Uporov, I., Lukashev, E., Kononenko, A., and Rubin, A. (1991). Photoconformation transition causes temperature and light effects during charge recombination in reaction centers of photosynthesizing objects. *Photosynth. Res.*, 25:560–569.

Shimazaki, K. and Sugahara, K. (1980). Inhibition site of the electron transport system in lettuce chloroplasts by fumigation of leaves with SO_2. *Plant Cell Physiol.*, 21:125–132.

Slemrod, M. (1978). Stabilization of bilinear control systems with applications to nonconservative problems in elasticity. *SIAM J. Contr. Optimiz.*, 10:131–141.

Smythe, W. (1950). *Static and Dynamic Electricity.* McGraw-Hill, New York.

Sontag E. (1988). Bilinear realizability is equivalent to existence of a singular affine differential I/O equation. *Syst. Contr. Lett.*, 11:181–187.

Sontag E. (1990). *Mathematical Control Theory: Deterministic Finite Dimensional Systems.* Springer–Verlag, New York.

Sorenson, H. (1988). Recursive estimation for nonlinear dynamic systems. In Spall, J.C., ed., *Bayesian Analysis of Time Series and Dynamic Models.* Marcel Dekker, New York.

Sragovich, V. (2006). *Mathematical Theory of Adaptive Control.* World Scientific, Singapore.

Stocker, J. (1950). *Nonlinear Vibrations*, Interscience, New York (reissued by John Wiley & Sons, 1993).

Susmann, H. (1972). The bang-bang problem for linear control systems in GL(n, R). *SIAM J. Contr. Optim.*, 10:470–476.

Susmann, H.J. (1976). Some properties of vector fields not altered by small perturbation. *J. Diff. Eqs.*, 20:292–315.

Susmann, H. (1983). Lie brackets and local controllability. *SIAM J. Contr. Optimiz.*, 21:686–713.

Susmann, H. and Jurdjevic, V. (1972). Controllability of non-linear systems. *J. Diff. Eqs.*, 12:95–116.

Susmann, H. (1998). Geometry and optimal control. In Baillieul, J. and Willems, J., eds., *Mathematical Control Theory*, pp. 140–198. Springer-Verlag, New York.

Swerling, P. (1957). Detection of fluctuating pulsed signals in the presence of noise. *IRE Trans.*, 6:269–308.

Takens, F. (1981). Detecting strange attractors in turbulence. In Rand, D.A. and Young, L.S., eds., *Dynamical Systems and Turbulence*, Lecture Notes in Mathematics, 898, pp. 366–381. Springer-Verlag, Berlin,

Tam, L., Wong, W., and Yau, S. (1990). On a necessary and sufficient condition for finite dimensionality of estimation algebras. *SIAM J. Contr. Optim.*, 28(1):173–185.

Traub, R., Miles, R., and Jeffreys, J. (1993). Synaptic and intrinsic conductances shape picrotoxin-induced synchronized after discharges in the quinca pig hippocampal slice. *J. Physiol. (London)*, 461:525–547.

Tsu, R. (2005). *Superlattice to Nanoelectronics*. Elsevier, Amsterdam.

Udilov, V. (1974). Application of methods of abstract algebra to investigation of multidimensional automatic control systems. *Kibernetica i Vichisl. Technika*, 23:20–27.

Udilov, V. (1999). Construction of Versal Models of Dynamic Systems with Variable Parameters . *J. Automat. Inf. Sci.*, 31(6):33–44.

Van der Shaft, A. (1982). Controllability and observability for affine nonlinear Hamiltonian systems. *IEEE Trans. Aut. Cont.*, AC–27:490–494.

Van der Shaft, A. (1987). Equations of motion for Hamiltonian systems with constraints. *J. Phys. A.: Math. Gen.*, 11:3271–3277.

Van Kampen, N. (1983). *Stochastic Processes in Physics and Chemistry*. North–Holland, Amsterdam.

Vapnik, V. (1995). *The Nature of Statistical Learning Theory*. Springer, New York.

Vapnik, V. (1998). *Statistical Learning Theory*. Wiley, New York.

Varga, A. (2001). Model reduction software in SLICOT library. In Datta, B., ed., *Applied and Computational Control, Signals, and Circuits: Recent Developments*, p. 239–251. Kluwer Academic, Boston.

Vick, D. (1970). The problem of measurement. *Usp. Fiz. Nauk*, 101(2):303-329.

Wagner, G. (1983). Characterization of the distribution of internal motions in the basic pancreatic trypsin inhibitor using a large number of internal NMR probes. *Q. Rev. Biophys.*, 16:1–57.

Wayland, R., Bromley, D., Pickett, D., and Passamante, A. (1993). Recognizing determinism in a time series. *Phys. Rev. Lett.*, 70:580–582.

Wei, J. and Norman, E. (1963). Lie algebraic solution of linear differential equations. *J. Math. Phys.*, 4(5):575– 581.

Wei, J. and Norman, E. (1964). On the global representation of the solutions of linear differential equations as a product of exponentials. *Proc. Amer. Math. Sci.*, 15:327–334.

White, D. and Woodson, G. (1959). *Electromechanical Energy Conversion*. John Wiley, New York.

Whitney, H. (1936). Differentiable Manifolds. *Ann. Math.*, 37:645–680.

Willems, J. (1998). Path integral and stability. In Baillieul, J. and Willems, J., eds., *Mathematical Control Theory*, pp. 1–32. Springer-Verlag, New York.

Williamson, D. (1977). Observation of bilinear systems with application to biological control, *Automatica*, 13:243–254.

Willsky, A. (1973). Some estimation problems on Lie groups. In Mayne, D.O. and Brockett, R.W., editors, *Geometric Methods in System Theory*, pp. 305–314. Reidel, Boston.

Wong, W. 1987. On a new class of finite-dimensional estimation algebras. *Syst. and Contr. Lett.*, 9:79–83.

Wong, W. (1998). The estimation algebra of nonlinear filtering systems. In Baillieul, J. and Willems, J., eds., *Mathematical Control Theory*, pp. 33–65. Springer-Verlag, New York.

Xue, G., Zall, A. and Pardalos, P. (1996). Rapid evaluation of potential energy functions in molecular and protein conformation. In *DIMACS Series*, Vol. 23, pp. 237–249. Amer. Math. Soc., Provedence, RI.

Yakovenko, G. (1972). Trajectory synthesis of optimal control. *Aut. Remote Contr.*, 6:5–12.

Yatsenko, V. (1984). Dynamic equivalent systems in the solution of some optimal control problems. *Avtomatika*, 4:59–65.

Yatsenko, V. (1985). Control systems and fiber bundles. *Avtomatika*, 5:25–28.

Yatsenko, V. (1987). A mathematical model of a controlled mechanical displacement sensor. In *Control of Distributed Systems*, pages 26–30. IK AN UkrSSR, Kiev (in Russian).

Yatsenko, V. (1989). Estimating the signal acting on macroscopic body in a controlled potential well. *Kibernetika*, 2, 81–85.

Yatsenko, V. (1993). Quantum mechanical analogy of Belman optimal principle for control dynamical processes. *Cybern. Comput. Techn.*, 99:43–49.

Yatsenko, V. (1995). Hamiltonian model of a transputer type quantum automaton. In *Quantum Communications and Measurement*. Plenum Publishing Corporation, New York.

Yatsenko, V. (1996). Determining the characteristics of water pollutants by neural sensors and pattern recognition methods. *J. Chromatography A*, 722(1–2):233–243.

Yatsenko, V. (2003). Functional structure of the cryogenic optical sensor and mathematical modeling of signal. In *Cryogenic Optical Systems and Instruments, Proc. of SPIE*, 5172:97–107.

Yatsenko, V.A. and Knopov, P.C. (1992). Parameter estimation of almost periodic signal via controllable bilinear observations. *Aut. Remote Contr.*, 3:65–70.

Yatsenko, V. and Rakitina, N. (1994). The decomposition of the controlled dynamical model of information transform system with local and global symmetry. *Avtomatica*, 3-4;61–70.

Yatsenko, V., Titarenko T., and Kolesnik Yu. (1994). Identification of the non-Gaussian chaotic dynamics of the radioemission back scattering processes. In *Proc. of the 10th IFAC Symposium on System Identification SYSID'94*, Kobenhavn, 4–6 July, 1994, 1:313–317.

Yruela, I., Churio, M., Geusch, T., Braslavsky, S., and Holzwarth, A. (1994). Opto-acoustic and singlet oxygen near-IR emission study of the isolated Dl-D2-cyt b-559 reaction center complex of photosystem. *J. Phys. Chem.*, 98:12789–12806.

Zakharov, V., Manakov, S., Novikov, S. and Pitaevski, L. (1980). *Soliton Theory: an Inverse Problem Theory*. Nauka, Moscow (in Russian).

Zagarii, G.I. and Shubladze, A.M. (1981). Adaptive control methods for industrial application. Differentiation and filtering of signals. *Avtomatika*, 4:50–60.